AF598036

METHODS IN MOLECULAR BIOLOGY™

For further volumes:
http://www.springer.com/series/7651

Brain Development

Methods and Protocols

Edited by

Simon G. Sprecher

Department of Biology, University of Fribourg, Fribourg, Switzerland

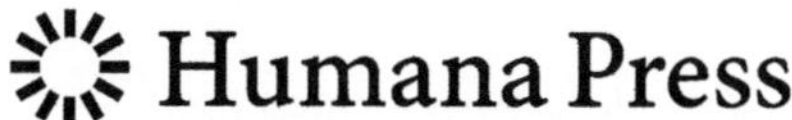

Editor
Simon G. Sprecher
Department of Biology
University of Fribourg
Fribourg, Switzerland

ISSN 1064-3745 ISSN 1940-6029 (electronic)
ISBN 978-1-62703-654-2 ISBN 978-1-62703-655-9 (eBook)
DOI 10.1007/978-1-62703-655-9
Springer New York Heidelberg Dordrecht London

Library of Congress Control Number: 2013946845

Printed on acid-free paper

Humana Press is a brand of Springer
Springer is part of Springer Science+Business Media (www.springer.com)

Preface

The brain is without any doubt the most complex organ. Particularly, cellular diversity, connectivity among neurons, formation of neuronal networks, the use of distinct neurotransmitter system, and the underlying function for behavior raise the question of how this highly interconnected organ develops. It is therefore not surprising that the intersection between developmental biology and neuroscience provides an exceptional field to address and investigate impacting biological questions. Complementing findings of an array of distinct animal model systems provide the basis of brain development research. Our current understanding is based on widely used genetic model systems including the fruit fly, zebra fish, chicken, and mouse. These animal models are impacting in particular since they allow elaborate genetic manipulations including transgenic expression systems and conditional knockout or knockdown of developmental genes. Genetic developmental studies are further complemented by several non-genetic animal models, which further substantiate general principles and mechanisms in brain development. Questions that can be investigated often depend on the methodological accessibility. Therefore, progress and developments in the constantly improving laboratory technologies provide an essential foundation for the advancement in the field.

This book aims to provide a broad overview and introduction of widely used leading-edge techniques in genetic model systems as well as some of the complementing animal models. The main focus lies on two key technical aspects of developmental neurobiology: Detection of gene expression and functional characterization of developmental control genes. The basic principle of expression and function studies are shared between different model systems. This includes in situ hybridization, reporter gene expression, and immunohistochemical staining methods, as well as RNA interference, Morpholino, or transgenic techniques. However the experimental procedure, such as, for instance, tissue treatment, fixation, dissection, genetic manipulation, and imaging, often differs substantially between animal models or even distinct stages in the same species. The collection of protocols aims to provide precise technical protocols but also allows for comparing a wide range of protocols in different tissues and species.

Fribourg, Switzerland *Simon G. Sprecher*

Contents

Contributors

IRWIN ANDERMATT • *Institute of Molecular Life Sciences and Neuroscience Center Zurich, University of Zurich, Zurich, Switzerland*
JENNIFER E. BESTMAN • *The Dorris Neuroscience Center, The Scripps Research Institute, La Jolla, CA, USA*
GEORGE BOYAN • *Developmental Neurobiology Group, Biocenter, Ludwig-Maximilians-Universität, Martinsried, Germany*
THOMAS G. BRACEWELL • *Department of Cell and Developmental Biology, University College London, London, UK*
HOLLIS T. CLINE • *The Dorris Neuroscience Center, The Scripps Research Institute, La Jolla, CA, USA*
MICHAEL J. CRAWFORD • *Biological Sciences, University of Windsor, Windsor, ON, Canada*
DANIELLE C. DIAPER • *Department of Neuroscience, MRC Centre for Neurodegeneration Research, Institute of Psychiatry, King's College London, London, UK*
BORIS EGGER • *Zoology Unit, Department of Biology, University of Fribourg, Fribourg, Switzerland*
ADÈLE FAUCHERRE • *Département de Physiologie, Institut de Génomique Fonctionnelle, CNRS UMR 5203, INSERM U661, Universtités Montpellier 1 & 2, Montpellier, France*
QIZHI GONG • *Department of Cell Biology and Human Anatomy, University of California, Davis, CA, USA*
FRANÇOIS GUILLEMOT • *Division of Molecular Neurobiology, MRC National Institute for Medical Research, London, UK*
DAFNI HADJIECONOMOU • *Division of Molecular Neurobiology, MRC National Institute for Medical Research, London, UK*
GISELBERT HAUPTMANN • *Department of Biosciences and Nutrition, NOVUM, Karolinska Institutet, Huddinge, Sweden*
THOMAS A. HAWKINS • *Department of Cell and Developmental Biology, University College London, London, UK*
RONG-QIAO HE • *State Key Laboratory of Brain and Cognitive Science, Institute of Biophysics, Chinese Academy of Sciences, Beijing, China*
FRANK HIRTH • *Department of Neuroscience, MRC Centre for Neurodegeneration Research, Institute of Psychiatry, King's College London, London, UK*
LARA HOOKER • *Biological Sciences, University of Windsor, Windsor, ON, Canada*
DAVID JUSSEN • *Institute of Genetics, University of Mainz, Mainz, Germany*
KAROLINE F. KRAFT • *Institute of Genetics, University of Mainz, Mainz, Germany*
CLAUDIUS F. KRATOCHWIL • *Friedrich Miescher Institute for Biomedical Research, Basel, Switzerland*
GILBERT LAUTER • *Department of Biosciences and Nutrition, NOVUM, Karolinska Institutet, Huddinge, Sweden*
ZI-LONG LI • *State Key Laboratory of Brain and Cognitive Science, Institute of Biophysics, Chinese Academy of Sciences, Beijing, China*

AIMIN LIU • *Department of Biology, Eberly College of Science, Center for Cellular Dynamics, Huck Institute of Life Sciences, The Penn State University, University Park, PA, USA*

JINLING LIU • *Department of Biology, Eberly College of Science, Center for Cellular Dynamics, Huck Institute of Life Sciences, The Penn State University, University Park, PA, USA*

KAI-LI LIU • *State Key Laboratory of Brain and Cognitive Science, Institute of Biophysics, Chinese Academy of Sciences, Beijing, China*

YING LIU • *State Key Laboratory of Brain and Cognitive Science, Institute of Biophysics, Chinese Academy of Sciences, Beijing, China*

YU LIU • *Developmental Neurobiology Group, Biocenter, Ludwig-Maximilians-Universität, Martinsried, Germany*

HERNÁN LÓPEZ-SCHIER • *Unit of Sensory Biology & Organogenesis, Helmholtz Zentrum München, Munich, Germany*

EMILIE PACARY • *Neurocentre Magendie, Physiopathologie de la Plasticité Neuronale, U862, Institut National de la Santé et de la Recherche Médicale, Université de Bordeaux, Bordeaux, France*

BENJAMIN PERRUCHOUD • *Zoology Unit, Department of Biology, University of Fribourg, Fribourg, Switzerland*

FILIPPO M. RIJLI • *Friedrich Miescher Institute for Biomedical Research, Basel, Switzerland*

SAQIB S. SACHANI • *Biological Sciences, University of Windsor, Windsor, ON, Canada*

IRIS SALECKER • *Division of Molecular Neurobiology, MRC National Institute for Medical Research, London, UK*

NANA SHIMOSAKO • *Division of Molecular Neurobiology, MRC National Institute for Medical Research, London, UK*

CRISTINE SMOCZER • *Biological Sciences, University of Windsor, Windsor, ON, Canada*

IRIS SÖLL • *Department of Biosciences and Nutrition, NOVUM, Karolinska Institutet, Huddinge, Sweden*

CLAUDIO D. STERN • *Department of Cell & Developmental Biology, University College London, London, UK*

ESTHER T. STOECKLI • *Institute of Molecular Life Sciences and Neuroscience Center Zurich, University of Zurich, Zurich, Switzerland*

ANDREA STREIT • *Department of Craniofacial Development & Stem Cell Biology, King's College London, London, UK*

RICHARD P. TUCKER • *Department of Cell Biology and Human Anatomy, University of California, Davis, CA, USA*

KATHERINE J. TURNER • *Department of Cell and Developmental Biology, University College London, London, UK*

ROLF URBACH • *Institute of Genetics, University of Mainz, Mainz, Germany*

GUDRUN VIKTORIN • *Biozentrum, University of Basel, Basel, Switzerland*

XIU-MEI WANG • *State Key Laboratory of Brain and Cognitive Science, Institute of Biophysics, Chinese Academy of Sciences, Beijing, China*

ANDREAS WANNINGER • *Faculty of Life Sciences, Department of Integrative Zoology, University of Vienna, Vienna, Austria*

TIM WOLLESEN • *Department of Integrative Zoology, Faculty of Life Sciences, University of Vienna, Vienna, Austria*

Part I

Drosophila Protocols

Chapter 1

Immunostaining of the Developing Embryonic and Larval *Drosophila* Brain

Danielle C. Diaper and Frank Hirth

Abstract

Immunostaining is used to visualize the spatiotemporal expression pattern of developmental control genes that regulate the genesis and specification of the embryonic and larval brain of *Drosophila*. Immunostaining uses specific antibodies to mark expressed proteins and allows their localization to be traced throughout development. This method reveals insights into gene regulation, cell-type specification, neuron and glial differentiation, and posttranslational protein modifications underlying the patterning and specification of the maturing brain. Depending on the targeted protein, it is possible to visualize a multitude of regions of the *Drosophila* brain, such as small groups of neurons or glia, defined subcomponents of the brain's axon scaffold, or pre- and postsynaptic structures of neurons. Thus, antibody probes that recognize defined tissues, cells, or subcellular structures like axons or synaptic terminals can be used as markers to identify and analyze phenotypes in mutant embryos and larvae. Several antibodies, combined with different labels, can be used concurrently to examine protein co-localization. This protocol spans over 3–4 days.

Key words *Drosophila*, Embryo, Larva, Brain, Immunostaining, Fluorescence immunocytochemistry, Dissection, Antibody

1 Introduction

Similar to mammalian brain development, the *Drosophila* brain derives from a monolayered epithelium called the neuroectoderm. Subsequent neurogenesis is characterized by two neurogenic periods: one during embryogenesis and another during larval and pupal stages. The precursor cells of the developing brain, termed neuroblasts (NBs), derive from the embryonic procephalic neurogenic region to form proliferative clusters; they divide repeatedly and asymmetrically in a stem cell mode to generate a new NB and a smaller daughter cell, called ganglion mother cell (GMC). Each GMC is a transient intermediate progenitor cell that generally divides once to produce two lineage-specific postmitotic cells, either neuron or glia. Both neuron and glia subsequently initiate their differentiation processes, finally shaping neural segments and circuits along the major body axes of the embryo [1, 2].

Simon G. Sprecher (ed.), *Brain Development: Methods and Protocols*, Methods in Molecular Biology, vol. 1082, DOI 10.1007/978-1-62703-655-9_1, © Springer Science+Business Media, LLC 2014

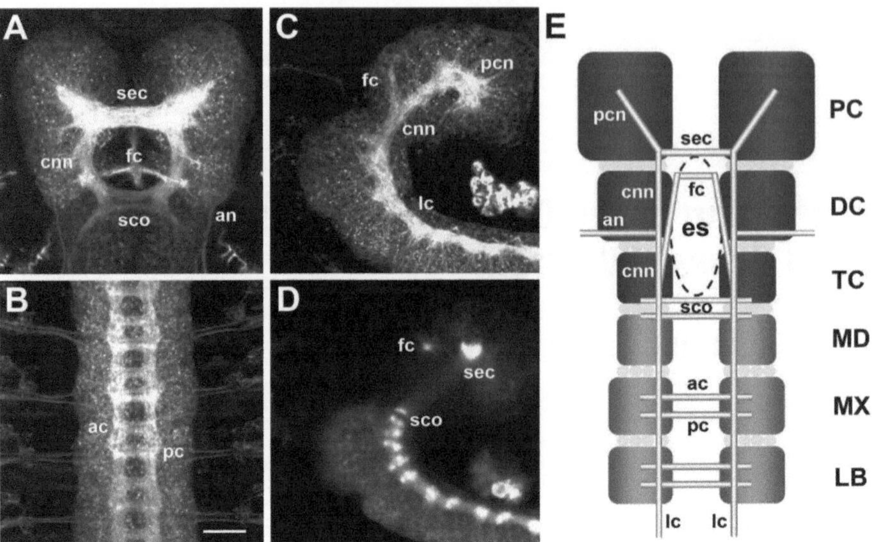

Fig. 1 The embryonic brain of *Drosophila*. (**a–d**) Laser confocal microscopy, superimposition of optical sections of stage 14 embryo immunolabeled with anti-HRP-FITC. (**a**) Frontal view of brain; (**b**) ventral view of ventral nerve cord; (**c**) lateral view of brain; (**d**) optical section along the midline of (**c**); (**e**) cartoon summarizing major neural segments and axon scaffolding which together constitute the embryonic *Drosophila* brain. Abbreviations: *sec* supraesophageal commissure, *fc* frontal connective, *cnn* circumesophageal connective, *sco* subesophageal commissure, an antennal nerve, *ac* anterior commissure, *pc* posterior commissure, *pcn* protocerebral connective, *lc* longitudinal connective, *PC* protocerebrum, *DC* deutocerebrum, *TC* tritocerebrum, *MD* mandibular neuromere, *MX* maxillary neuromere, *LB* labral neuromere, *es* esophagus. Scale bar: 25 μm

The central brain of *Drosophila* derives from 106 embryonic brain NBs that can be identified based on their positional relationships and NB-specific gene expression [3]. Towards the end of embryogenesis, most NBs stop proliferating and enter a period of quiescence. During the first larval instar stage, quiescent NBs reenter the cell cycle in a characteristic spatiotemporal pattern to perform a second round of neurogenesis, resulting in postembryonic neural progeny. These two phases of brain development are followed by an extensive morphological transformation during metamorphosis that ultimately leads to the adult fly brain [2].

Although approximately 95 % of the 200,000 neurons in the adult *Drosophila* brain are generated postembryonically, the main architecture of the adult brain is already laid down during embryonic neurogenesis [3, 4]. Thus, the *Drosophila* brain is composed of an anterior (supraesophageal) and a posterior (subesophageal) part, both of which are interconnected by an axon scaffold surrounding the gut; the anterior part comprises the protocerebral, deutocerebral, and tritocerebral neuromeres, whereas the posterior part comprises the mandibular, maxillary, and labial neuromeres (Fig. 1).

Immunostaining of the developing embryonic and larval *Drosophila* brain is used for two primary purposes. First, once a gene of interest has been cloned and antibodies raised against its encoded protein product(s), immunostaining can be used to visualize the spatiotemporal pattern of protein expression during embryonic and larval brain development. Second, antibody probes that recognize defined tissues, cells, or subcellular structures like axons or synaptic terminals can be used as markers to identify and analyze phenotypes in mutant embryos and larvae. Thus, following protein localization throughout *Drosophila* development reveals insights into gene regulation, cell-type specification, neuron and glial differentiation, and posttranslational protein modifications underlying the patterning, specification, and neuronal connectivity of the maturing brain.

In this respect, whole-mount immunohistochemistry (IHC), which is the process of using antibody probes to detect antigens (i.e., proteins), is among the most valuable of tools for analysis. It is a relatively cheap and reliable way to visualize discrete structures or cell types within the developing embryonic and larval brain of *Drosophila*. With the use of multiple antibodies, or when used in conjunction with genetic labeling, such as mosaic analysis with a repressible cell marker (MARCM) [5], it is possible to follow several different proteins, examining co-localization, or identify specific cells and follow their neuronal projections from the cell body to its terminal dendritic arborizations [6–8]. Immunostaining of *Drosophila* brain tissue is often used to help deconstruct the complexities of neural circuit development or neurodegeneration [9, 10], in many cases with the aim of understanding human disease pathogenesis [11, 12]. The availability of *Drosophila* antibodies to human homologues can pave the way for follow-up studies in mammals or human tissue [13, 14].

There is a growing supply of *Drosophila*-specific antibodies available that bind to all manner of cellular proteins, from organelle components to synaptic vesicle markers [15]. Immunostaining can be done in basically two ways, which are defined by way of visualization method. One option visualizes proteins/antigens with secondary antibodies that carry labels, which, upon enzymatic reactions, lead to precipitates that are visible under the light microscope. Typical examples are alkaline phosphatase or horseradish peroxidase reactions that lead to brown, black, or purple precipitates (for details, *see* ref. 16). These histochemical staining methods have obvious limitations: enzymatic reactions do not penetrate well into tissue, and resulting precipitates do not allow 3D reconstructions unless the tissue is embedded in plastic, microdissected, and subsequently scanned, which is laborious and time-consuming. The advent of laser confocal and super-resolution microscopy together with computer algorithms nowadays allows rapid and reproducible 3D reconstructions of optical sections derived from fluorescence immunostaining, which has thus become the method of choice. It is based on secondary antibodies that are conjugated with various fluorochromes that, following excitation, emit at specific wavelengths that are detected by the various optical imaging techniques. Our protocol outlines the basics for this method.

First, the tissue must be carefully dissected from the whole organism and fixed; the fixation process described here uses paraformaldehyde; however, other fixation processes may be used depending on the antibody used or the structures being visualized (*see* Subheading 1.1). In the case of the embryo, the two protective membranes, an outer shell called the chorion and a thick inner vitelline membrane [17], must first be removed. This vitelline barrier becomes permeable to the fixative agent when treated with a fixative-heptane mix. The membrane is then removed by methanol to allow the diffusion of the antibodies. Following fixation, the tissue is blocked, usually with normal serum. This step reduces binding of the antibody to nonspecific reactive sites of *Drosophila* proteins. Blocking with normal serum from the species used to generate the secondary antibodies is preferred. Either monoclonal or polyclonal primary antibodies can be used for immunohistological staining of *Drosophila* tissue.

Antibodies are raised against specific antigens, and their most effective concentration should be determined by first carrying out a dilution series on embryos. Occasionally, primary antibodies are already conjugated to a fluorophore, meaning that secondary antibodies are not needed to visualize the antigen's location. When choosing fluorophores for multiple antibody stainings, you should bear in mind what filters/lasers are available for imaging the brains and any crossover in the excitation or emission wavelengths [18]. Separating each incubation step is a series of washes. Thorough washing is essential to prevent antibodies interacting with surplus fixative and to clear the sample of residual antibodies. This immunohistological protocol can also be applied to adult brain tissue as well as to larval imaginal discs.

1.1 Troubleshooting

Although now a relatively standard technique, elements of this protocol may benefit from tweaking, depending on what you are hoping to visualize and the antibodies you use. If you have persistent trouble in visualizing your proteins, you may want to try the following suggestions.

1.1.1 Alter the Fixation Technique

Formaldehyde is suitable for deep penetration of the tissue; it forms strong cross-links between proteins and is suitable for long-term storage [19] and preserving chromosome morphology [20]. You may want to consider replacing the paraformaldehyde with freshly prepared formaldehyde solution. Glutaraldehyde may also be used [21, 22]; however, as it is a larger molecule, it does not diffuse as well through deep tissue. Its cross-links span a larger distance, so it can stably fix proteins that are further apart. Glutaraldehyde fixation is not ideal for immunohistochemistry as further treatment of the tissue is necessary to avoid aldehyde groups binding to the antibody [23]. Embryos may be fixed with methanol; while suitable for studying embryo morphology, this method shows poor preservation of cytoplasmic proteins [24]. A more complex technique is cryofixation [25]. With all these techniques, inadequate or over-fixation may occur, so it may be useful to slightly increase or decrease the fixation period.

1.1.2 Antibody Optimization

Too high or low concentrations of antibody may cause either antibody aggregation on the surface of the tissue or poor penetration of the tissue [26]; therefore, carrying out a dilution series will help to determine the most effective titer. To improve the signal-to-noise ratio, try using affinity-purified antibodies or altering the temperature, e.g., room temperature instead of 4 °C, or length of antibody incubation, e.g., 3 h instead of overnight [22].

2 Materials

Make all stock buffers in a sterile, sealable bottle. Prepare solutions with distilled H_2O (dH_2O). To avoid contamination of the stock solutions, keep individual 50 ml aliquots of PBS, PBL, and PBT and use these for dissections. These solutions are prone to contamination; check for wispy or cloudy cultures that may form after around 1 month of storage and dispose of contaminated solutions. All solutions should be stored at 4 °C unless stated otherwise.

2.1 Solutions for Embryo Preparations

1. PEM: 100 mM PIPES, 2 mM EGTA, 1 mM $MgSO_4$. To 800 ml dH_2O, add 34.63 g PIPES, 0.76 g EGTA, and 0.12 g $MgSO_4$. Adjust to pH 7 with HCl. Make up to final volume of 1 l with dH_2O.
2. PEM-FA: Add 1 ml 37 % formaldehyde solution to 9 ml PEM. Make fresh. Do not store.
3. PBT 0.1 %—for embryo and early larval stages (L1 and L2): Add 0.2 g BSA and 0.2 ml Triton X-100 to 200 ml PBS. Store at 4 °C (*see* **Note 1**).
4. 5 % PBT-NGS: Add 1 ml normal goat serum (Invitrogen) to 19 ml PBT 0.1 %.
5. 50 % sodium hypochlorite.
6. Heptane.
7. Methanol.

2.2 Solutions for Larval Preparations

1. PBL: Dissolve 1.8 g of lysine HCl in 50 ml dH_2O, adjust pH to 7.4 by adding *x* ml of 0.1 M NaH_2PO_4 (*see* **Note 2**), and adjust volume to 100 ml 0.1 M PBS. Store at 4 °C.
2. PBS 0.1 M buffer: Make 0.1 M Na_2HPO by dissolving 8.52 g in 600 ml dH_2O. Make 0.1 M NaH_2PO_4 by dissolving 2.4 g in 200 ml dH_2O. Take 500 ml of 0.1 M Na_2HPO and adjust

pH to 7.4 by adding *x*ml of 0.1 M NaH_2PO_4 (*see* **Note 2**). Store at 4 °C.

3. PBT 0.5 %—late larval stage (L3): Add 1 ml Triton X-100 to 200 ml PBS. Store at 4 °C (*see* **Note 1**).
4. 10 % PBT-NGS: Add 1 ml normal goat serum (Invitrogen) to 9 ml PBT 0.5 %.
5. 8 % PFA: Caution—make in a fume hood. In a sealable 50 ml container, dissolve 1.6 g paraformaldehyde in 20 ml dH_2O and add 140 μl 1 M NaOH. Place in 37 °C water bath, vortexing occasionally until completely dissolved. Aliquot 400 μl into 2 ml tubes and store at −20 °C (*see* **Note 3**).
6. PLP: Mix three parts PBL (1.2 ml) to 1 part 8 % PFA (400 μl). Make fresh. Do not store (*see* **Note 3**).

2.3 Mounting Medium and Antibodies

1. VECTASHIELD mounting medium with or without DAPI (*see* **Note 4**) (Vector Laboratories, Burlingame, CA).
2. Antibodies: Developmental Studies Hybridoma Bank supplies a large range of *Drosophila*-specific primary antibodies. Secondary antibodies should be selected based on the animal in which the primary antibody was raised, i.e., if you are using rat-Elav, then you will need something like a goat anti-rat 488 secondary antibody.

2.4 Equipment

1. Microscope (larval preps).
2. Nylon mesh (embryo preps).
3. Vertical rotator.
4. Glass pipette (embryo preps).
5. Small paintbrush (embryo preps).
6. Rocker.
7. Fine tip forceps and sharpening stone (Dumont no. 5 tweezers, superfine, straight tip) (larval preps).
8. Glass microscope slides, SuperFrost (Thermo Scientific Gerhard Menzel).
9. Cover slips: 22 × 50 mm, 0.13–0.17 thickness (embryo preps); 22 × 22 mm, 0.13–0.17 thickness (larval preps).
10. Silicone grease or petroleum jelly (larval preps).
11. Modeling clay (such as Plasticine) (embryo preps).
12. Dissection watch glass (larval preps).
13. Pin holder and stainless steel minutien pins 0.2 mm diameter (InterFocus Ltd, Fine Science Tools) (larval preps).
14. Timer.

3 Methods

To be carried out at room temperature unless otherwise stated.

3.1 Embryo Fixation

1. Recover plates from egg collection and remove any remaining yeast or dead flies with a spatula or brush being careful not to damage the agar (*see* **Note 5**).
2. Dechorionate the embryos by adding 50 % hypochlorite and agitate for 2–5 min until dechorionated embryos float to the surface (*see* **Note 6**).
3. Pour the embryos and hypochlorite through the nylon mesh and rinse thoroughly with dH_2O (*see* **Note 7**) (*see* Fig. 2).
4. Transfer embryos to a 2 ml tube containing 1 ml heptane and 1 ml PEM-FA (*see* **Note 8**).
5. Agitate on a rotator at high speed for 10–30 min (no longer than 30 min).
6. Allow embryos to settle—they should be at the interface between the two phases.
7. Remove the lower phase, then the upper phase (*see* **Note 9**).

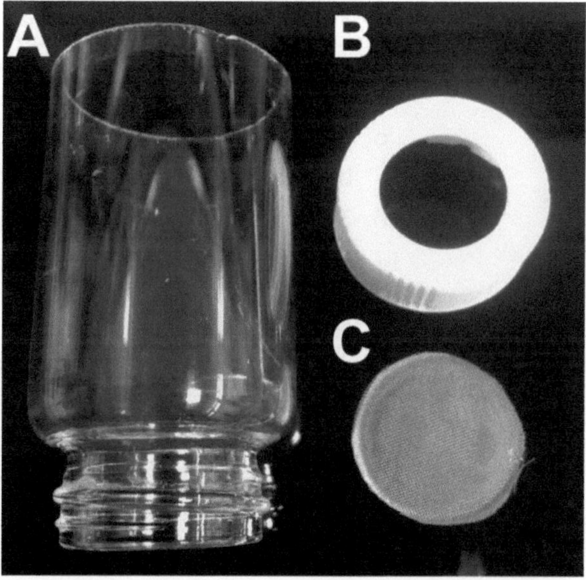

Fig. 2 Embryo collection apparatus. An open-bottom container (**a**) has a hole cut in the lid (**b**); the hole is closed by a fine mesh (**c**), which is included into the lid when screwed on top of the container. A 50 ml plastic container can also be adapted in this way

8. Replace with another 1 ml heptane. Then add 1 ml 100 % methanol and shake vigorously for 1–2 min. The devitellinized embryos will fall to the bottom of the tube.
9. Remove both phases making sure to remove all vitelline debris (*see* **Note 10**).
10. Quickly add 1.5 ml of 100 % methanol and agitate on rotator for 2 × 5 min, then 1 × 30 min (*see* **Note 11**).
11. Replace with fresh methanol and store at −20 °C or rehydrate for immunostaining.

3.2 Embryo Immunostaining

1. Transfer embryos into a 1.5 ml tube (*see* **Note 12**).
2. Rehydrate the embryos by removing the methanol and washing 2 × 5 min, then 1 × 30 min with PBT 0.1 %.
3. Incubate embryos for 30 min in 5 % PBT-NGS.
4. Remove PBT-NGS and add appropriate amount of primary antibody diluted in 5 % PBT-NGS to a reaction volume of 100 μl (*see* **Note 13**). Incubate overnight at 4 °C.
5. Remove antibody solution and wash 1 × 1 min, 3 × 5 min, and 4 × 30 min with PBT 0.1 % (*see* **Note 11**).
6. Incubate embryos for 30 min in 5 % PBT-NGS.
7. Remove PBT-NGS and add appropriate amount of secondary antibody diluted in 5 % PBT-NGS to a reaction volume of 100 μl. Incubate at 4 °C over night in the dark (*see* **Note 14**).
8. Remove antibody solution and wash 3 × 5 min and 4 × 30 min with PBT 0.1 % (*see* **Note 11**).
9. Add 1 drop of VECTASHIELD mounting medium and incubate at 4 °C over night in the dark.
10. Mount on glass slide with 22 × 50 mm cover slip (*see* **Note 15**).
11. Your samples are now ready for image acquisition (*see* **Note 16**).

3.3 Larval Immunostaining

1. Collect larvae and place into a watch glass containing cold PBS.
2. Fill a 0.5 ml tube with cold PBS and keep on ice with the lid open.
3. Roughly remove the larval CNS (*see* **Note 17**), placing dissected brains into the PBS-containing 0.5 ml tube.
4. After 30 min of dissection, remove PBS and add 500 μl of PLP. Agitate on a rotator for 1 h at room temperature.
5. Remove PLP and wash 3 × 10 min in PBT (*see* **Note 18**).
6. Block by incubating the brains in 5 % PBT-NGS (for L1 and L2) or 10 % PBT-NGS (for L3) on the rotator for 15 min.
7. Remove PBT-NGS and add appropriate amount of primary antibody diluted in 5 or 10 % PBT-NGS to a reaction volume

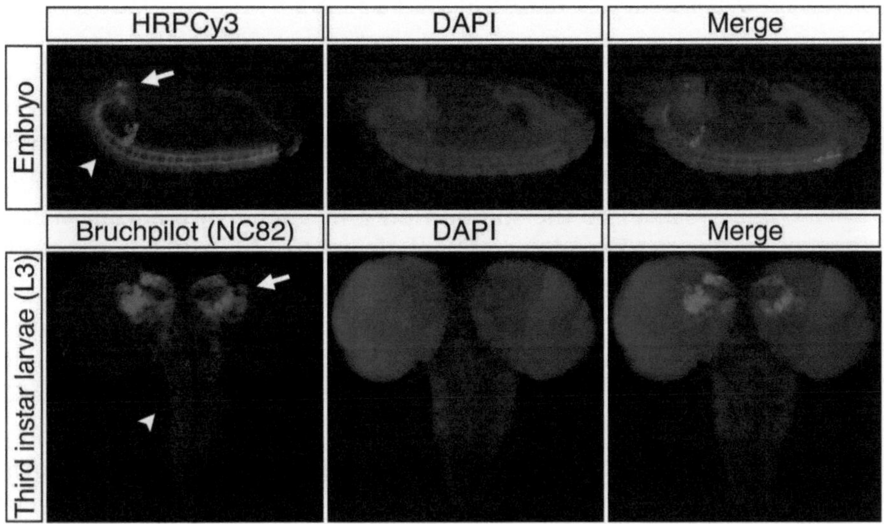

Fig. 3 Immunostaining of whole mount embryonic CNS and 3rd instar larval CNS. *Top row*: The embryonic nervous system is staining with horseradish peroxidase Cy3 (HRP-cy3; *red*) and visualized under a standard fluorescence microscope. The anterior brain is visible (*arrow*) as well as the ladderlike axon tracts of the ventral nerve cord (*arrowhead*); compare to Fig. 1. The nuclei of all embryonic cells are visualized with DAPI. Merge shows co-labeling of HRP-cy3 and DAPI. *Bottom row*: 3rd instar larval CNS immunostained with anti-Bruchpilot counterlabeled with goat anti-mouse 568, and nuclei are highlighted with DAPI; merge shows co-labeling of nc82 and DAPI. The nc82 antibody recognizes the Bruchpilot protein which is specifically enriched in the active zone of synapses

of 100 μl (*see* **Note 13**). Incubate overnight (unless alternative duration is specified by antibody manufacturer) at 4 °C.

8. Remove antibody solution and wash 1 × 1 min, then 3 × 20 min in PBT.
9. Add appropriate amount of secondary antibody diluted in 5 or 10 % PBT-NGS to a reaction volume of 200 μl. Incubate for 2–3 h on a rocker in the dark (*see* **Note 14**).
10. Wash brains 2 × 15 min in PBT, then 2 × 15 min in PBS (*see* **Note 19**).
11. Add ~2 drops of VECTASHIELD mounting medium and incubate overnight at 4 °C in the dark.
12. Fine dissect in mounting medium on microscope slide under microscope (*see* **Note 20**). Remove debris, arrange brains, and make a small dab of silicone grease or petroleum jelly at four corners where the cover slip will go. Gently lower the 22 × 22 mm cover slip in to place and seal with nail varnish (*see* **Note 21**).
13. Store in a microscope slide box in the dark at 4 °C (*see* **Note 22**).
14. Your samples are now ready for image acquisition (*see* **Note 16**).

Examples of successful fixation, labeling, and imaging under a standard fluorescence microscope are given in Fig. 3.

4 Notes

1. Triton X-100 is very viscous, so pipette slowly to make sure you take up the right amount. Mix by using a slow setting on a magnetic stirrer to avoid an abundance of bubbles.
2. Place the solution on a magnetic stirrer while measuring the pH and use a 10 ml pipette to add small amounts of NaH_2PO_4 to the solution.
3. Before you start the dissections, place a frozen aliquot of PFA in a 37 °C water bath. By the time you have finished the dissection and are ready to fix the tissue, the PFA should be fully dissolved and you can add the PBL directly to it.
4. DAPI (4,6-diamidino-2-phenylindole) is a nuclear stain that binds to double-stranded DNA and fluoresces when excited with a mercury arc lamp or UV. Even when not examining the nuclei, it is useful to use the consistent DAPI staining pattern to help orientate around the embryonic and larval brains.
5. Embryos can be collected by placing fruit agar plates (12.5 g sugar, 21.25 g agar, 750 ml dH_2O; autoclave in large flask; add 250 ml apple juice, 2 g nipagen; stir and pour into 55 mm petri dishes; allow to cool; store at 4 °C for 1 month) that have a smear of yeast paste in the middle, on top of a bottle of flies. Keep the bottle, which should have holes pierced in the bottom for air circulation, inverted in the dark and replace the agar plate as necessary.
6. You can use 50 % household bleach instead of hypochlorite but make sure that it hasn't gone off—yellow bleach is still in date.
7. Wash agar plates with water and pour this through the mesh ×2. Hold the nylon mesh over a waste beaker and use a water bottle to rinse thoroughly. You can test for remaining bleach by dabbing it on a colored paper towels (e.g., blue roll); if any bleach is present, it will change the paper color. Residual bleach will interfere with the fixation process. If you plan on doing a lot of embryo stainings, it is worth trying to make a straining tool out of a small glass jar with a screw lid and the bottom removed (*see* Fig. 2).
8. Hold the mesh with forceps and dip into the 2 ml tube to transfer the embryos and use the brush to transfer any embryos stuck to side of straining tool (if you have one).
9. Use a glass pipette as embryos are less likely to adhere to glass than to a plastic pipette. Remove as much of the solutions as possible—this may mean pipetting off some of your embryos too.
10. Remove the layers by pipetting from the interface of the two phases.

11. For each wash, remove the old solution and add the same amount of fresh solution. Washes should be agitated on a rotator at full speed.
12. Pipette as fast as you can; as soon as the liquid stops being turbulent, the embryos tend to stick to the sides of the pipette and are very difficult to remove. This means you should have your two tubes ready, with lids open and suck up the liquid and embryos and transfer as quickly as possible.
13. Depending on what you are attempting to visualize, you can add several primary antibodies at once. Take care to avoid using similar species that may cross-react, for example, if you use a mouse and rat primary antibody, the secondary anti-mouse will bind to the rat antibody and the secondary anti-rat will bind to the mouse antibody. If you are visualizing an endogenous GFP signal (e.g., using the GAL4>UAS system), you may get a better signal by also using an antibody against GFP and a secondary with a 488 (green) fluorophore. Some primary antibodies are already conjugated to a fluorophore and do not need a secondary antibody, such as horseradish peroxidase Cy3 (HRP-cy3), which specifically labels the *Drosophila* nervous system.
14. You will need a secondary antibody that targets the animal species in which the primary antibody was raised. You can use up to four separate channnels by picking fluorophores that are unlikely to interfere with each other, e.g., DAPI, UV; 488, green; 568, red; and 647, far red. Of course, the fluorophore you use will depend on what filters, lamps, or lasers are available for image capture.
15. Suck up VECTASHIELD and embryos and pipette out in a wide zigzag line on the labeled microscope slide (*see* Fig. 7a). Roll 4 small balls of modeling clay and place in the corners where the cover slip will go. Lower one short end of the cover slip in place and gently lower the cover slip, avoiding air bubbles. Gently press cover slip at the 4 corners.
16. To avoid bleaching the brains, first use a fluorescent microscope to select your best brains or embryos. When imaging using a laser scanning microscope, scan each channel sequentially to avoid interference between fluorophores. If you have issues with the noise-to-signal ratio, check that you are using the antibody at the appropriate dilution, ensure you are completing all wash steps, and check the pH of the buffer solutions.
17. This bit takes practice! There are several ways to pull the CNS from the larvae, but the thing that will help the most is being able to differentiate the CNS from the imaginal discs and fat bodies (*see* Figs. 4 and 5). I find it easier to gently clamp the body of the larvae about two thirds the way towards the mouth hooks; then when the larvae extends its mouth hooks, grab

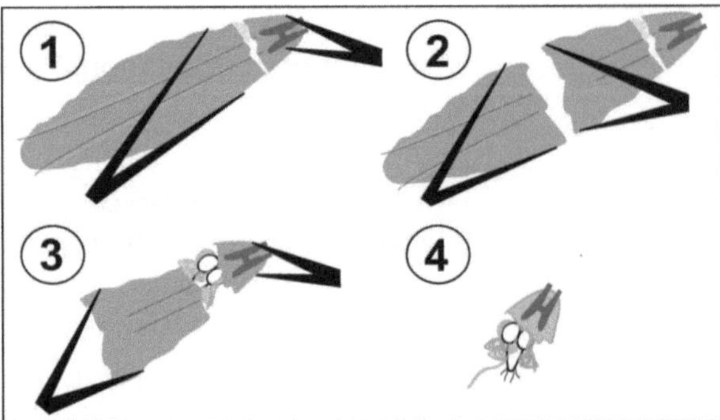

Fig. 4 Cartoon illustrating a rough larval CNS dissection. (**1**) First grab the extended mouth hooks and make a gentle tear in the cuticle. (**2**) Rip the larvae in half. (**3**) Holding the mouth hooks again, slowly pull off the remaining cuticle until (**4**) you are left with the larval brain still attached to the head, imaginal discs, and fat bodies

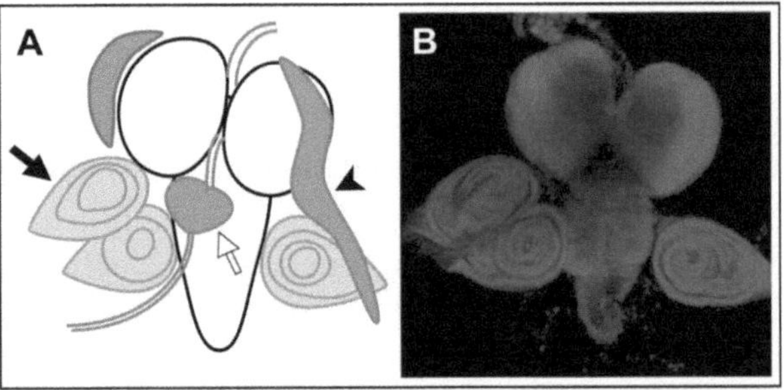

Fig. 5 Identifying the 3rd instar larval brain and CNS. (**a**) Several imaginal discs (*filled arrow*), fat bodies (*arrowhead*), and digestive components (*empty arrow*) will be attached to the 3rd instar larval brain and CNS. Identifying the brain among these other tissues will aid successful dissections. The larval brain and CNS are outlined with a *black line*. (**b**) DAPI labeled 3rd instar larval brain and CNS visualized under the fluorescence microscope after successful dissection; note that wing and leg imaginal discs are still attached

them with the forceps (*see* Fig. 4). Gently pull the mouth hooks until the cuticle rips and the internal matter is exposed. The round optic lobes of the CNS may be visible. Place both forceps on the body and rip the lower half of the larvae off. Grab the mouth hooks and body again and gently pull until the mouth hooks and CNS come away from the rest of the larvae. Holding onto the mouth hooks, transfer to the 0.5 ml tube that is sat on ice. If you do not have much luck this way, or find that the ventral nerve cord is lost using this method, you can also pinch both forceps just below the mouth hooks (by the scruff of its neck!) and rip the cuticle open and then bit by bit remove the lower part of the larvae, the digestive tract, fat, and some imaginal discs.

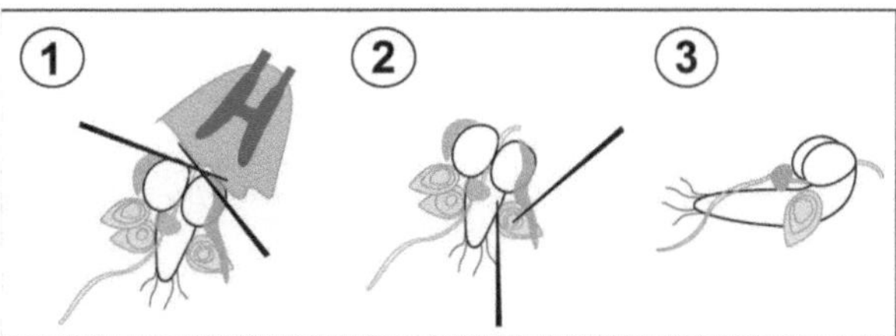

Fig. 6 Cartoon illustrating fine dissection of the larval brain and CNS. (**1**) Using the dissecting pins like a knife and fork, remove the mouth hooks from the brain and then (**2**) gently remove all other tissues taking care (**3**) not to damage or puncture the brain

18. When washing it is best to leave brains to settle for 15 s and then pipette off the solution. Eject the pipetted solution into a clean watch glass, check under the microscope, and rescue any brains by pipetting them back into the tube. If carrying out several dissections, you can leave samples washing for an extended period of time until all preparations have reached the same stage.
19. As the secondary antibodies are photosensitive, you should keep your preparation in the dark as much as possible to avoid bleaching the fluorophores. Cover the tubes with tinfoil or repurpose a small container to hold your tubes and attach to the rotator using Velcro strips.
20. Cut around 2 cm off the end of a P200 pipette tip and transfer brains and VECTASHIELD to a microscope slide. Label the microscope slide with a dissection code (for cross-reference with your records, e.g., Initials_01), genotype, the primary antibody and secondary fluorophore used, and the date. Remove the majority of the VECTASHIELD, so that the brains are no longer floating, and pipette back into your dissection tube. Use the pin holder tools to remove unwanted imaginal discs, fat bodies, and mouth hooks. You can use them like a knife and fork to slice away tissue, or pin an imaginal disc into place with one needle and cut the connecting tissue with the other (*see* Fig. 6). Take care not to pin or damage the CNS itself.
21. Arrange brains in concentric circles with the ventral nerve cord pointing outwards (*see* Fig. 7b). Orientate the brains in a mixture of dorsal or ventral side up. Add a small amount of VECTASHIELD in a circle around the brains—not too much or they will float out of position. Once the cover slip is in place, gently press on the four corners to flatten the brains slightly, and be careful as they are easily ruptured by squashing. Add more VECTASHIELD if necessary by pipetting small amounts at the edge of the cover slip. Remove excess VECTASHIELD with cotton buds or tissues dipped in ethanol—again, be very careful not to dislodge the cover slip as this will destroy your brains!

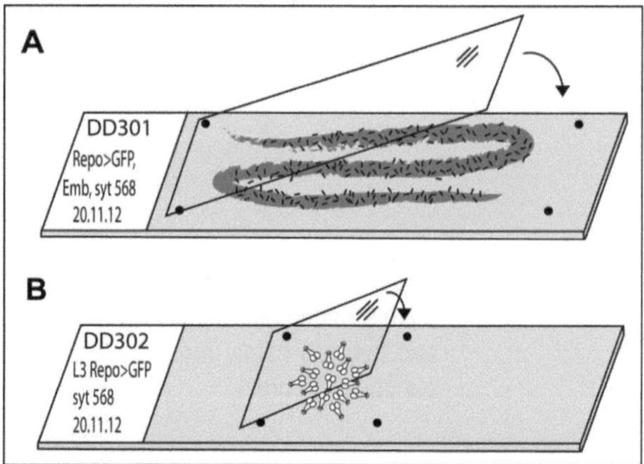

Fig. 7 Cartoon illustrating slide preparation. (**a**) Pipette embryos in VECTASHIELD mounting medium in a wide zigzag line onto labeled microscope slide, place small balls of modeling clay where the corners of the cover slip will be, and gently lower the cover slip (*arrow*) onto the embryos in VECTASHIELD; make sure to avoid air inclusions. (**b**) Arrange the larval brains/CNS in a rosette onto labeled microscope slide, place small dabs of silicone grease (or petroleum jelly) at the corners of where the cover slip will be, and gently lower the cover slip (*arrow*); make sure to avoid air inclusions

22. A standard 4 °C fridge is a good place to store your slide box. Cold rooms often have a damp atmosphere, which can cause the nail varnish to peel off.

Acknowledgements

This work was supported by grants from the UK Medical Research Council (G070149), Royal Society (Hirth2007/R2), Parkinson's UK (G-0714), Motor Neurone Disease Association (Hirth/Mar12/6085, Hirth/Oct07/6233), Alzheimer Research UK (Hirth/ARUK/2012), and the Fondation Thierry Latran (DrosALS to F.H.).

References

1. Skeath JB, Thor S (2003) Genetic control of Drosophila nerve cord development. Curr Opin Neurobiol 13:8–15
2. Kim DW, Hirth F (2009) Genetic mechanisms regulating stem cell self-renewal and differentiation in the central nervous system of Drosophila. Cell Adh Migr 3:402–411
3. Urbach R, Technau GM (2004) Neuroblast formation and patterning during early brain development in Drosophila. Bioessays 26: 739–751
4. Hirth F, Reichert H (1999) Conserved genetic programs in insect and mammalian brain development. Bioessays 21:677–684
5. Lee T, Luo L (2001) Mosaic analysis with a repressible cell marker (MARCM) for Drosophila neural development. Trends Neurosci 24: 251–254

6. Grueber WB, Ye B, Yang CH, Younger S, Borden K, Jan LY, Jan YN (2007) Projections of *Drosophila* multidendritic neurons in the central nervous system: links with peripheral dendrite morphology. Development 134: 55–64
7. Selcho M, Pauls D, Han KA, Stocker RF, Thum AS (2009) The role of dopamine in *Drosophila* larval classical olfactory conditioning. PLoS One 126:e5897
8. White KE, Humphrey DM, Hirth F (2010) The dopaminergic system in the aging brain of *Drosophila*. Front Neurosci 4:205
9. Muqit MK, Feany MB (2002) Modelling neurodegenerative diseases in *Drosophila*: a fruitful approach? Nat Revs Neurosci 3:237–243
10. Koizumi K, Higashida H, Yoo S et al (2007) RNA interference screen to identify genes required for Drosophila embryonic nervous system development. Proc Natl Acad Sci USA 104:5626–5631
11. Stochmanski SJ, Therrien M, Laganière J et al (2012) Expanded ATXN3 frameshifting events are toxic in Drosophila and mammalian neuron models. Hum Mol Genet 21:2211–2218
12. Dunlop J, Morin X, Corominas M et al (2004) glaikit is essential for the formation of epithelial polarity and neuronal development. Curr Biol 14:2039–2045
13. Tsuji T, Higashida C, Yoshida Y et al (2011) Ect2, an ortholog of Drosophila's pebble, negatively regulates neurite outgrowth in neuroblastoma × glioma hybrid NG108-15 cells. Cell Mol Neurobiol 31:663–668
14. Pandey UB, Nichols CD (2011) Human disease models in Drosophila melanogaster and the role of the fly in therapeutic drug discovery. Pharmacol Rev 63:411–436
15. Developmental Studies Hybridoma Bank developed under the auspices of the NICHD and maintained by The University of Iowa, Department of Biology, Iowa City, IA 52242 http://dshb.biology.uiowa.edu/
16. Patel N (1994) Imaging neuronal subsets and other cell types in whole mount Drosophila embryos and larvae using antibody probes. In: Goldstein LSB, Fryberg E (eds) Methods in Cell Biology, Vol 44. Drosophila melanogaster: Practical Uses in Cell Biology. Academic, New York, NY, For an amended and updated version, follow the link: http://patelweb.berkeley.edu/Images/Protocols/pdf%20files/Antibody%20Methods%202006.pdf
17. Ashburner M (1989) *Drosophila*: a laboratory manual. Cold Spring Harbor Laboratory Press, New York, NY
18. Hoffman, G. (2008) Seeing is believing: Use of antibodies in immunohistochemistry and in situ hybridization. In: Short course II of SfN's 38 annual meeting: 15–19 November 2008; Washington, DC. Society for Neuroscience
19. Rothwell WF, Sullivan W (2000) Fluorescent analysis of *Drosophila* embryos. In: Sullivan W, Ashburner M, Hawley RS (eds) *Drosophila* protocols. Cold Spring Harbor Laboratory Press, New York, NY, p 141
20. Bonaccorsi S, Giansanti MG, Cenci G, Gatti M (2012) Formaldehyde fixation of *Drosophila* testes. Cold Spring Harb Protoc. doi:10.1101
21. Heimbeck G, Bugnon V, Gendre N, Häberlin C, Stocker RF (1999) Smell and taste perception in *Drosophila melanogaster* larva: toxin expression studies in chemosensory neurons. J Neurosci 19:6599–6609
22. Stocker RF, Heimbeck G, Gendre N, de Belle JS (1997) Neuroblast ablation in *Drosophila* P[GAL4] lines reveals origins of olfactory interneurons. J Neurobiol 32:443–456
23. Hassell J, Hand AR (1974) Tissue fixation with diimidoesters as an alternative to aldehydes. I. Comparison of cross-linking and ultrastructure obtained with dimethylsuberimidate and glutaraldehyde. J Histochem Cytochem 22: 223–229
24. Wieschaus E, Nüsslein-Volhard C (1998) Looking at embryos. In: Roberts DB (ed) *Drosophila*, a practical approach. Oxford University Press Inc, New York, NY, p 205
25. Ripper D, Schwarz H, Stierhof YD (2008) Cryo-section immunolabelling of difficult to preserve specimens: advantages of cryofixation, freeze-substitution and rehydration. Biol Cell 100:109–123
26. Rebay I, Fehon R (2000) Generating antibodies against *Drosophila* proteins. In: Sullivan W, Ashburner M, Hawley RS (eds) *Drosophila* protocols. Cold Spring Harbor Laboratory Press, New York, NY, p 400

Chapter 2

Non-fluorescent RNA In Situ Hybridization Combined with Antibody Staining to Visualize Multiple Gene Expression Patterns in the Embryonic Brain of Drosophila

David Jussen and Rolf Urbach

Abstract

In *Drosophila*, the brain arises from about 100 neural stem cells (called neuroblasts) per hemisphere which originate from the neuroectoderm. Products of developmental control genes are expressed in spatially restricted domains in the neuroectoderm and provide positional cues that determine the formation and identity of neuroblasts. Here, we present a protocol for non-fluorescent double in situ hybridization combined with antibody staining which allows the simultaneous representation of gene expression patterns in *Drosophila* embryos in up to three different colors. Such visible multiple stainings are especially useful to analyze the expression and regulatory interactions of developmental control genes during early embryonic brain development. We also provide protocols for whole mount and flat preparations of *Drosophila* embryos, which allow a more detailed analysis of gene expression patterns in relation to the cellular context of the early brain (and facilitate the identification of individual brain neuroblasts) using conventional light microscopy.

Key words *Drosophila*, Embryonic brain, Neuroectoderm, Neuroblast identification, In situ hybridization, Antibody staining

1 Introduction

In situ hybridization and immunohistochemistry are fundamental and widely used methods in the field of developmental biology. They serve multiple purposes, such as the identification of a specific cell type or tissue by the visualization of marker genes, or the study of genetic interactions by the comparative analysis of gene expression patterns in different genetic backgrounds. Huge progress has been made in that field in the last decades. The invention of fluorescence-imaging techniques allows the detection of gene products in deeper layers of tissue, at consistently increasing resolutions. This also led to the development of staining protocols involving different fluorescent dyes, which allow the detection of multiple gene products at once [1].

Simon G. Sprecher (ed.), *Brain Development: Methods and Protocols*, Methods in Molecular Biology, vol. 1082, DOI 10.1007/978-1-62703-655-9_2,

However, fluorescence-imaging methods have limitations. Those concern, for example, the relatively low stability of fluorescence dyes hampering the long-term storage of specimens or the need of costly documentation systems.

Here, we present a protocol for non-fluorescent double in situ hybridization combined with antibody staining on *Drosophila* whole mount embryos, which allows the simultaneous visualization of transcripts and proteins in up to three different colors. This method combines the benefits of simultaneously labeling multiple factors at once, like it is popular in fluorescence microscopy, while allowing the easy handling of traditional, non-fluorescent labeling techniques. That way, a high amount of specimens can be analyzed under a conventional light microscope, without any need of fluorescent detection systems. Using differential interference contrast (i.e., Nomarski), histochemically labeled cells are visualized in context of the surrounding (uncolored) tissue, which often simplifies interpretation of gene expression patterns and their relation to specific cells without the need of additional markers (as, e.g., membrane markers used in fluorescence microscopy to visualize cell boundaries).

This protocol is based on the sequential detection of differentially labeled RNA probes via the alkaline phosphatase (AP)-catalyzed turnover of different chromogenic compounds. The precipitates formed in those reactions are insoluble in aqueous solution, which allows the elution of AP-bound antibodies after the chromogenic reaction without affecting the staining per se. By that, the AP-coupled antibody directed against the first probe can be eluted after the first chromogenic reaction, with the subsequent introduction of an AP-coupled antibody directed against the second probe. Here, we use NBT/BCIP and Vector Red as chromogenic compounds to sequentially detect DIG- and FITC-labeled probes, in combination with a subsequent antibody staining which is visualized by the peroxidase-catalyzed turnover of DAB. That way, three different gene products are visualized by blue, red, and brown precipitates, respectively.

Such stainings are particularly useful to analyze the expression and regulatory interactions of developmental control genes during the early period of embryonic brain development. In organisms as diverse as insects and mammals, the brain arises from multipotent neural stem cells (in insects called neuroblasts) which originate from the neuroectoderm. Products of developmental control genes are expressed in spatially restricted domains in the neuroectoderm and provide positional cues that regulate the development of neuroblasts [2, 3]. Therefore, the expression of these genes has to be precisely controlled. Recently, we uncovered a novel gene regulatory network in *Drosophila*, in which "DorsoVentral" and "AnterioPosterior" patterning genes interact to precisely control their spatially restricted expression in the brain neuroectoderm, which is essential for the correct formation and fate specification of neuroblasts [4–6]. These patterning genes display not only a high degree of intersegmental modulation but also considerable changes in their expression patterns over time. The staining technique presented here has been used to visualize multiple patterning genes at once in order to analyze their dynamic expression patterns in relation to each other and to uncover their genetic interactions and cellular functions. Non-fluorescent multiple stainings are also suitable to identify embryonic brain neuroblasts. The brain of *Drosophila* is built up by about 100 neuroblasts, which develop in a stereotypic pattern during the first third of embryonic development [7]. Brain neuroblasts can be identified individually by position, time point of formation, morphology, and the specific combination of marker genes they express [8]. As the identified marker genes are usually not selective enough, it is often necessary to stain for a combination of those in order to identify individual brain neuroblasts. Using this protocol, marker gene expression can be analyzed using conventional Nomarski optics. That way, labeled as well as unlabeled neuroblasts are visible alongside each other, which allows the identification of individual neuroblasts with the use of only a few molecular markers.

2 Materials

2.1 Reagents

1. Anti-digoxigenin (DIG)-AP, Fab fragments from sheep (Roche Applied Science).
2. Anti-fluorescein (FITC)-AP, Fab fragments from sheep (Roche Applied Science).
3. Methanol (≥99.9 %, p.a.).
4. ssDNA (10 mg/μl, MB grade, from fish sperm) (Roche Applied Science).

2.2 Buffers

1. AP buffer (50 ml): 41.2 ddH_2O, 1 ml 5 M NaCl, 2.5 ml 1 M $MgCl_2$, 5 ml 1 M Tris–HCl. Adjust pH to 9.5. Add 50 μl Tween®20.
2. Blocking buffer (100 ml): 10 ml 1 M Tris–HCl, 3 ml 5 M NaCl. Adjust pH to 7.5. Add 0.5 g blocking reagent (delivered with TSA Biotin System, PerkinElmer). Store aliquots at −20 °C.
3. Glycine buffer (50 ml): 50 ml ddH_2O, 0.038 g glycine, 1.46 g NaCl. Adjust pH to 2.3. Add 0.05 g BSA and 50 μl Triton™ X-100.
4. Hybridization buffer (50 ml): 12.5 ml SSC (20×), 12.5 ml DEPC-H_2O, 25 ml formamide, 50 μl Tween®20.
5. PBS (20× stock solution, 500 ml): 500 ml ddH_2O, 75.97 g NaCl, 9.94 g Na_2HPO_4, 4.14 g NaH_2PO_4. Adjust pH to 7.4. Dilute with ddH_2O to obtain 1× working solution.

6. PBT (0.3 %, 50 ml): 47.5 ml ddH_2O, 2.5 ml PBS (20×), 150 μl Triton™ X-100.
7. PBTween (0.1 %, 50 ml): 2.5 ml PBS (20×), 47.5 ml ddH_2O, 50 μl Tween®20 (use DEPC-treated PBS and ddH_2O for PBTween-DEPC).
8. Vector Red buffer (100 ml): 100 ml distilled ddH_2O, 1.21 g Tris–HCl. Adjust pH to 8.2–8.5. Add 100 μl Tween®20.
9. Washing buffer (100 ml): 10 ml 1 M Tris–HCl, 7.5 ml 2 M NaCl, 82.5 ml ddH_2O. Adjust pH to 7.5. Add 50 μl Tween®20.

2.3 Solutions and Media

1. ABC solution (avidin-biotinylated peroxidase complex): 1 ml PBT, 4 μl solution A, 4 μl solution B (both are delivered with the Vectastain ABC Kit). Prepare solution 1 h before use and keep on a shaker (100 rpm) at room temperature (RT).
2. AP staining solution: 1 ml AP buffer, 3 μl NBT stock solution, 1.5 μl BCIP stock solution. Prepare freshly before use.
3. Apple juice agar: 1,000 ml apple (or grape) juice, 28 g agar. Heat until the apple juice agar solution starts boiling and becomes clear. Dispense in fly culture vials or petri dishes (the latter requires additional fly cages). Once the agar has cooled and hardened, store at –4 °C.
4. BCIP stock solution: 10 ml dimethylformamide (100 %), 500 mg BCIP. Store 1 ml aliquots at –20 °C.
5. Chlorine bleach (6 %): Dilute sodium hypochlorite solution (12 % Cl) 1:1 with H_2O.
6. 3,3′-Diaminobenzidine tetrahydrochloride (DAB) stock solution: Dissolve 1 DAB tablet (10 mg; Sigma-Aldrich) in 35 ml PBT. Store 400 μl aliquots at –20 °C.
7. DAB staining solution: 400 μl DAB stock solution, 600 μl PBT, 2 μl H_2O_2. Prepare freshly before use.
8. DEPC-H_2O: 1,000 ml H_2O, 1 ml DEPC. Autoclave before use.
9. Fixative: 450 μl PBT, 70 μl formaldehyde (37 %), 600 μl n-heptane.
10. Glycerol (70/90 %). Dilute glycerol with PBS to obtain the particular concentration.
11. NBT stock solution: 10 ml dimethylformamide (70 %), 500 mg NBT. Store 1 ml aliquots at –20 °C.
12. Vector Red (VR) staining solution: 1 ml VR buffer, 16 μl VR Substrate Solution 1, 16 μl VR Substrate Solution 2, 16 μl VR Substrate Solution 3 (VR Substrate Solutions are delivered with the Vector Red Alkaline Substrate Kit I). Vortex ~5 s after addition of each VR Substrate Solution. Prepare freshly before use.

2.4 Kits

1. TSA™ Biotin System (PerkinElmer).
2. Vectastain ABC Kit (Standard) (Vector Labs).
3. Vector Red Alkaline Phosphatase Substrate Kit I (Vector Labs).

2.5 Equipment for Staining, Preparation, and Mounting

1. Aluminum foil (thickness 0.02–0.03 mm).
2. Preparation forceps (e.g., Dumont no. 5, Fine Science Tools).
3. Concavity slide.
4. Cover slips (18 × 18 mm, 22 × 22 mm, and 24 × 60 mm).
5. Microscope slides.
6. Minutien pins (stainless steel, 0.1 mm diameter, Fine Science Tools).
7. Nail polish (transparent).
8. 2 nickel-plated pinholders (Fine Science Tools).
9. Nylon mesh (120 μm, e.g., from Merck Millipore).
10. Small scalpel.
11. Spot plate (white).
12. Weighing dishes (white, 41 × 41 mm).
13. Whetstone (e.g., Sharpening stone for Dumont forceps, Fine Science Tools; alternatively abrasive paper, grain size 600–1,000).

2.6 Equipment for Microscopy

1. Cold light source (halogen or LED) equipped with fiber light guides (e.g., KL1500 by Schott).
2. Digital microscope camera (e.g., ProgRes® series, Jenoptik).
3. Upright light microscope (equipped with differential interference contrast and 40×–100× objectives, e.g., Axioscope Zeiss).
4. Stereo microscope (e.g., MZ series by Leica Microsystems).

3 Methods

General remarks: Wash steps and incubations are carried out at RT upon shaking (100 rpm) with 1 ml of solution used, unless stated differently. Reactions are carried out in standard 1.5 ml reaction tubes (Eppendorf tubes).

3.1 Embryo Collection and Dechorionization

1. Place flies on apple juice agar for egg laying (*see* **Note 1**).
2. Dechorionize embryos after egg laying by covering the apple juice agar with chlorine bleach (6 %) for 3 min. Slightly rotate once in a while. Dechorionized embryos will float up.
3. Collect embryos by transferring the chlorine bleach into an egg basket (*see* **Note 2**).
4. Wash embryos with water.

3.2 Fixation

1. Open the egg basket. Take embryos from the nylon mesh with a small scalpel and transfer them into an Eppendorf tube with fixative.
2. Fix embryos by vigorously shaking for 20 min (900 rpm).
3. Let the two phases of the fixative separate.
4. Remove the lower phase of the fixative without removing too many embryos.
5. Add 500 μl methanol.
6. Remove the vitelline membrane by vortexing at maximum speed for 2 min.
7. Let the devitellinized embryos sink down and remove lower phase of the solution.
8. Add 500 μl methanol.
9. Vortex for 1 min.
10. Let the embryos sink down and remove as much of the solution as possible without removing too many embryos.
11. Rinse 4× with methanol.
    - Embryos may be stored in methanol at –20 °C (*see* **Note 3**).

3.3 (Double) In Situ Hybridization

Use gloves, filtered tips, and DEPC-treated PBTween for the following in situ hybridization procedure:

1. Rinse 5× with PBTween.
2. Incubate 5 min in PBTween/hybridization buffer (1:1).
3. Incubate 5 min in hybridization buffer.
4. Perform prehybridization by incubating 1 h in hybridization buffer + ssDNA (1:100) at 55 °C upon shaking (300 rpm).
5. Perform hybridization by incubating in hybridization buffer + ssDNA (1:100) containing your probe (*both* of your probes for double in situ hybridization) at the appropriate working dilution(s) (*see* **Note 4**).
6. Incubate overnight at 55 °C upon shaking (300 rpm).
7. Incubate 30 min in hybridization buffer at 65 °C upon shaking (350 rpm).
8. Incubate 30 min in PBTween/hybridization buffer (1:1) upon shaking (350 rpm).
9. Wash 4 × 20 min with PBTween at 65 °C upon shaking (350 rpm).

   Standard tips and untreated PBTween may be used from now on.
10. Wash 10 min with PBTween at RT.

3.4 Detection of the First Probe (via NBT/BCIP)

1. Incubate 1.5 h in anti-DIG-AP *or* anti-FITC-AP (1:1,000 in PBTween) (*see* **Note 5**).
   - The antibody can alternatively be incubated overnight at 4 °C upon slight shaking.
2. Rinse 3× and wash 3 × 10 min with PBTween.
3. Incubate 2 × 5 min in AP buffer.
   - Prepare NBT/BCIP staining solution during that step.
4. Incubate in AP staining solution and transfer the solution containing the embryos into a white weighing dish (41 × 41 mm). Keep it cool and dark.
5. Judge the chromogenic reaction from time to time under a stereo microscope (*see* **Note 6**).
6. When staining intensity has reached the desired level, transfer the embryos back into the tube (*see* **Note** 7).
7. Rinse 3× with PBTween and terminate the reaction by incubating 10 min in methanol.
8. Rinse 3× with PBTween.
   - For double in situ hybridization combined with antibody staining, continue with Subheading 3.5.
   - For single in situ hybridization combined with antibody staining, continue with Subheading 3.7.
   - For double in situ hybridization without further antibody staining, continue with Subheading 3.9.

3.5 Antibody Elution and Inactivation of Residual AP

1. Rinse 1× and incubate 3 × 10 min in glycine buffer.
2. Rinse 3× with PBT.
3. Incubate 5 min in PBT/formaldehyde (10:1).

3.6 Detection of the Second Probe (via Vector Red)

1. Rinse 3× and wash 3 × 10 min with PBT.
2. Incubate overnight in anti-FITC-AP or anti-DIG-AP (1:1,000 in PBT; depending on the choice of the first antibody) at 4 °C.
3. Rinse 3× and wash 3 × 10 min with PBT.
4. Incubate 10 min in Vector Red buffer. Prepare Vector Red staining solution during that step.
5. Incubate in Vector Red staining solution. Transfer the solution containing the embryos into a white weighing dish (41 × 41 mm). Keep it cool and dark. Judge the chromogenic reaction (in this case, a red colored staining) from time to time under a stereo microscope (*see* **Note 8**).
6. Transfer embryos back into the tube.
7. Rinse 3× and wash 3 × 10 min with PBT.
8. Incubate 5 min in PBT/formaldehyde (10:1) (*see* **Note 9**).

3.7 Incubation with Primary Antibody

1. Rinse and wash 3 × 10 min in PBT.
2. Incubate overnight with primary antibody (at the appropriate working dilution in PBT) at 4 °C (*see* **Notes 10** and **11**). Depending on the quality of the antiserum, 2–4 h incubation at RT may be suitable, too.
3. Rinse 3× and wash 3 × 10 min with PBT.

3.8 Detection of Primary Antibody (via DAB)

1. Incubate 2–3 h with biotin-coupled secondary antibody (1:500 in PBT) RT.
2. Optional: Continue with Subheading 3.10 for Tyramide Signal Amplification (*see* **Note 12**). If not, prepare ABC before the following step. ABC has to incubate for at least 1 h before use upon shaking (100 rpm).
3. Rinse 3× and wash 3× 10 min with PBT.
4. Incubate 1 h in ABC.
5. Rinse 3× and wash 3 × 10 min with PBT. Prepare DAB staining solution during the last wash step.
6. Incubate in DAB staining solution. Transfer the solution containing the embryos into a white weighing dish (41 × 41 mm). Judge the chromogenic reaction under a stereo microscope. The staining should emerge quickly (~ 1–15 min).
7. When the intensity of the brown DAB staining has reached the desired level, transfer the embryos back into the tube.

3.9 Final Washing Steps

8. Rinse 3× and wash 10 min with PBT.
9. Rinse 2× with PBS.
10. Store in 70 % glycerol.

3.10 Optional: Tyramide Signal Amplification

1. Rinse 1× and wash 1 × 10 min with PBT.
2. Incubate 1 × 5 min and 1 × 30 min in blocking buffer.
3. Incubate 30 min in Streptavidin-HRP (1:500 in blocking buffer).
4. Wash 3 × 10 min with TSA washing buffer. Prepare ABC before the following step. ABC has to incubate for at least 1 h upon shaking (100 rpm) before use (*see* **step 4** of Subheading 3.8).
5. Incubate 5 min in 140 μl TSA reagent (1:70 in amplification diluent) (*see* **Note 13**).
6. Continue with **step 3** of Subheading 3.8.

3.11 Manufacturing of Preparation Needles

Two different kinds of needles are needed for the preparation of embryos. One will be used for handling the embryos (i.e., moving in glycerol, transferring onto microscope slide, holding during preparation; termed “handling needle”), while the other is used for cutting the tissue during preparation (termed “cutting needle”) (Fig. 1a). For that, the pinpoint of both needles will be processed differently.

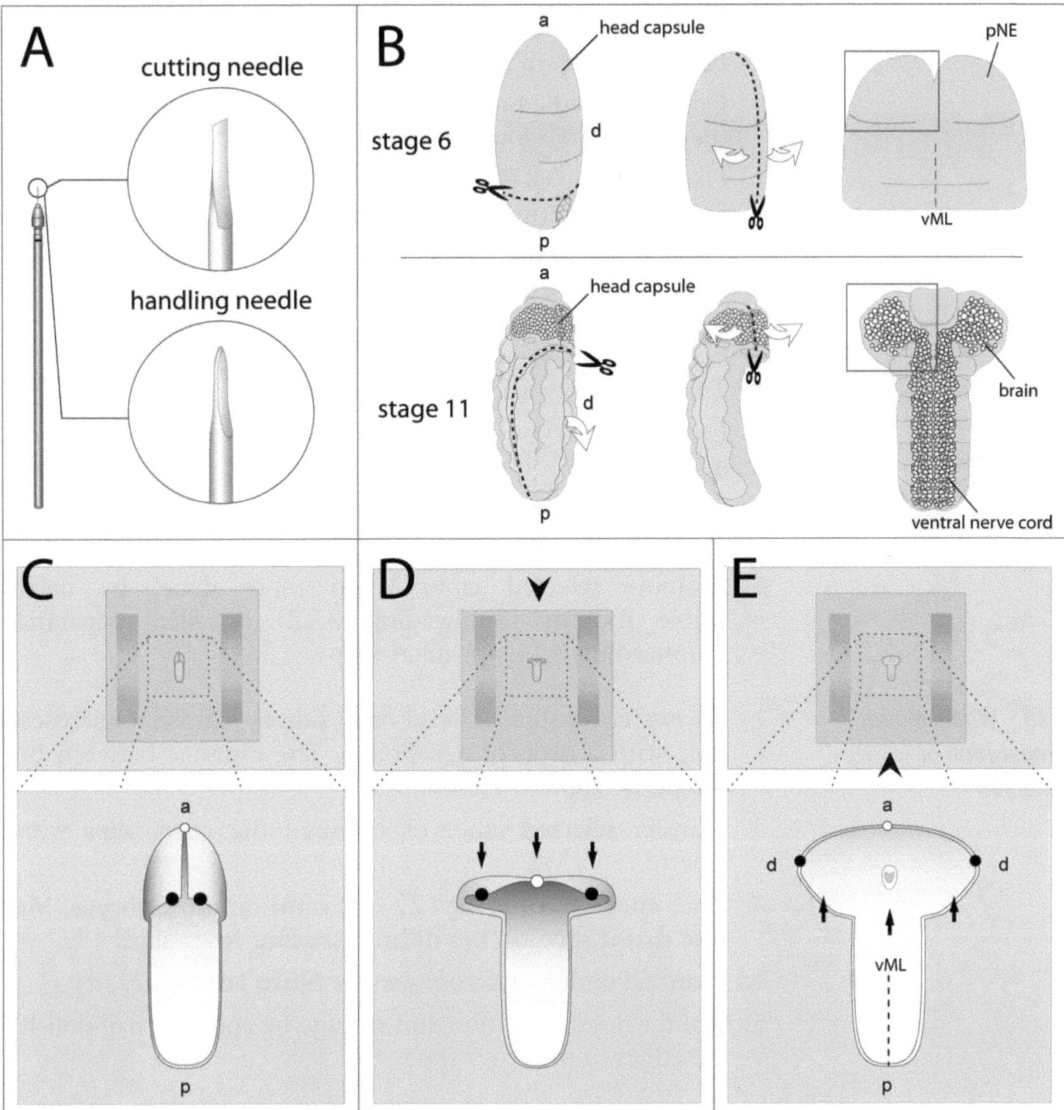

Fig. 1 Filet preparation of *Drosophila* embryos. (**a**) The cutting needle has a flattened tip with sharp edges, while the tip of the handling needle is rounded. (**b**) Main steps performing filet preparation of embryos at different stages. The posterior part of the abdomen is removed by cutting the embryo transversally. Then, the remaining ectoderm is cut along the dorsal midline towards the anterior tip of the embryo to open the head capsule. Finally, the ectoderm has to be flattened. (**c–e**) Instructions for flat preparations. *Black* and *white dots* indicate corresponding positions of the head capsule. *Arrowheads* indicate how to move the upper (small) cover slip. *Arrows* indicate the flow of glycerol caused by moving the cover slip. (**c**) An embryo with opened head capsule is positioned in glycerol between a large (24 × 60 mm) and a small (18 × 18 mm) cover slip (separated by aluminum spacers) with anterior facing forwards and ventral facing down. (**d**) The small cover slip is carefully moved in posterior direction (regarding the body axis of the embryo), which causes the hemispheres to erect. (**e**) The small cover slip is carefully moved back until the hemispheres are unfurled flatly on the cover slip. *a* anterior, *p* posterior, *d* dorsal, *vML* ventral midline, *pNE* procephalic neuroectoderm

1. Lock the dull end of a minutien pin tightly in a pinholder.
2. Process the tip of the needle under a stereo microscope.
3. Place the needle horizontally on a solid steel ground (e.g., on the flat end of a forceps).
4. Flatten the tip of the needle by placing the flat end of another forceps on the tip of the needle, slowly streaking away from it with some pressure (*see* **Note 14**).
5. Adjust both needles according to Fig. 1a with the aid of a whetting stone or fine abrasive paper (*see* **Note 15**).

3.12 Initial Examination of Embryos

1. Transfer embryos into a well of a white spot plate.
2. Examine embryos under a stereo microscope using maximum magnification.
3. Sort embryos with the handling needle according to staining quality, stage, and genotype of interest.
4. Examine selected embryos in more detail by making whole mounts (Subheading 3.13) or filet preparations (Subheading 3.14) for microscopy.

3.13 Whole Mount Preparation of Embryos

1. Fix two cover slips (22 × 22 mm) side by side on a microscope slide with a drop of nail polish. The distance between both should be approx. 15 mm.
2. Transfer selected embryos between the cover slips with a pipette.
3. Place another cover slip (22 × 22 mm) on the embryos. Make sure that it stays on top of both spacing cover slips.
4. Examine under a microscope (*see* **Note 16**; Fig. 2a, d)
5. Seal for documentation and storage by applying nail polish to the edges of the cover slip

3.14 Filet Preparation of Early to Mid-Stage Embryos

The following filet preparation is recommended for embryos until developmental stage 12 and is performed under a stereo microscope (*see* **Note 17**).

1. Fill the pit of a concavity slide with 70 % glycerol.
2. Transfer the selected embryo into the pit with the handling needle.
3. Place the embryo on the lateral side and hold it carefully with the handling needle.
4. Remove the posterior part of the abdomen and cut along the dorsal midline from posterior to anterior with a cutting needle to open the head capsule (Fig. 1b).
5. Spread the hemispheres slightly with both needles or a Dumont forceps.

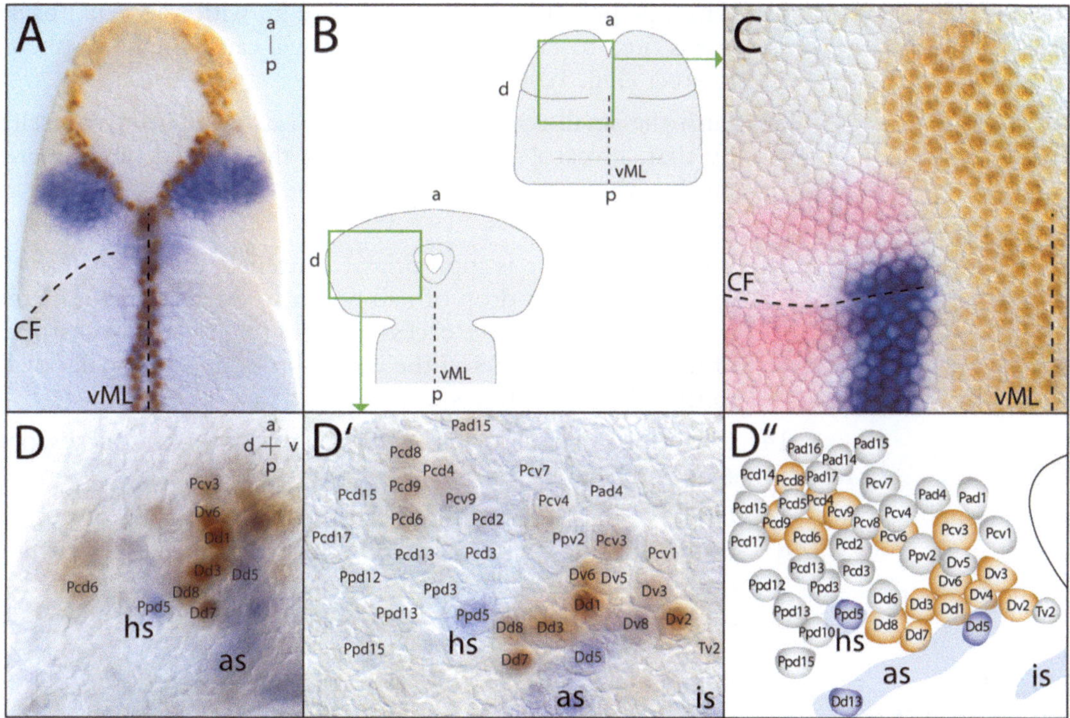

Fig. 2 Non-fluorescent multiple stainings of the head neuroectoderm and brain neuroblasts in whole mounts and flat preparations. (**a**) Ventral view on the head region of a stage 8 whole mount embryo (anterior is up) stained for *Nkx6*-mRNA (via NBT/BCIP, *blue*) and single-minded protein (via DAB, *brown*). (**b**) Schematic representation of the details presented in C/D′. (**c**) View on the head region (left hemisphere) of a filet prepared stage 6 embryo with almost all neuroectodermal cells in one plane: *blue* is *ind*-mRNA (via NBT/BCIP), magenta is *msh-mRNA* (via Vector *Red*) and *brown* is Vnd-Protein (via DAB). (D/D′) Early brain neuroblast pattern of a late stage 9 embryo as whole mount (**d**) and after filet preparation of the same embryo (D′), stained against Engrailed (via NBT/BCIP, *blue*) and *svp*-LacZ (via DAB, *brown*). Note that filet preparation allows a more clear view, and thus the identification of individual neuroblasts by marker expression and relative position (as illustrated in D″) according to [7, 8]. *a* anterior, *p* posterior, *v* ventral, *d* dorsal, *CF* cephalic furrow, *vML* ventral midline, *hs engrailed* head spot, *as engrailed* antennal stripe, *is engrailed* intercalary stripe. D-D″ Adapted from [7]

6. Remove excess yolk carefully with the cutting needle while holding the embryo with the handling needle.
7. Arrange two spacers (i.e., strips of aluminum foil, approx. 2 × 15 mm) in parallel at a distance of 10–12 mm in the middle of a large cover slip (24 × 60 mm).
8. Add 45 μl glycerol (90 %) between the aluminum spacers.
9. Transfer the prepared embryo with the handling needle into the glycerol drop and push it carefully to the ground, with anterior forwards and ventral downwards.
10. Place a small cover slip (18 × 18 mm) on the glycerol drop. And wait until the glycerol is spread between the two cover slips (Fig. 1c).

11. Carefully pull the small cover slip (e.g., with a forceps) backwards (i.e., towards yourself). This will cause the hemispheres to erect (Fig. 1d).
12. Push the small cover slip forwards (i.e., away from yourself). By that, glycerol will float towards the erected hemispheres causing them to unfurl flatly on the cover slip (Fig. 1e; *see* **Note 18**).
13. When the embryo is appropriately flattened, fix the small cover slip by applying a drop of nail polish to every corner. Let the nail polish dry, remove excess glycerol (e.g., with small stripes of tissue paper), and seal the cover slip by applying nail polish to the edges (two times).
14. Clean the cover slip from glycerol and fix it on a microscope slide with small stickers.
15. Under a microscope, the filet preparation can be examined from both sides (with a view from ventral or dorsal) by turning the cover slides including the filet preparation (Fig. 2c, D′; *see* **Note 19**).

3.15 CNS Preparation of Stage 17 Embryos

1. Fill the pit of a concavity slide with 70 % glycerol.
2. Transfer the selected embryo into the pit.
3. Place the embryo with the dorsal side up and hold it carefully at the anterior end with the handling needle.
4. Cut the epidermis from the posterior tip of the embryo along the dorsal midline towards the anterior tip with the cutting needle.
5. Open the embryo by moving the epidermis aside with the preparation needles or a Dumont forceps.
6. Remove the non-neural tissue that lies on top of the CNS (i.e., guts and fat body).
7. Carefully detach the CNS from the epidermis.
8. Continue as in **steps 7–10** of Subheading 3.14.
9. Push the small cover slip slightly forwards. The glycerol flow should push the brain hemispheres forwards in front of the ventral nerve cord (*see* **Note 20**).
10. When the CNS is appropriately aligned, fix the small cover slip by applying a drop of nail polish to each corner. Let the nail polish dry and seal the cover slip by applying nail polish to the edges (two times).
11. Clean the cover slip from glycerol and fix it on a microscope slide with small stickers.
12. Under a microscope, the CNS can be examined from both sides (with a view from ventral or dorsal) by turning the cover slips including the filet preparation.

3.16 Analysis of the Developing Pattern of Brain Neuroblasts

The development of brain neuroblasts occurs in a stereotypic spatiotemporal pattern, with every neuroblast expressing a unique combination of marker genes. Both have been described in detail for *Drosophila* [7, 8], which allows the identification of individual brain neuroblasts. The brain neuroblast pattern is best examined in filet preparation of embryos from late stage 8 to late stage 11 using 100× magnification and differential interference contrast. We recommend (combinatorial) antibody stainings for distinct molecular markers (particularly against Deadpan, *svp*-LacZ, and Engrailed protein) to analyze the patterns of brain neuroblasts in more detail. Deadpan (Dpn) is a general neuroblast marker. In early developmental stages (i.e., late stage 8–stage 9), Dpn antibody may even be sufficient to identify single neuroblasts according to their position within the entire brain neuroblast pattern [7]. From stage 9 onwards, identification of single brain neuroblasts becomes increasingly difficult with general markers. We recommend the use of *svp-LacZ* enhancer trap line (available at Bloomington Stock Center, Bloomington, Indiana, USA). The expression pattern of *svp*-LacZ lines is stable, well described, and persists throughout embryonic brain development. *svp*-LacZ marks a subset of (about 40) brain neuroblasts in a characteristic pattern, which covers all three main subdivisions of the brain (i.e., proto-, deuto-, and tritocerebrum). By that, it can be used to identify the *svp*-LacZ-labeled neuroblasts within the entire pattern of brain neuroblasts but also the adjacent (unlabeled) ones (Fig. 2D-D″) according to ref. 7. In any case, it is helpful to combine stainings with Engrailed, which is segmentally expressed in the posterior part of each subdivision of the brain neuroectoderm (*engrailed head spot*—protocerebrum, *engrailed antennal stripe*—deutocerebrum, *engrailed intercalary stripe*—tritocerebrum) and in distinct subsets of brain neuroblasts that emerge from the corresponding Engrailed neuroectodermal domains (Fig. 2D-D″). These Engrailed-positive brain neuroblasts additionally serve as reference points for orientation within the Dpn- or *svp*-LacZ-labeled neuroblast patterns [7, 8]. The (combinatorial) use of the markers mentioned above may in most cases be sufficient to identify single brain neuroblasts. However, further markers are listed in [8].

4 Notes

1. The desired embryonic stage, and thus time window of egg collections, depends on the process to be examined and has to be chosen according to [9]. For example, the analysis of patterning in the neuroectoderm requires early developmental stages (starting at stage 5), while the analysis of NB formation requires stages of mid-embryogenesis (stages 8–11). For the

former, we suggest egg collections of 3–4 h time windows, while for the latter, 8–16 h time windows are suitable (at 25 °C).

2. Egg baskets can easily be made by cutting a 50 ml falcon tube in half, cutting a hole into the cap and screwing a piece of nylon mesh between tube and cap (*see* ref. [10]).
3. Note that long-term storage of fixed embryos may affect the staining quality. Some antibodies yield best stainings with freshly fixed embryos.
4. For double in situ hybridization, one probe has to be labeled with DIG and the other one with FITC (the use of different labels like biotin and dinitrophenol may be suitable, too). The dilution at which the probes work best must be tested before. We suggest testing probes by performing a standard in situ hybridization procedure (as described above) for a serial dilution of probes (e.g., 1:50, 1:500, 1:2,500) on wild-type embryos. Probes should be detected via NBT/BCIP, since this method is very sensitive and easy. The optimal dilution is reached when the specific signal develops clearly without background staining (which may take up to 2 h or longer). If this is not the case, dilute the probe further. Higher dilutions may prolong the staining reaction. However, a clear signal should be preferred, since background staining of the different colorimetric reactions adds up and can decrease the overall staining quality significantly. Probes may be reused multiple times when stored at –20 °C. Often, the best results are reached when probes have been used before, since less unspecific bindings occur.
5. The choice of the first antibody (anti-DIG-AP or anti-FITC-AP) depends on the probe to be detected first. The first detection step is the generally less demanding and more sensitive NBT/BCIP reaction. It results in a deep blue precipitate, which is per se richer in contrast alongside the whitish tissue of the embryo. For those reasons, the inferior probe should be detected first. Here, only AP-dependent reactions are used for probe detection. Other enzymes used for colorimetric detection are peroxidases (using DAB as chromogen) or beta-galactosidase (using X-beta-D-Gal, Salmon-beta-D-Gal, or Magenta-beta-D-Gal as chromogen). However, we found that the peroxidase-dependent reactions produce relatively poor signal/background ratio with this protocol, while the beta-Gal-dependent reactions have been reported to be little sensitive (*see* ref. [11]); this report also contains information on the use of further AP substrates.
6. The AP reaction may take from ~5 min up to several hours, depending on the probe used and the amount of transcript present. The staining solution should be replaced every hour. Usually, staining patterns start to emerge during the first 15 min.

The reaction should be terminated as soon as a) the relative amount of background vs. specific staining rises markedly or b) the specific signal becomes too strong. The optimal staining intensity also depends on the context. When co-labeling of certain cells is expected (e.g., when a general NB marker is combined with an NB subset marker), the staining should be terminated earlier, since strong stainings may mask the presence of further dyes. Apart from that, it is recommended to terminate stainings relatively late to assure the representation of low transcript levels. Note that the NBT/BCIP staining will appear slightly lighter after the incubation in methanol and changes to a blue color since the red component of the initially purple reaction product becomes washed out.

7. Embryos can be transferred into the tube by slightly tilting the dish and flushing the embryos into one corner with PBTween. From there, they can be sucked up with a pipette.
8. Vector Red staining reactions may take up to several hours. The staining solution should be replaced every hour. The expression pattern usually starts to emerge during the first 30 min. The reaction can be slowed down by incubating at 4 °C. This may allow overnight incubation. Note that none of the dye will be washed away during subsequent steps of this protocol. Therefore, the staining reaction should be terminated as soon as the staining intensity is judged optimal.
9. This fixation step inactivates AP. Methanol is not suitable for terminating the Vector Red reaction, because the red precipitate is (at least partially) soluble in methanol. Efficient AP inactivation is especially important if further NBT/BCIP stainings are planned (e.g., for the detection of balancer expression) to prevent cross-reactions.
10. Many primary antibodies can be used in solutions containing 1 % sodium azide; this prevents contamination with fungi or bacteria and is advantageous as an antibody can be reused several times often with increasingly better staining results. However, keep in mind that some antibodies can be impaired or do not work at all when using sodium azide; therefore we recommend to compare stainings in which the antibody has been incubated in PBT and in PBT with sodium azide.
11. Additional antibodies for balancer detection can be added to the primary antibody solution. Balancers can be detected via DAB in parallel to the actual primary antibody. However, we recommend detecting balancers via AP and NBT/BCIP, since this method is more sensitive.
12. Tyramide Signal Amplification increases the staining intensity of DAB stainings. This amplification step may not be necessary with good antisera.

13. The parameters of this signal amplification step are particularly critical and may have to be adapted. Prolonged incubation and inappropriately high concentrations of TSA reagent quickly lead to high background levels during the DAB staining reaction.
14. The cutting needle has to be flattened more intensely than the handling needle to obtain a sharp edge for preparation.
15. Add a drop of glycerol on the whetstone/abrasive paper. Prepare the needle under the stereo microscope by gently streaking it over the glycerol-covered surface of the stone/paper. The cutting needle has to be sharpened to obtain an even blade with somewhat rectangular edges, while the handling needle should be whetted in a way to obtain a thicker tip with rounded edges to prevent damaging of embryos during preparation.
16. Embryos can be examined from different angles by carefully moving the upper cover slide over the spacers, which will cause the embryos to rotate in the glycerol. By that, different regions of the embryonic brain can be examined in one specimen.
17. Flat preparations, though time consuming in production, allow the most detailed examination of gene expression pattern, even on the level of single cells (e.g., neuroblasts) [7, 8]. Yolk and other tissue which often accumulate unspecific staining are removed, and the whole neuroectoderm is presented in one plane, which significantly improves differential interference contrast. That way, neuroectodermal cells and neuroblasts, including the unstained, can be easily detected.
18. This step may need some exercise. The position of the embryo may be corrected by sliding the small cover slip in different directions. Avoid squeezing the embryo by pushing the cover slip down.
19. We recommend 40× or 63× objectives for the examination of gene expression in the neuroectoderm and 63× or 100× objectives for the analysis of the NB pattern. Stainings can be documented on a microscope equipped with a standard digital (CCD/CMOS) camera. For digital documentation, we recommend recording selected regions of flat preparations. Depending on the magnification used, a stack of 5–10 pictures along the *z*-axis is usually enough to cover all cells of the embryonic brain (i.e., neuroectoderm and all underlying neuroblasts of one hemisphere).
20. The hemispheres of the late embryonic brain lie on top of the ventral nerve cord, which hampers the view during microscopy. After this step, the brain hemispheres should lie in front of the ventral nerve cord, so that the whole embryonic CNS is arranged in a straight line.

Acknowledgments

The authors thank Janina Seibert and Dagmar Volland for sharing protocols and considerable expertise, Janina Seibert for Fig. 2c, and Karoline F. Kraft for critically reading the manuscript. This work was supported by grants from the Deutsche Forschungsgemeinschaft (UR163/2-1 and UR163/3-1).

References

1. Kosman D, Mizutani CM, Lemons D, Cox WG, McGinnis W, Bier E (2004) Multiplex Detection of RNA Expression in Drosophila Embryos. Science 305:846
2. Skeath JB, Thor S (2003) Genetic control of Drosophila nerve cord development. Curr Opin Neurobiol 13:8–15
3. Urbach R, Technau GM (2004) Neuroblast formation and patterning during early brain development in Drosophila. Bioessays 26: 739–751
4. Urbach R, Volland D, Seibert J, Technau GM (2006) Segment-specific requirements for dorsoventral patterning genes during early brain development in Drosophila. Development 133:4315–4330
5. Seibert J, Volland D, Urbach R (2009) Ems and Nkx6 are central regulators in dorsoventral patterning of the Drosophila brain. Development 136:3937–3947
6. Seibert J, Urbach R (2010) Role of en and novel interactions between msh, ind, and vnd in dorsoventral patterning of the Drosophila brain and ventral nerve cord. Dev Biol 346: 332–345
7. Urbach R, Schnabel R, Technau GM (2003) The pattern of neuroblast formation, mitotic domains and proneural gene expression during early brain development in Drosophila. Development 103:3589–3606
8. Urbach R, Technau GM (2003) Molecular markers for identified neuroblasts in the developing brain of Drosophila. Development 103: 3621–3637
9. Campos-Ortega J, Hartenstein V (1997) The embryonic development of Drosophila melanogaster. Springer, Berlin Heidelberg
10. Rothwell WF, Sullivan W (2000) Fluorescent analysis of Drosophila embryos. In: Sullivan W, Ashburner M, Hawley RS (eds) Drosophila protocols, 1st edn. CSHL Press, Cold Spring Harbor, New York, pp 143–145
11. Hauptmann G (2001) One-, two-, and three-color whole-mount in situ hybridization to Drosophila embryos. Methods 23:359–372

Chapter 3

Analysis of Complete Neuroblast Cell Lineages in the Drosophila Embryonic Brain via DiI Labeling

Karoline F. Kraft and Rolf Urbach

Abstract

Proper functioning of the brain relies on an enormous diversity of neural cells generated by neural stem cell-like neuroblasts (NBs). Each of the about 100 NBs in each side of brain generates a nearly invariant and unique cell lineage, consisting of specific neural cell types that develop in defined time periods. In this chapter we describe a method that labels entire NB lineages in the embryonic brain. Clonal DiI labeling allows us to follow the development of a NB lineage starting from the neuroectodermal precursor cell up to the fully developed cell clone in the first larval instar brain. We also show how to ablate individual cells within a NB clone, which reveals information about the temporal succession in which daughter cells are generated. Finally, we describe how to combine clonal DiI labeling with fluorescent antibody staining that permits relating protein expression to individual cells within a labeled NB lineage. These protocols make it feasible to uncover precise lineage relationships between a brain NB and its daughter cells, and to assign gene expression to individual clonal cells. Such lineage-based information is a critical key for understanding the cellular and molecular mechanisms that underlie specification of cell fates in spatial and temporal dimension in the embryonic brain.

Key words *Drosophila*, Embryonic brain, Neural stem cell, Neuroblast lineage, DiI labeling, Antibody staining

1 Introduction

Proper functioning of the brain relies on an enormous diversity of neural cells generated by neural stem cells, called neuroblasts (NBs) in insects. In *Drosophila*, the central nervous system (CNS) including the brain develops from a bilaterally symmetrical sheet of neuroectodermal cells. It gives rise to a fixed number of NBs that segregate to the interior of the embryo. Upon segregation, a NB typically undergoes rounds of asymmetric cell division, budding off smaller intermediate precursor cells, which usually divide once to produce two postmitotic cells. Each NB produces specific neural cell types in defined time periods and, by that, generates a nearly invariant and unique cell lineage, suggesting stereotyped patterns of lineage development [1]. The fate of an individual NB and,

Simon G. Sprecher (ed.), *Brain Development: Methods and Protocols*, Methods in Molecular Biology, vol. 1082,
DOI 10.1007/978-1-62703-655-9_3,

accordingly, the main features of its developing cell lineage are specified by positional cues within the neuroectoderm [2, 3] and, thereafter, by temporal cues as well as by a combination of further developmental control genes that it expresses [1, 4–7]. Many studies have shown that knowledge about the precise lineage relationships between a NB and its daughter cells is a critical key to investigate the molecular mechanisms that underlie specification of cell fates in spatial and temporal dimension [5, 8–13]. Such lineage-based information on the developmental origin of a NB and the neuron types it develops will not only decipher the sophisticated circuitry of the brain but also elucidate how a complex brain develops. Therefore, analysis of NB lineages with high resolution is needed, permitting a systematic identification of neural cell types by resolving the development of each single cell in a NB lineage.

To unravel NB lineages in the fly brain, especially in the postembryonic period of development, very straightforward genetic clonal labeling techniques have been established, such as FLP/FRT-based (e.g., [14, 15]), Gal4-based G-TRACE [16], or MARCM-based methods [17–20]. However, due to technical limitations, these genetic labeling techniques seem to be less applicable to the elucidation of the entire NB lineages in the embryo. As the production of FLP/FRT clones depends on heat shock-flippase necessary to induce recombination, during embryogenesis critical levels of heat shock-flippase may not become enriched before the early-born part of a lineage has developed [21]. Also, MARCM fails to disclose the entire NB lineages in the embryo, since after recombination (which depends also on a critical level of heat shock-flippase), the clonal reporter expression additionally relies on the loss of the GAL80 repressor, which seems to persist over the entire embryonic development [22]. To our knowledge, only clonal DiI labeling undoubtedly reveals the entire lineage of embryonic brain NBs. Clonal DiI labeling is highly selective and noninvasive. It is based on the application of a lipophilic fluorescent dye onto early neuroectodermal cells. The dye is easily absorbed by the membrane of the cell and diffuses through the lipids quickly without being transmitted to neighboring cells (*see* also ref. [23]). There are several different versions of carbocyanine dyes available (e.g., the most commonly used DiI, DiO, and DiD) with different fluorescent properties. DiI, for example, has similar excitation properties to rhodamine; excited by green light, it fluoresces red. Depending on requirement, a number of alternative versions of the classical lipophilic carbocyanine dyes have been developed over the last few years, e.g., a chloromethylbenzamino derivate of DiI (CM-DiI), which shows a better staining persistence after standard tissue fixation procedures. As lipophilic carbocyanine dyes are entirely atoxic [24], the labeled cell can develop completely unaffected by the dye according to its position-dependent fate.

Clonal DiI labeling allows in vivo to study the development of all the progeny cells derived from a single-labeled neuroectodermal precursor cell (e.g., processes of cell proliferation and differentiation) and, after fixation, to analyze the cellular composition (i.e., glial or neuronal cells) and overall morphology of the fully differentiated cell clone. By using this technique, the embryonic lineage of all neuroblasts in the ventral nerve cord (VNC) has been described previously [25–27]. Recently, clonal DiI labeling in combination with molecular markers and cell ablation experiments allowed us to analyze the embryonic development of a prominent central brain structure, the mushroom bodies (MBs). We unraveled the origin of the four mushroom body neuroblasts (MBNBs), their mode of formation, and could clarify the spatiotemporal development and individual cellular composition of their embryonic lineages [28].

Here we describe a protocol that was basically developed by Bossing and Technau [29] and modified by us to allow targeted DiI labeling of brain NB clones. Our methodical modifications describe also how to selectively ablate cells within a labeled NB clone, which, for example, reveals informations about the temporal succession at which daughter cells are generated. Finally, we describe how to combine clonal DiI labeling with fluorescent antibody staining that permits relating protein expression to individual cells within a DiI labeled NB clone.

2 Materials

2.1 Reagents

1. DiI (1,1′-dioctadecyl-3,3,3′,3′-tetramethylindocarbocyanine perchlorate; DiI, CM-DiI, DiD, and other carbocyanine tracers can be obtained at Molecular Probes).
2. DiI/oil (5 mg/ml).
3. CM-DiI/oil (3 mg/ml).
4. DiD/oil (10 mg/ml).
5. Apple juice agar: Boil 29 g agar in 1 l apple juice until mixture is translucent. Decant into Petri dishes as long as the mixture is hot; let cool.
6. PBS (1 l 20× stock solution): Solve 151.92 g NaCl, 19.88 g Na_2HPO_4 (dehydrated), and 8.28 g NaH_2PO_4 in ddH_2O, pH 7.4.
7. PBTween (0.1 % Tween20 in PBS).
8. Heptane glue: Mix Cello 31.39 (Tesa) with heptane in a 2:1 ratio.
9. DAB solution: Dissolve 2–3 mg/ml DAB in 100 mM Tris–HCl, pH 7.4.

10. Fixatives: For antibody staining after DiI labeling, mix 200 μl 37 % formaldehyde in 900 μl 1× PBS. For photoconversion, mix 260 μl 37 % formaldehyde in 900 μl 1× PBS.
11. Glycerol 70 %: Dissolve 70 ml 100 % glycerol in 30 ml 1× PBS.
12. Fluorocarbon oil Voltalef 10S.
13. Fetal calf serum.

2.2 Equipment

1. Inverse microscope (e.g., Leitz Fluovert FU) equipped with different fluorescent filters (Fluorescein FITC, Rhodamin, etc.).
2. 100 W halogen lamp.
3. 100 W mercury lamp.
4. Stereomicroscope.
5. Cold light source.
6. Capillary grinder (e.g., Bachofer Type 462).
7. Capillary puller (e.g., Sutter P97).
8. Micromanipulator (e.g., Leitz M).
9. Bunsen burner.
10. Cover slips (22 × 22 mm, 24 × 60 mm).
11. Glass slide (76 × 26 × 1 mm).
12. Plasticine.
13. Borosilicate glass microcapillaries (Hilgenberg normal "Meterware").
14. Glass cutter.
15. Wet chamber (e.g., Petri dish containing wet pieces of filter paper).
16. Petri dish (60 × 20 mm).
17. Scalpel.
18. Preparation needle.
19. Fine forceps (e.g., Dumont 5SF).
20. Black block dish.
21. Single-use syringe (5 or 10 ml).
22. Polyethylene tube (6 × 4 × 1 mm).
23. Transparent nail polish.
24. Tape (single sided, double sided).
25. Dissecting knife.
26. Cage (60 × 70 mm).
27. Weighing tray (41 × 41 × 8 mm).

3 Methods

3.1 The Workplace

1. We recommend performing clonal DiI labelings, if possible, on a microscopic stage with controllable thermocouple element, or alternatively, in a temperature-controlled room (18–20 °C), so that development of the embryos is decelerated. This prolongs the time period in which successful labeling can be obtained. At warmer temperatures, the viability of treated embryos declines and precise timing of DiI labeling becomes more difficult.
2. Labeling and observation are carried out on an inverted microscope.
3. Ideally, to minimize vibration, the microscope and micromanipulator carrying the DiI-filled capillary are posed on a heavy-weight balance table or comparable equipment.
4. The microscope needs to be equipped with exchangeable halogen and mercury lamps and appropriate filter settings capable of exciting the respective dye (for detailed informations on different dyes and filter settings, see "Molecular Probes Handbook of Fluorescent Probes and Research Chemicals") and GFP.

3.2 Preparation of Fluorescent Dye/Oil Mixture

1. The fluorescent, lipophilic dye is mixed at a given concentration (*see* Subheading 2.1) with any commercially available vegetable oil (e.g., canola or soya oil).
2. Shake slowly (ca. 200 rpm) for 3–5 h in the dark at room temperature. The aliquots (~10 μl) can be stored at −20 °C for several years.

3.3 Preparation of Polyethylene Tube

1. Take a piece of polyethylene tube (ca. 8 cm) and heat it briefly over a small flame of a Bunsen burner (until the polyethylene starts "bubbling").
2. Pull apart slowly from both ends until the tube reaches a length of about 50 cm.
3. Cut one end of the tube at appropriate position so that the inner diameter fits the capillary (Fig. 1c).

3.4 Manufacturing Labeling Capillaries

1. Capillaries most suitable for DiI labeling have the following characteristics: (a) a relatively long shank and (b) a relatively small tip diameter. The outer diameter at the distal tip of the capillary should be around 3–5 μm (Fig. 1d) (*see* **Note 1**).
2. Grind the capillary on a commercial capillary grinder in an angle of ca. 30°; the resulting capillary will be sharp enough to easily penetrate the vitelline membrane. For best results, the capillary should be wet-ground. In the following, this capillary type will be called "labeling capillary," whereas a pulled but unground capillary, needed for transfer and preparation of embryos, will be called "flat capillary."

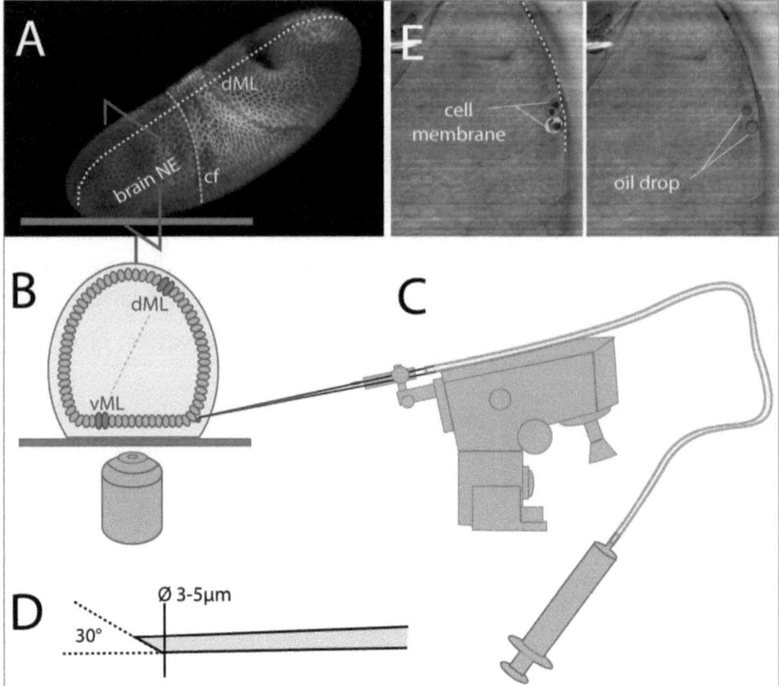

Fig. 1 Set up for DiI labeling. (**a**) Gastrula-stage embryo (at stage 7) glued to the cover slip for DiI labeling in the brain neuroectoderm. *cf* cephalic furrow; *dML* dorsal midline. (**b**) Schematic cross section through the head region boxed in (**a**); the embryo is fixed on a cover slip in appropriate orientation to label cells in the ventral or intermediate neuroectoderm (NE) under an inverse microscope. *vML* ventral midline. Modified from [29]. (**c**) Micromanipulator equipped with a labeling capillary coupled with a polyethylene tube which is connected to a syringe. (**d**) The angle of the bevel of the labeling capillary should be around 30°, and the outer diameter at the distalmost tip between 3 and 5 µm. (**e**) Labeling of two neighboring neuroectodermal cells with DiI (*magenta*) and DiD (*green*). *Dashed line* indicates the adhesive border (color figure online)

3. Under visual control through a stereomicroscope, slowly lower the capillary until its tip touches the abrasive wheel of the grinder. The tip is open when you see water climbing up into the capillary. Note, grinding does not take longer than a few seconds.
4. Mark the untreated, upper end of the labeling capillary with a water-resistant pen while it is still clamped in the grinder holder. This will help you later to fix the capillary in the micromanipulator in the right orientation.
5. To evaluate the inner diameter of the capillary tip, couple a polyethylene tube (as prepared above, *see* Subheading 3.3) to the capillary. Insert the syringe at the other end of the tube and apply pressure while the tip of the capillary is under distillated water

filled in a black block dish: the capillary is useful for labeling if you can hardly detect the escaping small air bubbles under intense light. For ablation of cells (*see* Subheading 3.16), we recommend using capillaries with a slightly larger tip opening.

6. Using the syringe, draw acetone into the capillary and eject it. The labeling capillary is then dry and ready for storage.
7. Store the labeling capillaries dust-free in a Petri dish; fix each capillary on a thin strip of plasticine. After labeling or ablation, the capillary can be restored when before rinsed with acetone.

3.5 Preparing Heptane Glue-Covered Cover Slips

1. In order to prepare a glue-coated cover slip (24 × 60 mm), spread a drop of heptane glue with the help of a small cover slip (22 × 22 mm) so that the cover slip becomes coated homogenously with a thin film of glue, and let it dry for at least 10 min (Fig. 2a) (*see* **Note 2**).
2. Cut frames out of single-sided tape (width 25 mm). The border of a frame should be ~5 mm and the opening ~15 mm wide (Fig. 2a). Stick the frame onto the glue-coated cover slip (*see* **Note 3**).

3.6 Egg Collection and Staging

1. At least 2 days before you start collecting embryos for DiI labelings: Place the flies (between 4 and 14 days old) in a fly cage on an apple juice plate with yeast. Change the plate every 24 h (*see* **Note 4**).
2. For DiI labeling, precisely staged embryos within a range of 1 h can be obtained when changing the plate every hour at 25 °C.

3.7 Dechorionization and Mounting of Embryos for DiI Labeling

1. Fix a piece of double-sided tape to a cover slip (22 × 22 mm) (Fig. 2b).
2. Take cooled, fresh apple juice agar plates, and cut the agar into blocks of 2 × 2 cm. Place two blocks of agar in a distance of about 1.5 cm on a glass slide (76 × 26 mm), and the cover slip on top, leaving a part of each agar block exposed (Fig. 2b1).
3. Make 3 lines of 5–7 holes each, by pressing a preparation needle into the agar blocks at an angle of about 20° (Fig. 2b2).
4. Transfer and scatter a few embryos onto the tape with a dissecting knife (Fig. 2b3).
5. Wait about 3 min to let the chorion dry.
6. Collect eggs at the blastoderm stage and dechorionate by rotating the embryo slightly along the dorsoventral axis with the tip of a preparation needle. The chorion breaks and releases the embryo (*see* **Note 5**).
7. Carefully pick up each dechorionated embryo and transfer it to one of the agar blocks to prevent further drying (Fig. 2b4).

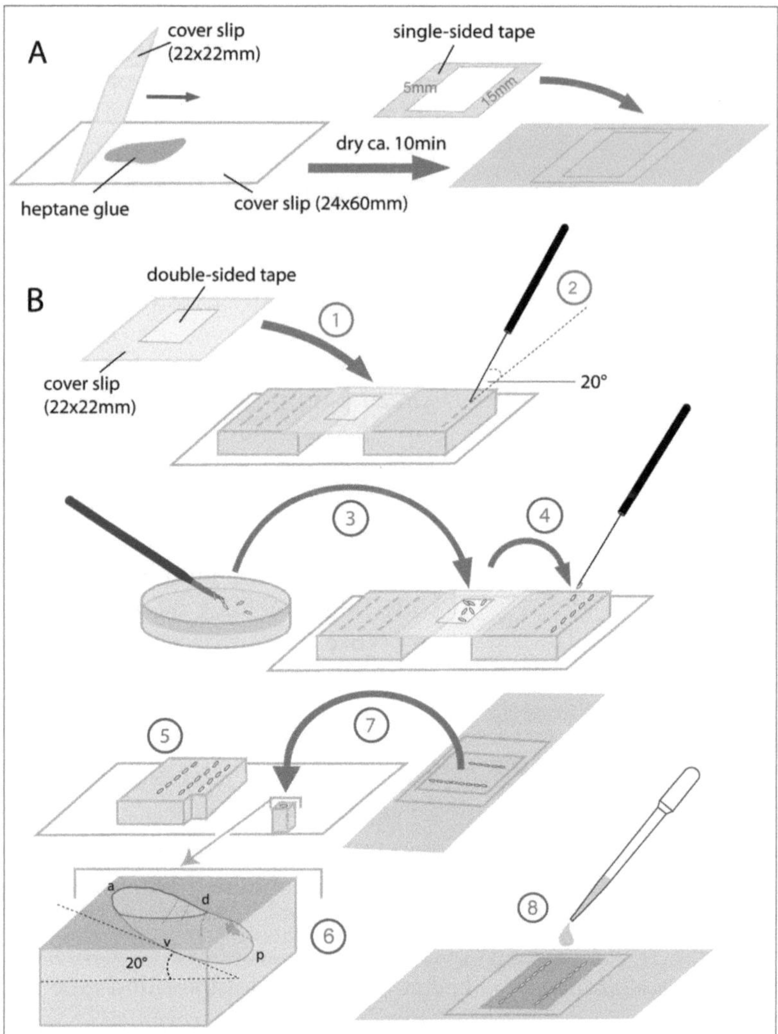

Fig. 2 Preparation of embryos for DiI labeling. (**a**) Spread a drop of heptane glue on a cover slip. Let the heptane glue dry for at least 10 min. Stick a single-sided tape frame on the glue-coated cover slip. (**b**) (1) Fix a piece of double-sided tape on a cover slip. Place the cover slip on top of two blocks of agar. (2) Make holes into the agar. (3) Transfer embryos onto the tape. Dechorionate embryos. (4) Transfer each embryo on an agar block and orientate them. (5) Cut up the agar blocks so that every embryo is on an individual block. (6) Check for optimal orientation. (7) Fix the embryo on the prepared cover slip. (8) When embryos are properly dried, cover them with Voltalef 10S oil. *a* anterior, *p* posterior, *d* dorsal, *v* ventral

8. For DiI labelings in the brain neuroectoderm, the anteroposterior axis of the embryo should be orientated perpendicular to the injection capillary. Transfer embryos into the prepared holes in the agar block, so that the posterior half of the embryo is sticking in the hole, and the procephalon is

facing you (Fig. 2b6). Each embryo has to be orientated accurately with respect to the dorsoventral axis depending on the area of the brain neuroectoderm you want to label in (Fig. 2a) (*see* **Note 6**).

9. Using a scalpel, separate embryos on smaller agar blocks, each embryo on a single block (Fig. 1b5).
10. Now, stick the part of the procephalon facing you on the cover slip. Carefully lower a prepared cover slip (*see* below and Subheading 3.5) onto a single block and press slightly in order to fix the embryo in proper orientation (Fig. 1b7) (*see* **Note 7**).
11. Repeat this process until all embryos are stuck next to each other (head-to-tail) to the prepared cover slip.
12. Let embryos desiccate for 5–12 min at room temperature (*see* **Note 8**).
13. As soon as the embryos are dried properly, cover them with a few drops of fluorocarbon oil Voltalef 10S (Fig. 2b8). Fluorocarbon oil (like halocarbon oil) allows oxygen to permeate to the embryo but prevents water from evaporating.
14. Until the beginning of injection, the cover slip is kept in a wet chamber at room temperature.

3.8 Filling the Capillary with Dye Solution

1. Put the unbeveled end of the capillary (as indicated by the pen mark, *see* Subheading 3.4) into the polyethylene tube, and a single-use syringe on the other end of the tube.
2. Fix the capillary with the topside up (as indicated by the pen mark) in the holder of a micromanipulator (Fig. 1c).
3. Place a drop of 1 μl dye/oil on a cover slip (24 × 60 mm) and bring both the dye drop and the capillary tip under the microscope in focus using a 10× objective.
4. Now change to a 50× objective, and slowly draw up the dye into the capillary.
5. As soon as the tip is filled with dye, the syringe has to be removed from the polyethylene tube, the tip of the capillary to be lifted up, and the syringe, with the plunger pulled out, to be inserted again into the tube.
6. To reuse the drop of dye later, you can store the cover slip in a dark, dry, and dust-free box.

3.9 Labeling of Individual Neuroectodermal Cells and Determination of Their Position

1. Place the cover slip with the embryos under the microscope.
2. If the embryos are properly desiccated, there should be a relatively sharp adhesive border around the area where each embryo is stacked to the glue and several rows of neuroectodermal cells should lie in one focal plane (Fig. 1b).

3. To identify the optimal time point for labeling and the position of the cell to be labeled, it is useful to have at least a 600× or higher magnification (e.g., using a 50× immersion oil objective and a 12.5× ocular magnification). Optionally, the labeling can be controlled on a monitor using a standard digital camera (CCD/CMOS).
4. Check under the microscope the developmental stage of the embryos. By stage 7 (staging according to [30]) you can detect, for example, the dorsal and ventral midline, the cephalic furrow (*see* **Note 9**). The latest time point for successful labeling seems to be around stage 9.
5. Determine the neuroectodermal cell to be labeled by making use of the above mentioned, morphological landmarks that help to recognize the position of individual cells (*see* **Note 9**).
6. Via the micromanipulator, lower the tip of the capillary until it is in focus, but still in a certain distance to the embryo. From now on, do not work with the micromanipulator anymore, as all movements should be effected via the microscope table.
7. Penetrate the vitelline membrane slowly and carefully with the tip of the capillary in the position where the neuroectodermal cell of interest is located, and approach the tip as closely as possible towards its cell membrane (*see* **Note 10**).
8. By slowly pushing the plunger of the syringe, deposit a small drop of dye next to this cell. The diameter of the applied drop is critical for successful labeling: about 1/2–1/5 of the diameter of a neuroectodermal cell seems to be most promising (Fig. 1e). Immediately after depositing the drop, the dye diffuses into the cell membrane leaving behind the nonfluorescent drop of oil solvent.
9. Slowly, pull the capillary out of the embryo.
10. For double labeling in the same embryo, place two capillaries filled with different dyes (each on a different micromanipulator) as close as possible to the embryo before labeling. If no additional micromanipulator is available, you can also quickly exchange the two capillaries between the labeling.
11. After labeling, briefly inspect the quality of the labeling under fluorescent light using a halogen lamp (*see* **Note 11**).
12. Finally, control the position of the labeled cell via DIC optics.
13. Take notes of each labeling concerning position and behavior of the labeled cell (*see* Subheading 3.10).
14. Store the labeled embryos on the cover slip in a wet chamber at 18–20 °C.

3.10 Control of Cell Behavior After Labeling

1. About 1 h after labeling (around late stage 8/early stage 9), judge the behavior of the labeled cell using the same fluorescence and magnification settings (*see* **Note 12**).
2. If the neuroectodermal cell has delaminated, the NB was successfully labeled. To follow the development of the NB lineage in more detail, it is possible to inspect the labeling in shorter time intervals (e.g., every 30 min) without disturbing the labeled cells.
3. The identification of a labeled NB clone is facilitated when labeling is done in embryos of appropriate Gal4 strains (most can be ordered in the Bloomington or Kyoto Stock Center) in which the Gal4 pattern is visualized via a GFP reporter. Alternatively, you can combine your labeling with an antibody staining (*see* Subheading 3.15).
4. For further development, the embryos are kept under fluorocarbon oil Voltalef 10S at 18 °C until they reach the desired stage. To obtain fully developed stage 17 NB clones, embryos are kept at 18 °C overnight.
5. Before further preparation, monitor the embryos and eliminate all those which do not show properly labeled cell clones.

3.11 Preparation of Embryos at Developmental Stage 17 (See Fig. 3)

1. For better handling, we recommend removing the parts of the cover slip outside of the single-sided tape frame using a glass cutter.
2. Tip the cover slip at an ~80° angle for several minutes so that the fluorocarbon oil can drain off.
3. Then, place the cover slip in a weighing tray (41 × 41 × 8 mm) and spread about 1 ml heptane on the leftover fluorocarbon oil in order to remove it.
4. Rock the cover slip slowly for about 10–20 s. Do not rock too long since the heptane will also solve the glue and the embryos will detach.
5. Quickly place the cover slip on a glass slide for better handling.
6. Pour away the remaining heptane carefully. Wait a few seconds until the heptane is almost completely evaporated and cover the embryos with PBS immediately to prevent drying.
7. Get the embryos, one by one, out of the vitelline membrane using a flat capillary.
8. Dissect the CNS out of the embryo by placing one fine forceps at the anterior end of the embryo and another one in its middle/posterior part. Carefully lacerate the embryo to make the CNS visible. Remove all remaining tissue from the CNS.

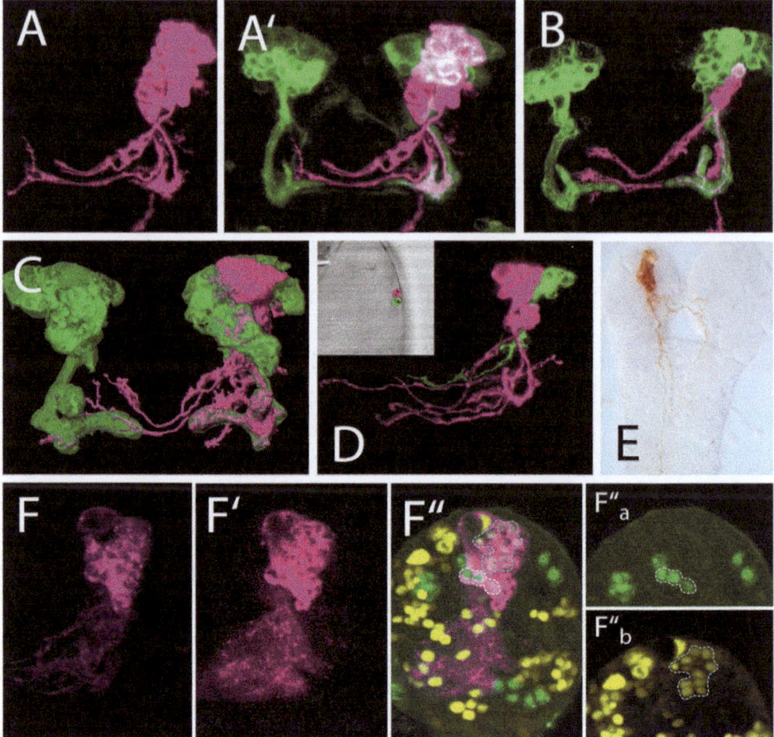

Fig. 3 Different applications of clonal DiI labeling. (**a**) Combined confocal images of a DiI labeled mushroom body NB clone (**a**) in a Ok107 (ey)-Gal4; UAS-CD8::GFP brain (**a′**) at mid/late stage 17. Note that some cells of the DiI labeled clone express GFP, indicating that this is a mushroom body NB clone. (**b**) DiI labeled mushroom body NB clone at late stage 17 after ablation of the NB at stage 11. Accordingly, only the early-born part of its lineage has developed. (**c**) 3D reconstruction of a DiI labeled mushroom body NB clone (*magenta*) in the early first larval instar in comparison to the structures of the entire mushroom body (*green*) as revealed by Ok107 (ey)-Gal4; UAS-CD8::GFP expression. (**d**) A mushroom body NB clone (*magenta*) and another protocerebral NB clone (*green*) at late stage 17 obtained after labeling of two neighboring cells with DiI (*magenta*) and DiD (*green*) (see inset). A–D Adapted from [28]. (**e**) Photoconverted protocerebral NB clone at early stage 17. (**f**) DiI labeled protocerebral NB clone (at early first larval instar) before (**f**) and after (**f′**) fluorescent antibody staining against the gene products Eyes absent (**f″**a) and Retinal homeobox (**f″**b). Note that antibody staining leads to partial loss of DiI signal. (**f″**) Merge

9. Use the flat capillary to transfer each CNS to a glue-coated cover slip (24 × 60 mm) prepared with a tape frame (as explained in Subheading 3.5) that is filled with PBS.
10. Stick the ventral nerve cord with its ventral side on the glue so that the brain hemispheres are orientated upwards.
11. The preparation can be stored in a wet chamber at 4 °C until fixation and documentation.

3.12 Preparation of First Instar Larvae

After about 30 h at 18 °C, embryos are checked for eclosion under a stereomicroscope in intervals of about 15 min. Take notes on which larva hatches at which position on the cover slip, as this information is necessary to make correlations to the notes you took during the labeling (*see* **Note 13**).

1. Immediately after eclosion, take the larvae out of the oil by using a flat capillary. Transfer each larva separately to a glue-coated cover slip prepared with a tape frame (as explained in Subheading 3.5) filled in with PBS.
2. To prepare the CNS, tear the larva in half with a pair of fine forceps. Then turn the cuticle "inside out" and the CNS will appear attached to the cuticle. Remove the CNS entirely from the tissue, transfer it with a capillary to another glue-coated cover slip prepared with a tape frame (as explained in Subheading 3.5) filled with PBS, and stick the ventral nerve cord with its ventral side on the glue.
3. The preparation can be stored in a wet chamber at 4 °C. If you want to photoconvert the clone continue as described in Subheading 3.17, otherwise go on with fixation (*see* Subheading 3.13).

3.13 Fixation

1. Remove the PBS under visual control using a stereomicroscope, and replace it immediately with fixation solution (*see* Subheading 2.1). Note, as fixatives are toxic, take precautions (e.g., place the stereomicroscope on a portable fume hood).
2. Fix 20 min for subsequent antibody staining (as described in Subheading 3.15), otherwise 10 min.
3. Thereafter, carefully replace the fixative with PBS.
4. Change the PBS 4–5 times.

3.14 Documentation

1. Fix the cover slip (24 × 60 mm) carrying the preparation on a glass slide with tape for better handling.
2. Remove excessive PBS so that the border of the liquid slightly overtops the single-sided tape frame (*see* **Note 14**).
3. Now carefully place a cover slip (22 × 22 mm) onto the single-sided tape frame.
4. By slightly moving the upper cover slip, the brain hemispheres that get in contact with this cover slip (as explained in Subheadings 3.11 and 3.12) can be brought into the best possible position for exposure.
5. Fix the cover slip to the underlying cover slip by applying nail polish at the corners.

6. Immediately document the labeled NB clone using a confocal microscope. Transport the specimen in a wet chamber (*see* **Note 15**). Optionally, 3D reconstructions can be generated from the LSM stacks using, for example, Amira or comparable software.

3.15 Optional: DiI Labeling Combined with Fluorescent Antibody Stainings

As far it was only possible to combine photoconverted DiI labelings with nonfluorescent histochemical antibody stainings (visualized, e.g., via the alkaline phosphatase reaction) with the disadvantage that the photoconverted cells themselves, which have accumulated strong levels of diaminobenzidine (DAB), could not be antibody stained. Now this limitation can be avoided by combining labeling, performed with the improved fixable DiI derivatives (e.g., CM-DiI), with fluorescent antibody stainings. To carry out fluorescent antibody stainings with fluorescent DiI labeled clones, another critical issue is the proper usage of detergents, needed for antibodies to penetrate the membranes of cell tissues. As the carbocyanine dyes are lipophilic, conventional detergents in widely used concentrations (such as 0.1–0.3 % Triton X-100 or 0.05 % saponin) result in extensive diffusion of the DiI label out of the tissue (own observations; *see* also ref. [31]). Presented below is a protocol in which we apply a low concentration of 0.1 % Tween20, and in addition, mechanically perforate the neurilemma that envelops the brain.

1. After fixation, to facilitate the penetration of the antibodies into the brain tissue, carefully perforate the neurilemma with a flat capillary. Perforate generously but do not damage the area containing the labeled clones.
2. Thereafter, cover the specimen for 48 h with primary antibody/antibodies dissolved in 0.1 % PBTween (~200 μl). Put the cover slip in a wet closed chamber at 4 °C.
3. Then take off the primary antibody/antibodies and replace with 0.1 % PBTween twice.
4. Cover three times for 20 min with 0.1 % PBTween.
5. Dissolve the secondary antibody/antibodies (take in consideration that you use appropriate fluorescent conjugates which do not interfere with applicated carbocyanine dyes) in 0.1 % PBTween and incubate for 18–20 h at 4 °C. In case of weak antibody stainings, a Tyramide Signal Amplification System (TSA) is recommended (according to the manufacturers' protocol of PerkinElmer, Waltham, USA).
6. After taking off the secondary antibody/antibodies, replace with 0.1 % PBTween twice and then cover for 15 min with 0.1 % PBTween twice.
7. Subsequently, wash twice with PBS.
8. Finally, put a cover slip on the tape frame, fix with nail polish, and follow up with documentation immediately as the DiI signal fades relatively fast (*see* Subheading 3.14).

3.16 Optional: Ablation of DiI Labeled NBs and/or Daughter Cells

1. For cell ablations, use embryos that have been labeled as described in Subheading 3.9 and have developed into the desired stage. Use a microscope equipped with a 100 W halogen lamp and appropriate filter settings. Bring the specimen in which cell ablations will be done under a sufficient magnification (600× or higher) in focus. Connect the ablation capillary (*see* Subheading 3.4) with the tube, then clamp the capillary into the micromanipulator and lower it into the Voltalef oil. Finally, connect the syringe to the other end of the tube to prevent Voltalef oil entering the tip.
2. Identify the labeled cell(s) to be ablated under the halogen lamp.
3. Using the micromanipulator, move the capillary towards the embryo under transmitted light.
4. Slowly and carefully move the cross table of the microscope towards the capillary tip until the capillary penetrates the vitelline membrane, and then carefully direct it to the cell(s) designated for ablation.
5. Once the tip of the capillary reaches the cell to be ablated, cautiously pull the syringe plunger in order to suck off the cell(s) (*see* **Note 16**).
6. Remove the capillary from the embryo by moving the cross table.
7. Store the treated embryos in a wet chamber at 18 °C until they have reached the desired developmental stage.
8. Then prepare, fix, and document the specimen as described in Subheadings 3.11–3.14.

3.17 Optional: Photoconversion of DiI Labeled Preparations

Photoconversion transforms the DiI label into a permanent reaction product via the Maranto reaction in which illumination releases a singlet oxygen that oxidizes DAB [32] (*see* also ref. [33]). DiI labeled photoconverted clones can be investigated using conventional light microscopy and preserved for many years.

1. Incubate for 1 h in calf serum.
2. Exchange the drop of PBS with a drop of DAB solution (that has been briefly centrifuged before). Note, DAB is a hazardous chemical—handle with appropriate precautions!
3. Photoconvert the labeled NB clone under the microscope with a 100 W mercury lamp and rhodamine filter using a water or oil immersion objective with about 50× magnification. Photoconversion is completed when all fluorescent signal is gone, this takes usually 10–20 min. Check regularly to make sure that the preparation is not over dyed.
4. After photoconversion, remove the DAB.
5. Rinse the preparation with PBS 3–4 times.

6. Fix in formaldehyde solution for 15 min.
7. Rinse 3–4 times with PBS.
8. Place a drop (20–25 μl) of 70 % glycerol on another cover slip (24 × 60 mm).
9. Using a flat capillary, transfer the preparation to the glycerol drop and bring it in appropriate orientation.
10. Carefully cover the glycerol drop containing the CNS with a small cover slip (18 × 18 mm). Control the orientation of the preparation under a stereomicroscope while the glycerol spreads between both glass slides. Excess glycerol can be removed with tissue paper.
11. Fix the cover slip by applying a drop of nail polish to all four corners. Let those drops dry and seal all sides two times with nail polish. The photoconverted NB clones can now be documented using Nomarski optics and a 63× (or 100×) immersion oil objective, as camera lucida drawings, or can be digitized. For example, with the help of a standard digital camera (CCD/CMOS), a motorized microscope table, and appropriate documentation software, it is useful to take a sequence of individual pictures in *z*-axis (the distance between different focal planes should be 1 μm). Saving each focal plane in Tiff format, you can merge all files to a film sequence using, e.g., QuickTime Player; this often facilitates understanding the spatial complexity of brain NB clones.

4 Notes

1. Glass capillaries for DiI labeling can be produced by any commercial pipette puller, e.g., Sutter P97 which allows both the heat and the strength to be varied. Change the settings of the puller according to the operating manual until you get ideal needles (*see*, e.g., the comprehensive guide to pulling capillaries on the website of Sutter instruments http://www.sutter.com/contact/faqs/pipette_cookbook.pdf).
2. Spread the heptane glue out carefully so that you have one thin but continuous layer of glue. If there is not enough glue, the embryos will not stick to the cover slip. However, if you apply too much glue, the embryo sinks into the glue and does also not stick properly. In addition, the capillary will be clogged by glue.
3. Make sure the frame is big enough so that there is a distance of at least 4 mm between the embryos and the inner edges of the frame. Otherwise the capillary cannot be placed near enough to the embryo. Fix the frame carefully without producing air bubbles. Otherwise the oil or PBS will later flow off the cover slip. After all embryos are fixed on the cover slip, place it onto a glass slide for better handling.

4. You do not need a vast amount of individuals, but as you will only work with embryos from a 1 h collection at 25 °C, you need to make sure that your flies lay enough eggs. In order to guarantee efficient labeling, it is important that you have sufficient numbers of precisely staged embryos. Of course, the collection schedule is temperature dependent.
5. We do not recommend chemical dechorionization as according to our experience this seems to cause misdevelopment or damage of embryos at higher rates.
6. The orientation of the embryo on the agar block is critical for the accessibility of the capillary to the neuroectodermal area one wants to label in. Orientate the embryos on the agar block by slightly moving them with the tip of the preparation needle. Consider that the adhesive border of the attached part of the neuroectoderm is close to the cell you want to label when you stick the embryo to the cover slip (Fig. 1b). For example, if you plan to label cells in the ventral neuroectoderm of the brain, the embryo should be placed sagittally so that the adhesive border is close to the ventral head midline (Fig. 1a). Accordingly, if you want to label in the dorsal neuroectoderm, orientate the embryo in the opposite direction. It is also important for the viability of the embryo that the ectodermal region that sticks to the glass surface is not too large, as otherwise the embryo does not develop any further.
7. While mounting the embryos, on the one hand, it is important to press carefully and vertically so that the embryos are neither damaged nor altered in position, on the other hand, you need to press slightly onto the cover slip so that several rows of neuroectodermal cells come to lie in one focal plane.
8. The average time for desiccation depends highly on temperature and humidity. Control the drying of the embryos under a stereomicroscope. Observe the embryos under a cold light source until you note two reflecting stripes on each side of the ventral midline. These stripes will disclose the level of desiccation: at the beginning, these stripes are well defined and smooth, while after a few minutes, they show tiny folds; once they are visible, the right level of desiccation has been achieved.
9. To perform targeted labeling of neuroectodermal cells in the brain, it is necessary to make use of morphological landmarks which help to recognize the position of individual neuroectodermal cells: easily identifiable are the cephalic furrow, the dorsal, and ventral midline. It is also helpful to use the stereotypical pattern of procephalic mitotic domains as morphological landmarks, in addition to the abovementioned landmarks. The procephalic ectoderm can be subdivided into several mitotic domains, which are characterized as discrete groups of cells that synchronously enter mitosis [34]. Each mitotic domain has a distinct time of entry into mitosis and discloses a specific shape.

Mitotic domains (except domain B) are only recognizable during their period of mitosis but not before or thereafter. Almost all procephalic mitotic domains have already completed mitosis (by early/mid-stage 8) before they give rise to first NBs. Recently, we were able to establish mitotic domain B, positioned in the central area of the brain neuroectoderm, as a useful landmark for a systematical DiI labeling of neuroectodermal cells, and could show that, among others, the NBs of the mushroom body originate there [28]. Mitotic domains can be visualized in vivo, for example, in *ubi-GFPnls* embryos: cells of the mitotically inactive domain B (continuously showing nuclear GFP) can be distinguished from the surrounding mitotically active domains which transiently loose nuclear GFP [28].

10. Once the tip of the dye-filled capillary has entered the embryo, you must completely avoid movements along the anteroposterior axis of the embryo. Otherwise you will destroy tissue or the labeled cell and risk misdevelopment of the cell clone.
11. You must use a halogen lamp for the inspection. Using a mercury lamp, due to the problem of phototoxicity, will lead to apoptosis of the labeled cells!
12. Check the behavior of the cell after labeling. The mode of brain NB formation reveals some differences depending on the neuroectodermal region a NB originates. For example, except in mitotic domain B, cells in all other parts of the brain neuroectoderm undergo mitosis before first NBs emerge. If a neuroectodermal cell divides in parallel to the ectodermal surface (typical for mitotic domain 1 or 5), then it may produce a NB and an epidermoblast, each generating a subclone, one becoming located in the later brain, the other in the epidermis (perhaps mapping in a sensory organ). If the neuroectodermal cell divides perpendicular to the ectodermal surface, then it may belong to mitotic domain 9. If the cell divides perpendicular (and asymmetrical) but in subectodermal position, then this is likely to be a NB producing a ganglion mother cell. It is also helpful to make notes on the time point of NB formation (indicating if it is an early or late developing NBs). For further details on different modes of brain NB formation, *see* [35].
13. According to our experience, DiI labelings can be followed until the first hours of the first larval instar since the lipophilic dye gradually dilutes, and thus fades, in the growing membrane of developing neuronal/glial cells.
14. Do not embed the preparations in pure glycerol nor in other glycerol-based mounting mediums (e.g., Vectashield). Lipophilic dyes, such as DiI, dissolve in glycerol. Embedding can only be done in PBS without glycerol. However, due to the evaporation of PBS, every supplement should be prepared individually and immediately prior to the exposure under microscope.

15. As DiI signal may become weaker over time, it is often useful to document a labeled NB clone before and after the antibody staining. Because of the relatively long procedure, you risk to lose the signal.
16. For cell ablations, halogen lamps with rhodamine filters and transmitted light at once are recommended. In order to recognize the DiI fluorescent signal of the cell(s) to be ablated, and simultaneously, the surrounding tissue, reduce the intensity of the transmitted light to an appropriate low level.

Acknowledgments

The authors thank Thomas Kunz, Martin Steimel, and Christof Rickert for sharing protocols and considerable expertise and David Jussen for critically reading the manuscript. We are grateful to Gerd Technau for his general support. This work was supported by grants from the Deutsche Forschungsgemeinschaft (UR163/2-1 and UR163/3-1) and by a research stipend to K.F.K. from the FTN of the University Mainz.

References

1. Technau GM, Berger C, Urbach R (2006) Generation of cell diversity and segmental pattern in the embryonic central nervous system of Drosophila. Dev Dyn 235:861–869
2. Skeath JB (1999) At the nexus between pattern formation and cell-type specification: the generation of individual neuroblast fates in the Drosophila embryonic central nervous system. Bioessays 21:922–931
3. Urbach R, Technau GM (2004) Neuroblast formation and patterning during early brain development in Drosophila. Bioessays 26:739–751
4. Jacob J, Maurange C, Gould AP (2008) Temporal control of neuronal diversity: common regulatory principles in insects and vertebrates? Development 135:3481–3489
5. Kao C-F, Lee T (2010) Birth time/order-dependent neuron type specification. Curr Opin Neurobiol 20:14–21
6. Lin S, Lee T (2012) Generating neuronal diversity in the Drosophila central nervous system. Dev Dyn 241:57–68
7. Skeath JB, Thor S (2003) Genetic control of Drosophila nerve cord development. Curr Opin Neurobiol 13:8–15
8. Baumgardt M, Karlsson D, Terriente J et al (2009) Neuronal subtype specification within a lineage by opposing temporal feed-forward loops. Cell 139:969–982
9. Jefferis GS, Marin EC, Stocker RF et al (2001) Target neuron prespecification in the olfactory map of Drosophila. Nature 414:204–208
10. Lai S-L, Awasaki T, Ito K et al (2008) Clonal analysis of Drosophila antennal lobe neurons: diverse neuronal architectures in the lateral neuroblast lineage. Development 135: 2883–2893
11. Pearson BJ, Doe CQ (2004) Specification of temporal identity in the developing nervous system. Annu Rev Cell Dev Biol 20:619–647
12. Yu H-H, Lee T (2007) Neuronal temporal identity in post-embryonic Drosophila brain. Trends Neurosci 30:520–526
13. Yu H-H, Kao C-F, He Y et al (2010) A complete developmental sequence of a Drosophila neuronal lineage as revealed by twin-spot MARCM. PLoS Biol 8
14. Golic KG, Lindquist S (1989) The FLP recombinase of yeast catalyzes site-specific recombination in the Drosophila genome. Cell 59: 499–509
15. Xu T, Rubin GM (1993) Analysis of genetic mosaics in developing and adult Drosophila tissues. Development 117:1223–1237
16. Evans CJ, Olson JM, Ngo KT et al (2009) G-TRACE: rapid Gal4-based cell lineage analysis in Drosophila. Nat Methods 6: 603–605

17. Lai S-L, Lee T (2006) Genetic mosaic with dual binary transcriptional systems in Drosophila. Nat Neurosci 9:703–709
18. Lee T, Luo L (1999) Mosaic analysis with a repressible cell marker for studies of gene function in neuronal morphogenesis. Neuron 22:451–461
19. Lee T (2009) New genetic tools for cell lineage analysis in Drosophila. Nat Methods 6: 566–568
20. Yu HH, Chen CH, Shi L et al (2009) Twin-spot MARCM to reveal the developmental origin and identity of neurons. Nat Neurosci 12:947–953
21. Larsen C, Shy D, Spindler SR et al (2009) Patterns of growth, axonal extension and axonal arborization of neuronal lineages in the developing Drosophila brain. Dev Biol 335:289–304
22. Luo L (2005) A practical guide: single-neuron labeling using genetic methods. In: Yuste R, Konnerth A (eds) Imaging in neuroscience and development. A laboratory manual. Cold Spring Harbor Laboratory, New York, NY, pp 99–110
23. Honig MG, Hume RI (1986) Fluorescent carbocyanine dyes allow living neurons of identified origin to be studied in long-term cultures. J Cell Biol 103:171–187
24. Honig MG, Hume RI (1989) Dil and DiO: versatile fluorescent dyes for neuronal labelling and pathway tracing. Trends Neurosci 12:333–340
25. Bossing T, Udolph G, Doe CQ et al (1996) The embryonic central nervous system lineages of Drosophila melanogaster. I. Neuroblast lineages derived from the ventral half of the neuroectoderm. Dev Biol 179: 41–64
26. Schmidt H, Rickert C, Bossing T et al (1997) The embryonic central nervous system lineages of Drosophila melanogaster. II. Neuroblast lineages derived from the dorsal part of the neuroectoderm. Dev Biol 189:186–204
27. Schmid A, Chiba A, Doe CQ (1999) Clonal analysis of Drosophila embryonic neuroblasts: neural cell types, axon projections and muscle targets. Development 126:4653–4689
28. Kunz T, Kraft KF, Technau GM, Urbach R (2012) Origin of Drosophila mushroom body neuroblasts and generation of divergent embryonic lineages. Development 139: 2510–2522
29. Bossing T, Technau GM (1994) The fate of the CNS midline progenitors in Drosophila as revealed by a new method for single cell labelling. Development 120:1895–1906
30. Campos-Ortega JA, Hartenstein V (1997) The embryonic development of Drosophila melanogaster. Springer, Berlin
31. Elberger AJ, Honig MG (1990) Double-labeling of tissue containing the carbocyanine dye DiI for immunocytochemistry. J Histochem Cytochem 38:735–739
32. Maranto AR (1982) Neuronal mapping: a photooxidation reaction makes Lucifer yellow useful for electron microscopy. Science 217:953–955
33. von Bartheld CS, Cunningham DE, Rubel EW (1990) Neuronal tracing with DiI: decalcification, cryosectioning, and photoconversion for light and electron microscopic analysis. J Histochem Cytochem 38:725–733
34. Foe VE (1989) Mitotic domains reveal early commitment of cells in Drosophila embryos. Development 107:1–22
35. Urbach R, Schnabel R, Technau GM (2003) The pattern of neuroblast formation, mitotic domains and proneural gene expression during early brain development in Drosophila. Development 130:3589–3606

Chapter 4

Flybow to Dissect Circuit Assembly in the *Drosophila* Brain

Nana Shimosako, Dafni Hadjieconomou, and Iris Salecker

Abstract

Visualization of single neurons within their complex environment is a pivotal step towards uncovering the mechanisms that control neural circuit development and function. This chapter provides detailed technical information on how to use *Drosophila* variants of the mouse Brainbow-2 system, called Flybow, for stochastic labeling of cells with different fluorescent proteins in one sample. We first describe the genetic strategies and the heat shock regime required for induction of recombination events. This is followed by a detailed protocol as to how to prepare samples for imaging. Finally, we provide specifications to facilitate multichannel image acquisition using confocal microscopy.

Key words *Drosophila*, Brainbow, Multicolor cell labeling, Genetics, Immunostaining, Confocal laser scanning microscopy

1 Introduction

Different neuron subtypes have elaborate axonal and dendritic processes with distinct shapes, which are indicative of their specific functions in neural circuits. Techniques labeling single neurons within the context of their complex surroundings are thus much needed additions to the neurobiologist's toolbox. In 2007, Livet et al. [1] pioneered the Brainbow system, a genetic multicolor cell labeling approach for mice, which makes it possible to visualize neurons and glia in different hues by the stochastic and combinatorial expression of three spectral variants of fluorescent proteins (FP). The *Drosophila* Brainbow [2] and Flybow [3] systems are adaptations of this approach for use in flies, containing distinct features that take advantage of genetic techniques available in this model organism. Here, we provide detailed information on how to use one of these approaches, Flybow, for studies in the brain.

Flybow (FB) is based on the Brainbow-2 strategy (*see* Fig. 1a). It relies on the *Gal4/UAS* system [4] to control transgene expression, while a FLP recombinase with altered specificity (mFLP5) mediates inversions and excisions of cassettes flanked by *mFRT71* sites [5].

Simon G. Sprecher (ed.), *Brain Development: Methods and Protocols*, Methods in Molecular Biology, vol. 1082,
DOI 10.1007/978-1-62703-655-9_4, © Springer Science+Business Media, LLC 2014

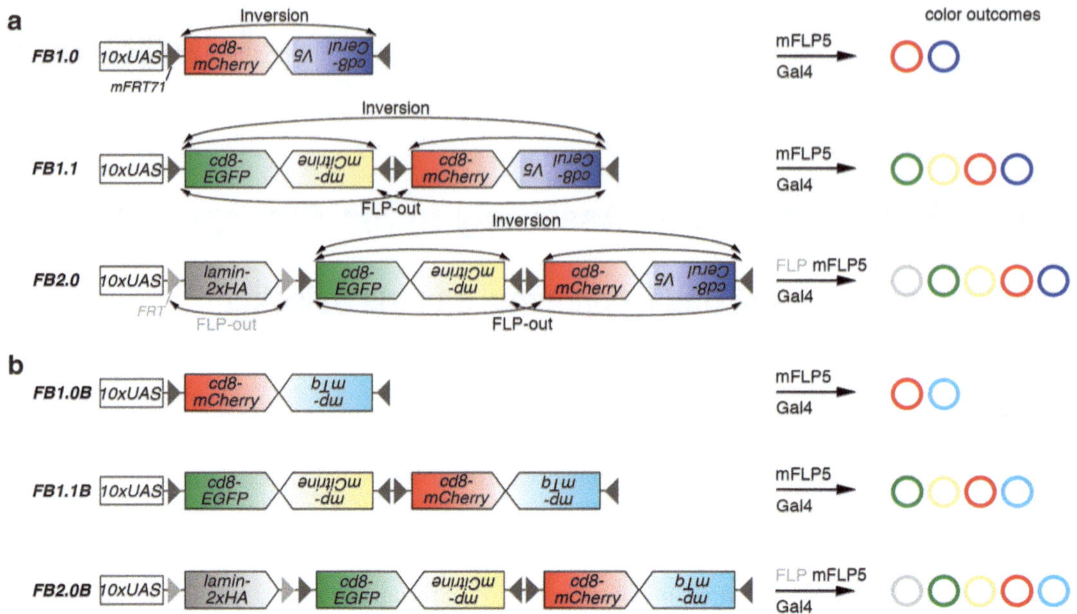

Fig. 1 Schematic representation of Flybow transgenes and color outcomes. Expression of the modified FLP recombinase mFLP5 induces inversions and excisions of cassettes each flanked by inward-facing *mFRT71* recognition sites (*black triangles*). Cassettes consist of two FP cDNAs with opposite orientations. In *FB2.0*, an additional FLP-out cassette flanked by canonical FRT sites facing in the same orientation (*grey triangles*) precedes the FP-containing cassettes. (**a**) A detailed description of *FB1.0*, *FB1.1*, and *FB2.0* transgenes is provided in [3]. (**b**) In *FB1.0B*, *FB1.1B*, and *FB2.0B* transgenes, Cerulean-V5 has been replaced by mTurquoise (mTq)

Each cassette contains two FP-encoding cDNAs in opposing orientations. FPs are membrane tethered by either Cd8a (cd8) [6] or the myr/palm (mp) sequence of Lyn kinase [7]. FP sequences are followed by *SV40* or *hsp70Ab* polyadenylation (pA) signals. mFLP5 is expressed under the control of the heat shock promoter (*hs-mFLP5*) and induces inversions of DNA cassettes by recombining *mFRT71* sites in opposing orientations, or excisions (FLP-out) by recombining *mFRT71* sites in the same orientation. Cassettes have been subcloned into a vector containing 10 *UAS* sites [8]; the FP closest to these sites will be expressed. *Flybow 1.0* consists of one cassette encoding two FPs (mCherry [9] and Cerulean-V5 [10]). *Flybow 1.1* contains two cassettes, each encoding two FPs (EGFP [11] and mCitrine [12]; mCherry and Cerulean-V5). *Flybow 2.0* features an additional stop cassette, flanked by canonical *FRT* sites facing in the same orientation, which can be excised by the canonical FLP recombinase. The stop cassette consists of *lamin* cDNA, followed by two HA tag sequences and *hsp70Aa* and *hsp27* pA signals. Since expression is only induced in cells, in which Gal4 and canonical FLP expression overlap, sparse labeling is achieved, thus facilitating the identification of single cells.

The first set of *FB* constructs relied on the cyan FP Cerulean, which requires detection with a V5 antibody because of its weak endogenous fluorescent signal in flies. To eliminate the necessity

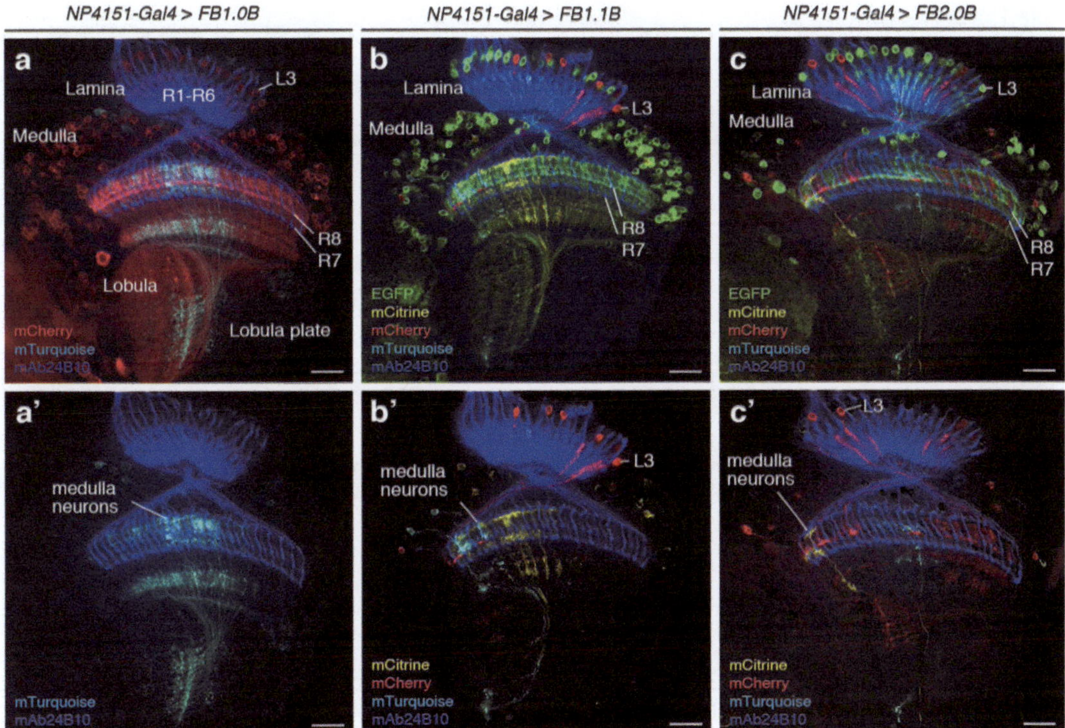

Fig. 2 Expression of B set of *FB* transgenes in the adult *Drosophila* optic lobe. Examples for original transgenes can be found in [3]. Photoreceptor cells (R1–R8) visualized with mAb24B10 (*blue*) extend axons into the optic lobe: R1–R6 axons innervate the lamina; R8 and R7 axons terminate in the medulla [15]. *NP4151-Gal4*, an enhancer trap insertion into the *Netrin B* locus [16, 17], drives expression of *FB* transgenes in lamina neurons L3 and medulla neuron subtypes, which extend axons into the lobula and lobula plate. (**a**, **a′**) Heat shock application induces expression of mFLP5 to trigger inversion events in *FB1.0B*^{260b} ($n = 14$), labeling individual neurons with mTurquoise instead of mCherry. (**b**, **b′**) In *FB1.1B*^{260b} ($n = 19$), excision and inversion events lead to expression of mCitrine, mCherry, or mTurquoise instead of the default marker EGFP. (**c**, **c′**) *FB2.0B*^{260b} ($n = 29$) facilitates sparse labeling of single neurons in a population because it relies on heat shock-induced expression of FLP for excision of the stop cassette flanked by FRT sites, and of mFLP5 for stochastic expression of the four fluorescent proteins. Confocal images represent single optic sections. Images in **b**, **b′**, **c**, and **c′** were processed using channel separation software. Heat shock exposure: 3 × 45 min (**a**–**b′**) and 3 × 30 min (**c**, **c′**) at 48, 72, and 96 h after puparium formation. Scale bars: 20 μm

for antibody detection, we generated a second set of transgenic lines, i.e., *FB1.0B*, *FB1.1B*, and *FB2.0B*, in which cd8-tethered Cerulean was replaced by myr/palm-tethered mTurquoise, a brighter cyan FP variant [13] (*see* Figs. 1b and 2). As a further improvement, we generated additional *hs-mFLP5* stocks for our toolkit. Three are homozygous viable transgene insertions on the X, second, and third chromosomes, while one line contains a homozygous lethal insertion on the second chromosome showing higher recombination efficiency (*see* Table 1).

Because Flybow is readily compatible with available genetic loss- and gain-of-function approaches, it can support anatomical and functional studies of any genetically accessible cell population

Table 1
Basic Flybow transgene toolkit

Flybow[a]	Chr.	Insertion	Related parental stocks		Reference
FB1.0^{260b}	2	Viable			[3]
FB1.0^{49b}	3	Viable			[3]
FB1.0B^{260b}	2	Viable			This study
FB1.0B^{49b}	3	Viable			This study
FB1.1^{260b}	2	Viable			[3]
FB1.1^{49b}	3	Viable			[3]
FB1.1B^{260b}	2	Viable			This study
FB1.1B^{49b}	3	Viable			This study
FB2.0^{260b}	2	Viable	*hs-FLP*^{1}*; FB2.0*^{260b}		[3]
FB2.0^{49b}	3	Viable	*hs-FLP*^{1}*; FB2.0*^{49b}		[3]
FB2.0B^{260b}	2	Viable	*hs-FLP*^{1}*; FB2.0B*^{260b}		This study
FB2.0B^{49b}	3	Viable	*hs-FLP*^{1}*; FB2.0B*^{49b}		This study
hs-mFLP	**Chr.**	**Insertion**	**Related parental stocks**	**Efficiency**[b]	
hs-mFLP5	2	Lethal	*hs-mFLP5/Gla Bc; TM2/TM6B*	**	[3]
hs-mFLP5	3	Lethal	*Gla Bc/CyO; hs-mFLP5/TM2*	**	[3]
hs-mFLP5^{MH15}	X	Viable		**	This study
hs-mFLP5^{MH1}	2	Lethal		***	This study
hs-mFLP5^{MH12}	2	Viable		**	This study
hs-mFLP5^{MH3}	3	Viable		**	This study

[a] *260b* and *49b* indicate the *attP* site-containing loci used for *FB* transgene insertion on the second and third chromosomes, respectively [3]

[b] Recombination efficiencies were estimated by monitoring excision events of a stop cassette flanked by *mFRT71* sites in 3rd instar larval eye imaginal discs

** good efficiency

*** very good efficiency

in the nervous system and other tissues [3]. While Flybow also can be applied to studies in embryos, in this chapter, we focus on its use in the larval, pupal, and adult nervous system.

2 Materials

Prepare all solutions with ultrapure deionized water (18.2 MΩ cm resistivity) from a water purification system.

2.1 Genetic Crosses and Clone Induction

1. Plastic vials containing standard cornmeal/agar medium.
2. Virgin and male flies with the proper genotype (*see* Table 1).
3. Water bath at 37 °C.

2.2 Buffers

1. Phosphate-buffered saline (PBS, 130 mM NaCl, 7 mM $Na_2HPO_4 \cdot 7H_2O$, 3 mM $NaH_2PO_4 \cdot H_2O$, pH 7.4). Prepare 10× PBS stock solution: 75.97 g NaCl, 18.76 g $Na_2HPO_4 \cdot 7H_2O$, 4.14 g $NaH_2PO_4 \cdot H_2O$. Add water to a volume of 990 mL. Dissolve crystals by stirring with a magnetic stirrer and adjust pH with drops of concentrated 12 N HCl and 10 M NaOH, and subsequently 1 M dilutions. Fill up to 1 L with water. Store this solution at room temperature in a glass bottle (*see* **Note 1**). To prepare 1 L PBS solution, mix 100 mL 10× PBS stock solution with 900 mL water and readjust pH using 1 N HCl or 1 M NaOH if necessary. Store at 4 °C.
2. Phosphate-buffered saline with Triton (PBT, 0.5 %): 100 mL PBS, 0.5 mL Triton®-X-100 (Sigma). Add Triton®-X-100 by using a 1 mL syringe and mix vigorously with a magnetic stirrer to dissolve the detergent. Store solution at 4 °C.
3. Phosphate buffer (PB, 0.1 M, pH 7.4): Weigh 1.73 g $NaH_2PO_4 \cdot H_2O$ into a glass beaker, add 125 mL distilled water, and stir until crystals are dissolved to make a 0.1 M NaH_2PO_4 solution. Weigh 13.4 g $Na_2HPO_4 \cdot 7H_2O$ into a second larger glass beaker, add 500 mL water, and stir to make a 0.1 M Na_2HPO_4 solution. Store 100 mL of 0.1 M Na_2HPO_4 solution in a separate glass bottle. Pour the 0.1 M NaH_2PO_4 solution into the remaining 0.1 M Na_2HPO_4 solution until pH reaches pH 7.4 to make 0.1 M PB. Store solutions at 4 °C.
4. Phosphate buffer with lysine (PBL): Dissolve 3.6 g L-Lysine-HCl (Sigma) in 100 mL water. Add approximately 40 mL 0.1 M Na_2HPO_4 solution until pH 7.4 is reached. Add 0.1 M PB to make up 200 mL. Filter-sterilize using a 250 mL filter unit (0.2 µm pore size, 50 mm membrane diameter, Nalgene, Cat. # 126-0020). Store solution at 4 °C for up to 3 months.
5. Blocking buffer (10 % NGS in PBT): 1 mL normal goat serum (NGS, Sigma), 9 mL PBT (*see* **Note 2**).

2.3 Fixative

1. Paraformaldehyde stock solution (8 %): 0.8 g paraformaldehyde (PFA, EM grade powder), 10 mL water, 70 µL of 1 M NaOH. Weigh PFA powder in 50 mL self-standing centrifuge tube (Corning). Wear a mask and gloves when weighing paraformaldehyde to minimize exposure, as the powder is classified as carcinogenic. Add 10 mL water and 70 µL of 1 M NaOH. Shake gently and microwave for 10 s with the cap slightly loosened. Fully dissolve in a 37 °C water bath for 30–60 min with occasional shaking of the tube. Filter into a fresh 15 mL centrifuge tube using a sterile hydrophilic syringe filter with 0.2 µm pore size (Sartorius). Store solution for up to 7 days at 4 °C (*see* **Note 3**).
2. Phosphate buffer with lysine and paraformaldehyde (PLP, 2 % PFA): 3 mL PBL, 1 mL 8 % PFA stock solution. Make up freshly prior to use.

2.4 Antibodies

1. Primary antibodies: mouse monoclonal antibody anti-V5 (1:500 dilution in blocking buffer; Invitrogen) to visualize Cerulean expression (*see* **Note 4**).
2. Secondary antibodies: goat anti-mouse $F(ab')_2$ fragments conjugated with Cy5 or Alexa Fluor®647 (Jackson ImmunoResearch Laboratories) used as a dilution of 1:200 in blocking buffer (*see* **Note 5**).

2.5 Dissections, Immunostaining and Mounting

1. Stainless steel No. 5 forceps.
2. Mesh baskets.
3. 24-well multidish (e.g., Nunclon™ Δ surface).
4. Terasaki plate (e.g., Nunc 60-well MicroWell™ MiniTray).
5. Parafilm®.
6. Sarstedt microtubes with O-ring screw caps.
7. Fine nylon mesh.
8. Rotating titer plate shaker (e.g., Thermo Scientific MiniMix microplate shaker).
9. Aluminum foil.
10. Vectashield (Vector Laboratories).
11. Microscope slide (SuperFrost®).
12. Cover slip (18 × 18 mm, No. 1 or 1.5).
13. Soft modeling mass (e.g., Fimo).
14. Nail polish.
15. Medical wipes.
16. Glas pipettes.
17. Sylgard (Dow Corning, 184 silicone elastomer kit).
18. Dissecting and fluorescence dissecting microscopes.

2.6 Image Acquisition and Analysis

1. Confocal laser scanning microscope.
2. Long-range objectives (e.g., Leica 20× (0.7 NA) air, 40× (1.25 NA), and 100× (1.46 NA) oil objectives).
3. Cargille immersion oil Type DF.
4. Confocal image acquisition software: (e.g. Leica LAS software).
5. Image analysis software (e.g. Volocity Improvision PerkinElmer; Image J; Fiji including Simple Neurite Tracer plug-in).

3 Methods

3.1 Genetic Crosses and Clone Induction Protocol

1. Build and expand the driver stock, which contains both an *hs-mFLP5* and a *Gal4* transgene active in a cell type, brain area, or other tissue of interest (*see* **Note 6**).

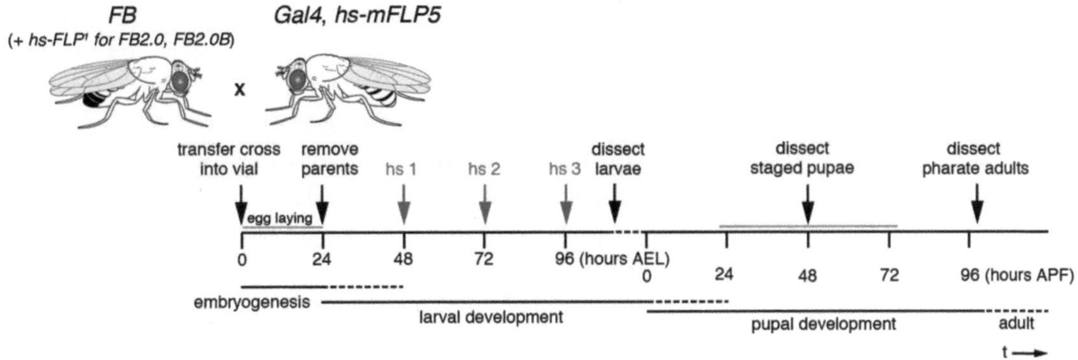

Fig. 3 Example of experimental strategy to induce recombination events. After setting up or transferring the cross into a fresh vial ($t=0$), flies are left to lay eggs for 1 day ($t=24$ h after egg laying (AEL)). Subsequently, parents are transferred into a fresh vial or removed. Progeny in vials are heat shocked up to three times at 48, 72, and 96 h AEL in a 37 °C water bath. Brains can be dissected from larvae, staged pupae, or adults. APF, after puparium formation

2. Collect male and unfertilized female flies from parental stocks for about 4 days (*see* **Note 7**). Set up crosses in plastic vials containing standard cornmeal/agar medium with about 10–12 females and 5–6 males per vial. The crosses should be designed so that progeny will have one *Gal4*, one *hs-mFLP5*, and one *FB1.0* or *FB1.1* transgene. Note that for *FB2.0* an additional transgene is required that expresses canonical FLP recombinase, such as *hs-FLP[1]* (*see* Fig. 3). *FB* and *hs-mFLP5* transgenes are available as insertions on different chromosomes to facilitate genetic crosses (*see* Table 1). We usually set up three to five parallel crosses for a given experiment. After a period of 24 h of egg laying, transfer parents into fresh vials. Repeat this process for maximally 1 week. All vials are kept in an incubator at 25 °C.
3. To induce recombination events (*see* Fig. 3), progeny of crosses are heat shocked by placing the vials into a water bath with the temperature set to 37 °C (*see* **Note 8**). The water level should be sufficiently high, so that vials, when weighed down, are immersed well above the food level to ensure that also larvae leaving the medium are exposed to heat.

3.2 Immunolabeling

1. Dissect 3rd instar larval, pupal, or adult brains (*see* **Note 9**) in drops of cold PBS using forceps on a dissecting pad. To maximize the number of dissected samples with recombination events, some *Gal4* drivers make it possible to preselect flies with clones under a fluorescence-dissecting microscope. Transfer the brains into a glass embryo dish filled with PBS using forceps or a glass pipette. Keep the dish on ice until all samples of a given genotype are dissected (*see* **Notes 10** and **11**).
2. Transfer the brains into a mesh basket (*see* **Note 12**) positioned in one well of a 24-well multidish filled with 1 mL PLP. Fix for 1 h at room temperature (*see* **Note 13**).

3. Wash the brains by transferring the mesh baskets sequentially into three wells filled with 1 ml PBT each. Keep baskets in the first two wells for a few seconds each (short washes) and in the third well for at least 15 min (long wash) (*see* **Note 14**).
4. Block tissue for at least 15 min by transferring the baskets into a new well containing blocking buffer (10 % NGS in PBT).
5. For incubation in primary antibody (*see* **Note 15**), transfer brains into the wells of a Terasaki plate, each filled with 10 μL of antibody diluted in blocking buffer. Distribute the brains over multiple wells, with each well not containing more than 10 brains (*see* **Note 16**). Add a stripe of moist filter paper on one side of the plate, close and seal tightly with Parafilm®. Place the plate into a humid chamber, such as a small closable box with a moist paper towel at the bottom. Keep on a gently horizontally rotating titer plate shaker overnight at 4 °C in a cold room.
6. Return the brains into the mesh baskets placed into 24-well plates filled with PBT. Wash for 15 min by transferring the baskets three times into the fresh wells for two short and one long washes.
7. For incubation in secondary antibodies, transfer the baskets into a new well containing the antibody diluted in blocking buffer. Wrap the 24-well plate with aluminum foil and keep on a titer plate shaker for 2.5 h at room temperature.
8. Transfer the baskets into the fresh wells: two filled with PBT for two short washes, and two filled with PBS for one short and one long wash. In total, wash at least for one hour. Immunolabeled brains should be stored in PBS at 4 °C, not in mounting media, to avoid softening of the tissue prior to mounting on slides.

3.3 Mounting Samples

1. Take immunolabeled brains with clones individually out of PBS and place in a drop of about 30 μL cold Vectashield on a pre-cleaned microscope slide.
2. Take a cover slip (18 × 18 mm, No. 1 or 1.5), place small pieces of soft modeling mass on the four corners of the cover slip to prevent squashing of the brains.
3. Using forceps, position the cover slip with the modeling mass pieces facing down on the sample. Gently press the corners. Then place the slide under a fluorescence dissecting microscope.
4. By pressing the forceps against the cover slip edges, move the cover slip gently up and down or left and right to roll at least one brain hemisphere into the correct orientation. Subsequently, brains can be gently flattened by pressing down the cover slip corners. Seal slide with clear nail polish. Store slides at 4 °C in the dark.

Table 2
Example of a confocal microscopy scanning method for imaging *FB* transgenes

Scan	Fluorescent protein[a]/dye	PMT[b] (e.g.)	Excitation maximum (nm)	Emission maximum (nm)	Excitation laser line (nm)	Emission range imaged (nm)
1	mCitrine	2	516	529	514	525–565
	Alexa Fluor®647	4	651	667	633	675–735
2	mCherry	3 (HyD)	587	610	561	572–630
3	EGFP	1	488	507	488	497–515
4	mTurquoise	3 (HyD)	434	475	458	460–495

[a]Detailed information about the FPs used in the Flybow approach can be found in these references: EGFP [11], mCitrine [12], mCherry [9], and mTurquoise [13]
[b]PMT, photomultiplier tube

3.4 Image Acquisition and Analysis

1. Collect images using a confocal laser scanning microscope equipped with high-quality long-range objectives. For oil objectives, use immersion oil suitable for immunofluorescence microscopy recommended by the confocal microscope provider. Set up a confocal microscope imaging method that combines sequential and simultaneous scan modes (*see* **Note 17**). For each FP or dye, select a laser line suitable for optimal excitation, as well as detection windows or acousto-optical beam splitter [AOBS] settings for collecting specific emission signals. An example of such a method is provided in Table 2. For the original set of *FB* transgenes, mCitrine and Cy5/Alexa Fluor®647 signals are collected using a simultaneous scan mode, followed by sequential scans of EGFP and mCherry signals. For set B of *FB* transgenes, a fourth sequential scan is added to image mTurquoise. To obtain optimal signals and minimal cross talk between channels, adjust the power of individual laser lines for each sample. The detection windows may also require fine-tuning depending on the experiment or microscope used. Assign a color to each channel, e.g., EGFP, green; mCitrine, yellow; mCherry, red; Cerulean-V5/Cy5, medium blue; and mTurquoise, cyan.
2. For samples displaying extensive bleed-through between channels, images can be processed using channel separation software. We use the Leica LAS AF suite channel separation tool for this purpose. In representative images of each detection channel, regions of interest with unambiguous strong but not saturated signals are manually selected for each FP or dye. Using these obtained values, the software algorithms subtract unspecific proportions of detected signals.

3. Confocal images can be further analyzed and processed using Volocity (Improvision PerkinElmer) and ImageJ (Fiji) software to perform z projections of selected sections. Fiji can be used for brightness and contrast adjustments of the four or five channels in color composite images, as well as for subsequent conversion into RGB images. The Simple Neurite Tracer plug-in is highly useful for the reconstruction of individual neurons from stacks.

4 Notes

1. We store most of our solutions in autoclaved standard screw cap laboratory glass bottles.
2. We use higher-quality goat serum to extend the lifetime of solutions.
3. For optimal fixation of tissue, we use EM grade paraformaldehyde powder (Polysciences, Cat. # 0030) in our protocol. Some investigators freeze paraformaldehyde aliquots. However, we prepare fresh solutions at a weekly basis to obtain best staining results.
4. Although Cerulean had been described as the best CFP derivative with respect to brightness, quantum yield, and oligomerization properties [10], consistent with other studies [2], we observed that this FP is less suitable for studies in *Drosophila* because of low endogenous emission signals. Cerulean is therefore visualized using anti-V5 primary and Cy5- or Alexa Fluor®647-conjugated secondary antisera.
5. $F(ab')_2$ fragment-based secondary antibodies ensure even immunolabeling throughout the entire brain, because their smaller size facilitates penetration deeper into the tissue.
6. To build stable stocks, we prefer to combine *hs-mFLP5* transgenes with the *Gal4* driver instead of *FB* transgenes, as continuous low levels of recombinase expression may cause transmittable recombination events in the germ line. Stocks containing both *hs-FLP*^{1} and *FB2.0* transgenes (*see* Table 1) should be continuously monitored. Note that the two *hs-mFLP5* transgenes described in [3] are homozygous lethal insertions and are therefore kept over a balancer.
7. As the occurrence of *hs-mFLP5*-induced recombination events decreases with the parental age of flies, maximally 4-day-old adult flies should be used for genetic crosses. The first emerging progeny of a cross should be prepared for analysis to maximize the number of samples with recombination events. In a 24-h egg collection, these correspond most closely to the progeny, at which the heat shock protocols were aimed.

8. Color outcomes are influenced by the time points, duration, and number of heat shocks. These parameters can be adjusted for each *Gal4* line, cell or brain area of interest, and experimental aim. While early heat shocks lead to recombination events in dividing progenitors and thus tend to label larger lineage-related groups of cells with the same FP, later or shorter heat shocks facilitate labeling of single cells. Repeated heat shocks further promote labeling of neurons or glia with different FPs. We observed that lineage-related cells born in a narrow time window are difficult to separate by the expression of different FPs using the FB approach. For studies in the optic lobe, the earliest time point, at which larvae are subjected to heat shock is 48 h after egg laying. The time point, at which crosses are transferred into a fresh vial is defined as t=0. In a regime requiring multiple heat shocks, this procedure was repeated at 72 and 96 h. The length of heat shocks ranges from 30 to 45 min. When combining the FB approach with mosaic analysis with a repressible cell marker (MARCM, [14]), it is useful to extend heat shocks to 90 min [3].
9. For the visual system, large late third instar larvae, which stopped wandering on the side of vials, are selected for dissections. To stage pupae, white pupae are collected from vials at 1-h intervals, placed on grape-juice agar plates to avoid desiccation, stored in a 25 °C incubator and dissected at specific time points after puparium formation. Pharate adult flies are dissected shortly before eclosion to avoid tracheal filling.
10. Brains are dissected on plates coated with Sylgard® (Dow Corning, 184 silicone elastomer kit). We use the lids of Terasaki plates for this purpose. When transferring the brains with a glass or plastic pipette, keep samples in the tip of the pipette, filled with PBS, to avoid losing them during the process.
11. Keep unfixed brains for not longer than 30 min on ice, as extended storage can cause connectivity defects. Fixed brains can be stored in PBT while completing all dissections for the day.
12. Mesh baskets are handmade using conical 1.5 mL Sarstedt microtubes with O-Ring Screw Caps and a fine nylon mesh. Cut microtubes at the bottom border of the ridged part using a pair of sharp scissors, so that the 24-well plate can still be closed with the lid. Carefully, cut off the top of the screw cap below the O-Ring using a sharp razor blade. Cut out a 2 × 2 cm square of nylon mesh. Place the mesh on top of the cut tube, and screw the cap ring in place while tightening the mesh. Remove all fabric on the outside of the baskets using a razor blade. Newly made baskets should be extensively washed in PBT followed by deionized water to avoid sticking of brains. After each use, baskets should be thoroughly washed by soaking and rinsing them in deionized water. They can be reused for years.

13. Fixation for 1 h in PLP at room temperature is central to obtaining good signals both from FPs and fluorophore-conjugated secondary antibodies. We observed that fixation using 4 % PFA even for 30 min significantly quenched FP signals.
14. Even if solely endogenous FP signals are collected and immunolabeling steps are not required, brains should still be washed extensively in PBT after fixation before storage in PBS. Otherwise the tissue will shrink when transferred into mounting media.
15. For the second set of *FB* transgenes, immunolabeling with anti-V5 is not required. Instead, additional antibody stainings, e.g., mAb24B10 (1:75 dilution in blocking buffer; Developmental Studies Hybridoma Bank) can be performed and visualized with a Cy5 or Alexa Fluor®647-coupled secondary antibody.
16. To avoid damaging brains during the transfer into the Terasaki plate wells, position brains in the remaining liquid between slightly held apart and curved forceps tips. To minimize the transfer of excessive PBT, gently blot up the liquid held between the forceps arms using a rolled up corner of a medical wipe (e.g., Kimberly-Clark Professional).
17. We optimized imaging conditions for a Leica TCS SP5 II upright laser confocal microscope equipped with a resonant scanner and four photomultiplier tubes (PMTs). However, all confocal microscope models are suitable, which have the five listed laser lines and whose detection windows can be adjusted. We typically collect image stacks using 1,024 × 1,024 pixels image size, 200 Hz line speed, and 5-line or 4-frame averages. To accelerate image acquisition, a bidirectional scan mode is used. Although not essential, the use of the resonant scanner provides an alternative way to increase the speed of image acquisition of large z stacks and thus to minimize photobleaching. When using the resonant scanner, images are collected at the fixed speed of 8 kHz and averaged 96 times. If the microscope is equipped with hybrid detection (HyD) GaAsP technology, scan the weakest fluorescent signals using the HyD PMT(s).

Acknowledgements

We thank J. Goedhart for sharing the *mTurquoise* cDNA. D.H. developed the original *FB* transgenes, while N.S. generated the second set of *FB* transgenes and validated the new *hs-mFLP5* insertions. The original Flybow approach was developed in collaboration with B.J. Dickson, S. Rotkopf, C. Alexandre, and D.M. Bell. We are grateful to H. Apitz, B. Richier, F. Rodrigues, and S. Schrettenbrunner for critical reading of this manuscript. This work is supported by the Medical Research Council (U117581332).

References

1. Livet J, Weissman TA, Kang H, Draft RW, Lu J, Bennis RA et al (2007) Transgenic strategies for combinatorial expression of fluorescent proteins in the nervous system. Nature 450:56–62
2. Hampel S, Chung P, McKellar CE, Hall D, Looger LL, Simpson JH (2011) Drosophila Brainbow: a recombinase-based fluorescence labeling technique to subdivide neural expression patterns. Nat Methods 8:253–259
3. Hadjieconomou D, Rotkopf S, Alexandre C, Bell DM, Dickson BJ, Salecker I (2011) Flybow: genetic multicolor cell labeling for neural circuit analysis in Drosophila melanogaster. Nat Methods 8:260–266
4. Brand AH, Perrimon N (1993) Targeted gene expression as a means of altering cell fates and generating dominant phenotypes. Development 118:401–415
5. Voziyanov Y, Konieczka JH, Stewart AF, Jayaram M (2003) Stepwise manipulation of DNA specificity in Flp recombinase: progressively adapting Flp to individual and combinatorial mutations in its target site. J Mol Biol 326:65–76
6. Liaw CW, Zamoyska R, Parnes JR (1986) Structure, sequence, and polymorphism of the Lyt-2 T cell differentiation antigen gene. J Immunol 137:1037–1043
7. Zacharias DA, Violin JD, Newton AC, Tsien RY (2002) Partitioning of lipid-modified monomeric GFPs into membrane microdomains of live cells. Science 296:913–916
8. Dietzl G, Chen D, Schnorrer F, Su KC, Barinova Y, Fellner M et al (2007) A genome-wide transgenic RNAi library for conditional gene inactivation in Drosophila. Nature 448: 151–156
9. Shaner NC, Campbell RE, Steinbach PA, Giepmans BN, Palmer AE, Tsien RY (2004) Improved monomeric red, orange and yellow fluorescent proteins derived from Discosoma sp. red fluorescent protein. Nat Biotechnol 22:1567–1572
10. Rizzo MA, Springer GH, Granada B, Piston DW (2004) An improved cyan fluorescent protein variant useful for FRET. Nat Biotechnol 22:445–449
11. Shaner NC, Steinbach PA, Tsien RY (2005) A guide to choosing fluorescent proteins. Nat Methods 2:905–909
12. Griesbeck O, Baird GS, Campbell RE, Zacharias DA, Tsien RY (2001) Reducing the environmental sensitivity of yellow fluorescent protein. Mechanism and applications. J Biol Chem 276:29188–29194
13. Goedhart J, van Weeren L, Hink MA, Vischer NO, Jalink K, Gadella TW Jr (2010) Bright cyan fluorescent protein variants identified by fluorescence lifetime screening. Nat Methods 7:137–139
14. Lee T, Luo L (1999) Mosaic analysis with a repressible cell marker for studies of gene function in neuronal morphogenesis. Neuron 22:451–461
15. Hadjieconomou D, Timofeev K, Salecker I (2011) A step-by-step guide to visual circuit assembly in Drosophila. Curr Opin Neurobiol 21:76–84
16. Hayashi S, Ito K, Sado Y, Taniguchi M, Akimoto A, Takeuchi H et al (2002) GETDB, a database compiling expression patterns and molecular locations of a collection of Gal4 enhancer traps. Genesis 34:58–61
17. Timofeev K, Joly W, Hadjieconomou D, Salecker I (2012) Localized netrins act as positional cues to control layer-specific targeting of photoreceptor axons in Drosophila. Neuron 75:80–93

Chapter 5

Immunofluorescent Labeling of Neural Stem Cells in the *Drosophila* Optic Lobe

Benjamin Perruchoud and Boris Egger

Abstract

The *Drosophila* visual system is an excellent model system to study the switch from proliferating to differentiating neural stem cells. In the developing larval optic lobe, symmetrically dividing neuroepithelial cells transform to asymmetrically dividing neuroblasts in a highly ordered and sequential manner. This chapter presents a protocol to visualize neural stem cell types in the *Drosophila* optic lobe by fluorescence confocal microscopy. A main focus is given on how to dissect, fix, immunolabel, and mount brains to reveal cellular morphology during early larval brain development.

Key words Neural stem cell, *Drosophila*, Brain, Optic lobe, Immunofluorescent labeling

1 Introduction

The developing *Drosophila* visual system or optic lobe is an attractive model system to study the transition from symmetric to asymmetric neural stem cell division. The larval optic lobe is comprised of an outer proliferation center that gives rise to the lamina and the outer medulla and an inner proliferation center that generates the inner medulla, lobula, and lobula plate [1]. In recent years the work of many research groups has shown that neurogenesis in the optic lobe resembles neural development in the mammalian brain and the retina [2–10]. In the optic lobe symmetrically dividing neuroepithelial cells progressively transform to asymmetrically dividing neuroblasts. This cell fate transformation is reminiscent of the transition of neuroepithelial cells to radial glial cells in the mammalian brain [11].

Here we present a detailed protocol to visualize optic lobe development by immunolabeling for fluorescent confocal microscopy. We specifically focus on early larval stages and describe how to dissect, fix, immunolabel, and mount 1st, 2nd, and early 3rd instar brains. In the described method, the dissected brains are attached to a cover slip during the entire procedure, which has

Simon G. Sprecher (ed.), *Brain Development: Methods and Protocols*, Methods in Molecular Biology, vol. 1082, DOI 10.1007/978-1-62703-655-9_5, © Springer Science+Business Media, LLC 2014

several advantages. Firstly, it prevents loosing the rather small brains during immunolabeling and washing procedures. Secondly, the brains can be orientated on the cover slips in a fixed position, which facilitates the imaging procedure. Thirdly, the method allows simultaneously assessing brains of various different stages and genotypes in rapid and efficient manner.

2 Materials

2.1 Staged Larval Collection

1. Fly cages.
2. Apple juice embryo collection plates with wet yeast.
3. Petri dishes 35 mm filled with fly food with wet yeast.

2.2 Preparation of Coated Cover Slips

1. Hot plate.
2. 22 × 22 cover slips thickness # 1.
3. 0.1 % w/v Poly-L-Lysine in water (Sigma).

2.3 Larval Brain Dissection

1. Fine forceps (Dumont # 5, Fine Science Tools).
2. Pipettes and sterile tips.
3. Paper tissue.
4. Two needles (25G or 30G) on cotton buds.
5. Petri dishes (plastic).
6. Phosphate-buffered saline, 1× PBS: 10 mM Na_2HPO_4, 2.68 mM KCl, 140 mM NaCl (PBS Tablets, Gibco, Invitrogen). pH is 7.45. Sterilize by autoclaving.

2.4 Fixation and Immunolabeling

1. Fixative: 4 % formaldehyde, 0.5 mM EGTA, 5 mM $MgCl_2$ in 1× PBS.
2. PBST: 0.3 % Triton X-100 in 1× PBS. Store at 4 °C.
3. Primary antibodies (e.g., anti-Discs large 4F3 (1:100), Developmental Studies Hybridoma Bank (http://dshb.biology.uiowa.edu/)).
4. Secondary antibodies (e.g., Goat anti Mouse Alexa488, Molecular Probes).
5. Columbia staining jars.
6. Rotating platform (shaker).
7. Humid chamber (e.g., empty Vectashield box).

2.5 Mounting and Visualization

1. Microscopy slides.
2. Vaseline.
3. Syringe with 18G needle.

4. Mounting media (Vectashield).
5. Transparent nail varnish.
6. Fluorescent (confocal) microscope.

3 Methods

3.1 Preparing Poly-L-Lysine-Coated Cover Slips

1. Prepare 0.05 % w/v Poly-L-Lysine in ddH_2O in a microfuge tube.
2. Place cover slips on a hot plate.
3. Add 10 μl of 0.05 % Poly-L-Lysine diluted in ddH_2O to each slide and let the drop dry (*see* **Note 1**).
4. Immediately after the first drop is dry, add another 10 μl of 0.05 % Poly-L-Lysine on top of the first drop and let it dry again.
5. Store slides in a dust-free environment (e.g., in a closed petri dish or cover slip holder).

3.2 Staging of Larvae

1. Collect embryos on apple juice plates for 4–6 h and let embryos develop at 25 °C (*see* **Note 2**).
2. 24 h after midpoint of egg collection, pick about 80–100 freshly hatched larvae and place them in a food plate containing a drop of wet yeast.
3. Let larvae develop at 25 °C to the appropriate larval stage (e.g., 24 h ALH, *after larval hatching* or 48 h ALH).

3.3 Dissection of Larval Brains

1. Pick the larvae from food plate and place them on a paper tissue soaked with PBS (*see* **Note 3**).
2. Place several drops of 1× PBS in a circle on the inside of a plastic petri dish lid. Add one larva to each drop (*see* Fig. 1a).
3. To dissect small larvae (0–48 h ALH), use two syringe needles on cotton buds (*see* Fig. 1a). Carefully hold the larva in place with one needle pressing down the middle of the larval body. Stick the tip of the other needle firmly into the very anterior part of the larva where the mouth hooks are located. Draw the body slowly away from the sticked down mouthpart. As the anterior head part breaks away from the rest of the body, the brain spills out. The brain might be still attached to the mouth hooks. Dissect off the mouth hooks and imaginal discs as necessary (*see* **Note 4**). To dissect larger larvae (72–96 h ALH), use fine forceps and the so-called "inverted sock" technique. Pull the larva in the middle apart and discard posterior part. Grab the larval mouth hooks at the anterior half with one forceps and invert the larval cuticle over the tip of the forceps in a manner like to invert a sock. The interior organs of the larva are now on the outside and the brain can be further dissected.

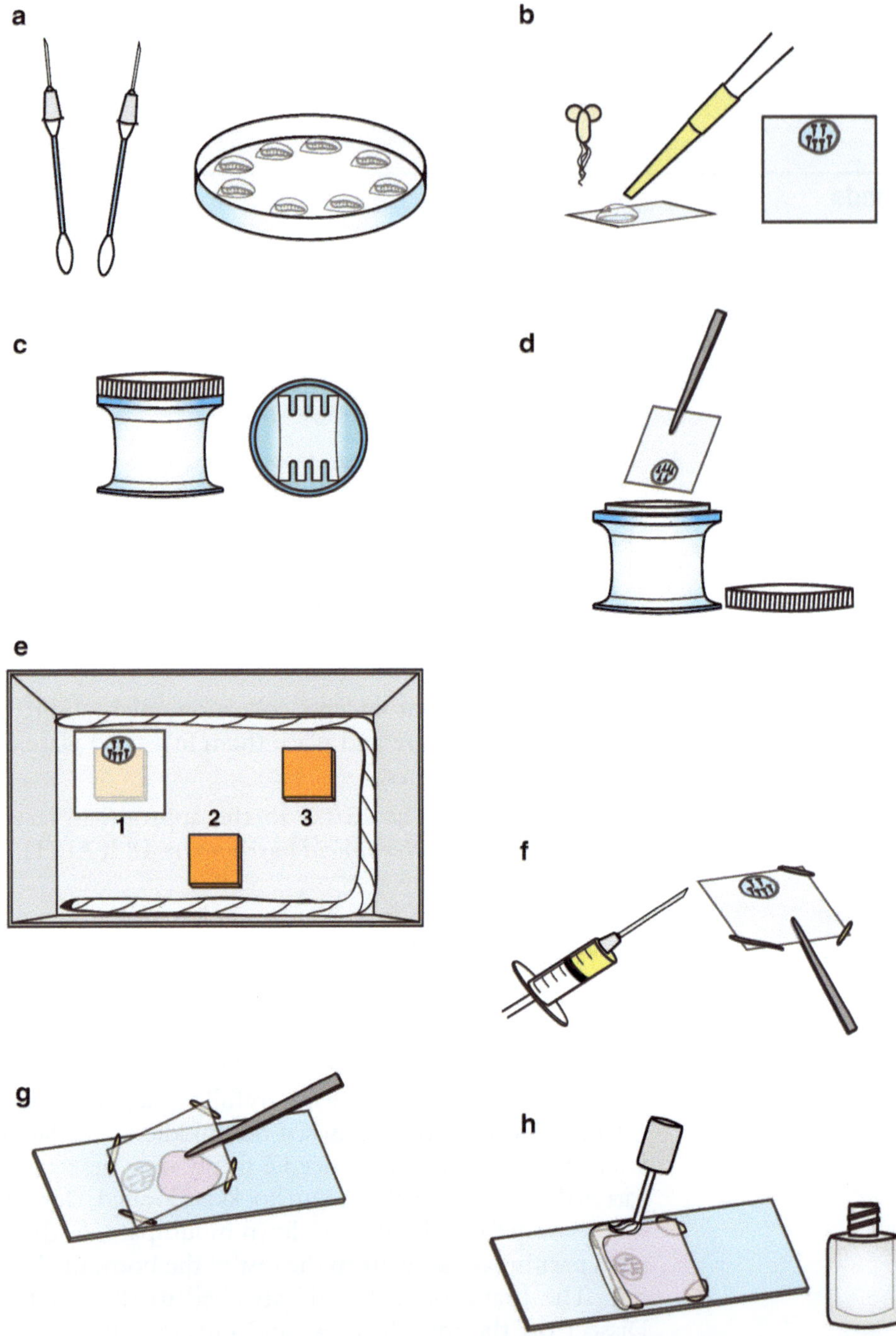

Fig. 1 (**a**) Larvae are placed in a drop of 1× PBS in a petri dish lid and dissected by using needles on cotton buds or fine forceps. (**b**) Dissected brains are transferred by using a pipette (small brains) or forceps (larger brains) and oriented in desired position on Poly-L-Lysine-coated cover slips. (**c**) Columbia jar holds maximal four cover slips. (**d**) Cover slips with attached brains are transferred to Columbia jars for fixation and washing steps. (**e**) A humid chamber is used for primary and secondary antibody incubation. (**f**) After the final washing step, a strip of Vaseline is added to each cover slip corner as a spacer. (**g**) Cover slip is carefully placed on a slide containing a drop of Vectashield and corners are gently pushed down. (**h**) Samples can be sealed with transparent nail varnish and stored at 4 °C

4. Place a drop of PBS onto the coated cover slip. For small brains (0–48 h ALH), cut off the very tip of a 200 μl pipette tip and transfer the brains to the cover slip (*see* **Note 5** and Fig. 1b). For larger brains (72–96 h ALH) use the forceps to transfer the brains.
5. Place brains to the Poly-L-Lysine-coated area in the desired orientation by using cotton bud needles or forceps (*see* **Note 6**).

3.4 Fixation and Staining

1. Rinse cover slips with attached brains in a Columbia jar (*see* Fig. 1c) containing 1× PBS.
2. Fix brains on cover slips in a Columbia jar containing 4 ml fixative by rotating on a platform for 18 min (*see* Fig. 1d).
3. Replace fixative by PBST ~6 ml and wash the brains 1 × 1 min, 2 × 5 min, and 1 × 15 min on a rotating platform.
4. Place the cover slips into a humid chamber and add primary antibodies (*see* **Note 7**) diluted in PBST (100 μl total volume) on top of slide (*see* Fig. 1e). Close lid of humid chamber and incubate overnight at 4 °C (*see* **Note 8**).
5. Wash the cover slips in a Columbia jar with PBST 1 × 1 min, 3 × 5 min, and 2 × 15 min.
6. Place the cover slips in a humid chamber and add secondary antibodies diluted in PBST (100 μl total volume) on top of the slide. Close the lid of humid chamber and incubate overnight at 4 °C (*see* **Note 9**).
7. Wash cover slips in a Columbia jar with PBST 1 × 1 min, 3 × 5 min, and 3 × 15 min.

3.5 Mounting

1. Transfer cover slips using a forceps from the Columbia jar and rinse off PBST with ddH_2O from top of cover slips (*see* **Note 10**).
2. Remove excess liquid on a paper tissue and add a strip of Vaseline to each corner as a space holder by using the syringe (*see* Fig. 1f).
3. Add a drop of Vectashield on a clean microscopy slide.
4. Place cover slips with attached brains onto microscopy slides and tap or gently press down each corner using cotton buds (*see* **Note 11** and Fig. 1g).
5. Seal cover slips with transparent nail varnish (*see* Fig. 1h).
6. Samples can be stored in slide folder at 4 °C.

4 Notes

1. In order to use smaller volumes of fixative (4 ml), add the Poly-L-Lysine drop towards one side of the cover slips (*see* Fig. 1b).
2. For more accurate staging, you may want to make a 1 h pre-collection for eggs that have been retained and developed in the female abdomen before starting the experimental collection.
3. Larvae clean themselves of yeast and fly food by moving around on a wet paper tissue.
4. Leaving some imaginal discs attached can help to stick the brains onto cover slips in the desired orientation.
5. Small brains often stick to the pipette tip. A trick is to coat the tip with cheek cells of your own saliva. Pipette saliva up and down and rinse once in 1× PBS before transferring the brains.
6. At this stage it helps to draw a brain orientation map of each cover slip. So the samples can be distinguished at a later time point. Alternatively, remember the position of each cover slip in the Columbia jar (e.g., wild type, position 1, mutant, position 2). Marker pen writings are washed off in PBST and are therefore not suitable to label cover slips.
7. We usually use a monoclonal mouse antibody against the protein Discs large (1:100), which is available from the Developmental Studies Hybridoma Bank (http://dshb.biology.uiowa.edu/). Discs large is localized to cell cortices in neuroblasts and septate junctions in neuroepithelial cells and is very suitable to visualize cell morphology in the larval optic lobe. To visualize optic lobe precursor cells, GAL4 lines such as $GAL4^{c855a}$ can be used to drive *UAS-GFP* constructs (*see* Fig. 2).
8. For most primary antibodies, best results are obtained with overnight incubation; however, primary antibody incubation for 3 h at room temperature can also give good results. Different conditions should be tested in cases where immunolabelings are unsatisfactory.
9. For better tissue penetration, secondary antibodies are also applied overnight at 4 °C.
10. PBST should be quickly washed off the top of the cover slip (not the side on which brains are attached) with ddH_2O because dried out PBST will leave a salt crust on cover slips.
11. Do press down gently only to leave morphology and shape of brains intact.

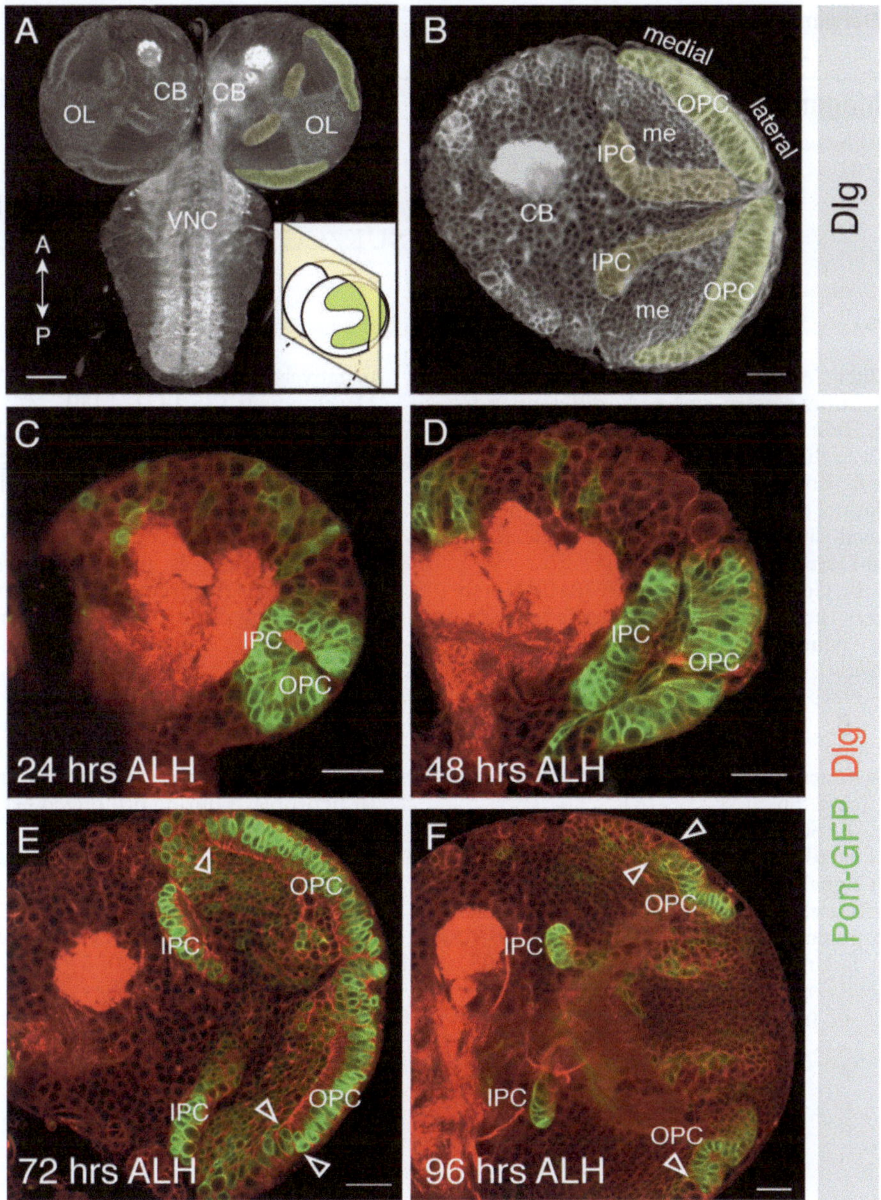

Fig. 2 Morphology and proliferation centers of the developing optic lobe revealed by immunofluorescent labeling. (**a**) A late third instar larval central nervous system (CNS): ventral nerve cord (VNC), central brain (CB), and optic lobes (OL). Subsequent images show frontal confocal sections, as shown in the inset diagram (OPC in *green*). Anterior and posterior refer to the neuraxis of the larval CNS. (**b**) A frontal section through a brain lobe at mid-third instar: the OPC (*green*), the inner proliferation center (IPC, *yellow*), and the medulla cortex (me). Discs large (Dlg; *gray*) outlines all cell cortices in the larval brain and highlights the morphology of the two optic lobe proliferation centers. (**c**) *GAL4*^{c855a} begins to drive expression of *UAS-pon-gfp* (*green*; Dlg in *red*) at first instar. At late first/early second instar (24 h ALH; after hatching), the OPC and the IPC can be distinguished as two closely associated epithelia. The cells belonging to the proliferation centers (*green*) are clearly distinguishable by their columnar shape, in contrast to the round, isolated central brain cells. (**d**) At the end of second/early third instar (48 h ALH), the epithelia of the OPC and IPC separate from each other and smaller progeny cells are located between the two epithelia. (**e**) As development progresses during second to mid-third instar (72 h ALH), the OPC cells at the medial edge of the epithelium lose their columnar shape (to the *left* of the *arrowheads*). (**f**) At late third instar (96 h ALH), the OPC epithelium decreases in size while the number of round neuroblast-like cells increases at the medial edges (to the *left* of the *arrowheads*). All images are single confocal sections, with anterior on top and lateral to the right. Scale bar is 50 μm (**a**) and 20 μm (**b–f**) (Fig. 2 is reproduced from ref. [2] originally published in *Neural Development*, BioMed Central)

Acknowledgments

We thank Mike Bate for his advice on how to dissect brains of 1st and 2nd instar larvae. B.P. and B.E. are funded by the Swiss University Conference (SUK/CUS) Project P-01 Bio-BEFRI.

References

1. Hofbauer A, Campos-Ortega JA (1990) Proliferation pattern and early differentiation of the optic lobes in *Drosophila melanogaster.* Roux's Arch Dev Biol 198:264–274
2. Egger B, Boone JQ, Stevens NR et al (2007) Regulation of spindle orientation and neural stem cell fate in the *Drosophila* optic lobe. Neural Develop 2:1
3. Egger B, Gold KS, Brand AH (2010) Notch regulates the switch from symmetric to asymmetric neural stem cell division in the *Drosophila* optic lobe. Development 137: 2981–2987
4. Egger B, Gold KS, Brand AH (2011) Regulating the balance between symmetric and asymmetric stem cell division in the developing brain. Fly 5(3):237–241
5. Yasugi T, Sugie A, Umetsu D et al (2010) Coordinated sequential action of EGFR and Notch signaling pathways regulates proneural wave progression in the *Drosophila* optic lobe. Development 137:3193–3203
6. Yasugi T, Umetsu D, Murakami S et al (2008) *Drosophila* optic lobe neuroblasts triggered by a wave of proneural gene expression that is negatively regulated by JAK/STAT. Development 135:1471–1480
7. Reddy BV, Rauskolb C, Irvine KD (2010) Influence of fat-hippo and notch signaling on the proliferation and differentiation of *Drosophila* optic neuroepithelia. Development 137:2397–2408
8. Ngo KT, Wang J, Junker M et al (2010) Concomitant requirement for Notch and Jak/Stat signaling during neuro-epithelial differentiation in the *Drosophila* optic lobe. Dev Biol 346:284–295
9. Orihara-Ono M, Toriya M, Nakao K et al (2011) Downregulation of Notch mediates the seamless transition of individual *Drosophila* neuroepithelial progenitors into optic medullar neuroblasts during prolonged G1. Dev Biol 351:163–175
10. Wang W, Liu W, Wang Y et al (2010) Notch signaling regulates neuroepithelial stem cell maintenance and neuroblast formation in *Drosophila* optic lobe development. Dev Biol 350:414–428
11. Gotz M, Huttner WB (2005) The cell biology of neurogenesis. Nat Rev Mol Cell Biol 6:777–788

Chapter 6

Using MARCM to Study Drosophila Brain Development

Gudrun Viktorin

Abstract

Mosaic analysis with a repressible cell marker (MARCM) generates positively labeled, wild-type or mutant mitotic clones by unequally distributing a repressor of a cell lineage marker, originally *tubP*-driven GAL80 repressing the GAL4/UAS system. Variations of the technique include labeling of both sister clones (twin spot MARCM), the simultaneous use of two different drivers within the same clone (dual MARCM), as well as the use of different repressible transcription systems (Q-MARCM). MARCM can be combined with any UAS-based construct, such as localized GFP fusions to visualize subcellular compartments, genes for rescue and ectopic expression, and modifiers of neural activity. A related technique, the twin spot generator, generates positively labeled clones without the use of a repressor, thus minimizing the lag time between clone induction and appearance of label. The present protocol provides a detailed description of a standard MARCM analysis of brain development that includes generation of MARCM stocks and crosses, induction of clones, brain dissection at various stages of development, immunohistochemistry, and confocal microscopy, and can be modified for similar experiments involving mitotic clones.

Key words *Drosophila*, MARCM, Twin spot generator, Flp-out, Clones, Somatic recombination, Protocol, Brain development, Cell labeling, Neuroblast lineage

1 Introduction

The generation of clones, composed of cells that are genetically different from the rest of the animal, is the genetic equivalent to the single-cell labeling and transplantation techniques of classical embryology that are possible to a limited extent in *Drosophila*. The resulting mosaic animals are essential in studying later functions of early lethal mutations, local tissue-specific effects of mutations, and non-cell-autonomous effects (reviewed in [1]). In *Drosophila*, mitotic clones are readily induced using the yeast Flp/FRT recombinase system [2, 3]. Expressing Flp recombinase from a heat-shock-inducible promoter or a tissue-specific enhancer in mitotic cells effects a crossing over at FRT sites that are located in the same position on homologous chromosome arms. The following cell division can segregate the recombined chromosomes into the two daughter cells such that each daughter cell is homozygous for one

Simon G. Sprecher (ed.), *Brain Development: Methods and Protocols*, Methods in Molecular Biology, vol. 1082, DOI 10.1007/978-1-62703-655-9_6,

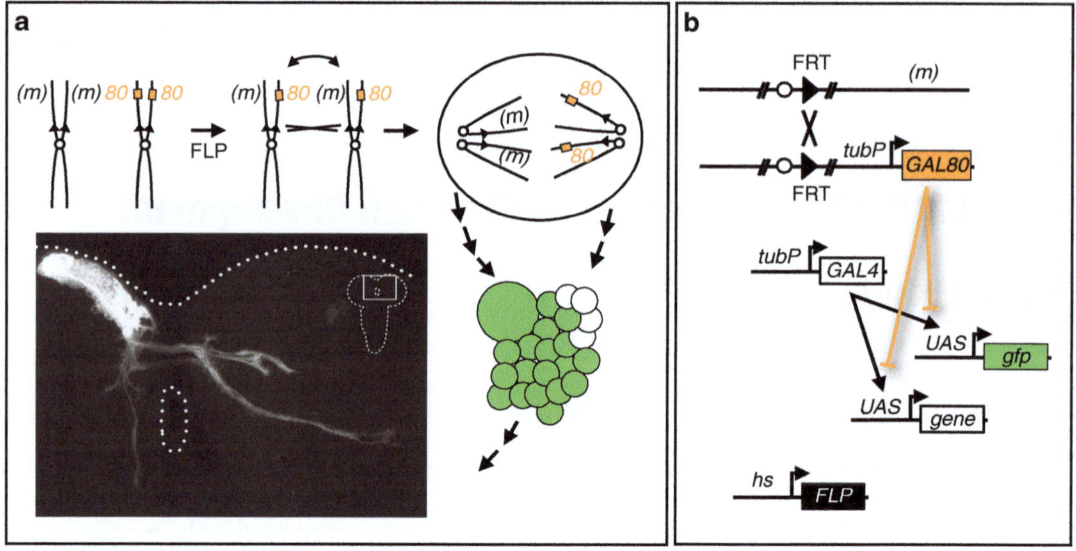

Fig. 1 (**a**) Schematic representation of unequal repressor (*orange*) distribution in MARCM by induced somatic recombination between homologous FRT sites, leading to positively labeled clones (*green*) and unlabeled sister clones. Only the recombining chromosome pair is shown, containing Flp recombinase target recognition (FRT) sites (*black triangles*) and repressor transgene (*orange*). Inset is a maximum projection of a confocal stack of a type II neuroblast clone that was induced around hatching, with wide secondary arborizations (Leica SP5, 63×/1.30 glycerol immersion objective). (**b**) Complete set of elements in cells of heterozygous offspring of a MARCM cross prior to recombination, and its interactions. Flp/FRT system, *black*; GAL4/UAS system, *white*; GAL80 repressor driven by ubiquitous promoter, *orange*; GFP marker, *green*. Some optional elements are included in *parentheses*: a mutation on the GAL80-free FRT chromosome (*m*), and an additional UAS-driven, GAL80-repressible transgene for rescue or ectopic expression of genes within the clone

of the recombined chromosome arms. Thus, if one chromosome arm contains a mutation, the mutation will become homozygous in one of the daughter cells and all its progeny, and the effects of the mutation can be studied in a heterozygous, phenotypically wild-type surrounding.

Positive labeling of clones using a genetically encoded marker such as GFP is not straightforward because any marker will be expressed throughout the heterozygous animal. Mosaic analysis using a repressible cell marker (MARCM) [4, 5] overcomes this limitation by asymmetrically distributing a repressor of marker expression, rather than the marker itself. Thus, within an animal that is heterozygous for the repressor transgene and therefore unlabeled, only clones that lost the repressor will express the marker (Fig. 1a). If a mutation is present on the other, homologous FRT-bearing chromosome, the labeled clone of cells will in addition be homozygous for that mutation. Therefore, MARCM generates positively labeled, wild-type or mutant clones of mitotic cells that allow the investigation of wild-type and mutant development with high temporal and spatial resolution.

As inducible, stochastic method of labeling single cells and their progeny, MARCM has enabled detailed morphological birth date analyses of single neurons as well as tracing of neuroblast lineages. For example, MARCM experiments identified neuronal subtypes produced from mushroom body and antennal lobe projection neurons, and revealed that these are produced by the corresponding neuroblasts in a strict temporal order [6, 7]. Comprehensive analyses of postembryonic neuroblast lineages [8, 9] and glial cells [10, 11] have also been performed. MARCM experiments identified a unique proliferation mode in type II lineages via neuroblast-like transit amplifying cells [12–14], and identified them as source of many larval brain glia [10, 15, 16]. MARCM experiments also revealed the existence of hemilineages that are formed as the two sibling neurons of ganglion mother cells adapt either of the two different fates, a choice mediated by Notch signaling [17, 18]. To the clonal analysis of mutant phenotypes, MARCM added not only a positive label but also the possibility for clone-specific rescue and ectopic expression of genes, for example the switching of cell fates by switching developmental control genes [19–21]. Using a behavioral assay together with clonal stimulation of neuronal activity via the thermosensitive TrpA1 channel, MARCM analysis has identified neuronal clusters that trigger male courtship behavior [22]. For clones that appear at higher frequency, or with directed expression of Flp [23], phenotypic screens for mutations on the FRT-bearing chromosome arm are feasible, and have for example led to the discovery of genes controlling the temporal identity of mushroom body neurons, branching of their axons, and cell type-specific axon pruning during metamorphosis [24–26].

The initially employed and most frequently used repressor in MARCM is the yeast GAL80 protein, driven by a ubiquitous *tubP* promoter, that can silence expression from a plethora of GAL4 driver and UAS responder stocks available for *Drosophila*, as well as the GAL80-repressible LexA-GAD fusion protein in dual MARCM [27]. Alternatively, other repressible transcription systems (Q MARCM, [28]) as well as UAS-marker-RNAi (twin spot MARCM [27]) have been adapted as MARCM systems and used for high-resolution lineage analysis [17, 29]. A limitation of MARCM is the long decay time of the repressor, 24–48 h for GAL80 [4]; this precludes its use for the embryo [30]. The repressor-free twin spot generator is more immediately visible as it recombines the 5′- and 3′- halves of two fluorescent marker genes, separated by an FRT-containing intron, to yield two different fluorescent proteins in sister clones. Thus, the time between clone induction and appearance of label is only limited by expression and fluorophore formation in the marker proteins [31]. Alternatively, flp-out clones, albeit not limited to mitotic cells, can be generated using *UAS>CD2,y>mCD8GFP* (G. Struhl in [32]; for neuroblast clones in the embryo, *see* [30, 33]), *Actin > CD2 > GAL4* [34–36], or *Actin > Draf+ > nlacZ* [37].

This protocol provides a detailed procedure of a standard MARCM experiment in brain development (*see* also [38–43]) that covers the generation of MARCM stocks, crosses and egg collection, clone induction, dissection of brains, immunohistochemistry, and confocal microscopy. The described tools and procedures can be adapted for related clonal techniques and other tissues.

2 Materials

2.1 Fly Stocks

A pair of MARCM stocks contains the following genetic elements (Fig. 1b), available through Bloomington [44]:

1. FRT chromosome with distally located *tubP-GAL80* transgene (*see* **Note 1**).
2. Homologous FRT chromosome, with or without distally located mutation.
3. *GAL4* driver (*see* **Note 2**).
4. *UAS* marker (*see* **Note 3**).
5. *hs-flp* or promoter-driven *flp* transgene (*see* **Note 4**).
6. Balancer chromosomes with markers visible at the time of dissection (*see* **Note 5**).

2.2 Preparing Crosses, Egg Collection, Heat-Shock Induction of Clones

1. Maintenance: Incubator at 25 °C (and 18 °C, optional).
2. Standard fly food [45] with a drop of fresh yeast paste (commercial live baker's yeast suspended in little sugar water), dried under a fan for a few minutes until the surface appears matte.
3. Egg collection: Grape juice plates [45] and matching containers covered with mesh, or food bottles, both with fresh yeast paste (*see* **Note 6**).
4. Heat shock: Water bath at 37 °C.
5. Parafilm to seal plates.

2.3 Dissection

1. Handling animals at different stages: Forceps or a wet brush for wandering larvae and pupae; 30% glycerol and spoon for floating and skimming off pre-wandering larvae; CO_2 fly station for anesthetizing adults.
2. Ice for keeping animals immobilized and for arresting development at exact stages.
3. Stereomicroscope with black working surface and adjustable side illumination, equipped with fluorescence for selecting genotypes prior to dissection.
4. Transparent glass or acrylic dishes for dissection and immunohistochemistry (*see* **Note 7**).
5. Dumont forceps (tip diameters 0.05 × 0.005–0.02 mm, Fine Science Tools).

6. 20 μl micropipette, or wide bore glass Pasteur pipette with 10 ml standard pipette pump (Bel-Art) for transferring brains (*see* **Note 8**).

2.4 Solutions for Dissection and Immunohistochemistry

1. PBS: Tablets (Sigma) or diluted from 10× PBS: 2.56 g/l NaH2PO4, 11.94 g/l Na2HPO4, 102.2 g/l NaCl, pH at 7.4 [46].
2. PBT: 0.1–0.5 % Triton X-100 in PBS.
3. PBL: [47] (*see* **Note 9**) (0.1 M lysine, 0.05 M phosphate buffer, pH 7.4—dissolve lysine in half the volume of water, adjust pH to 7.4 with 0.1 M Na_2HPO4, adjust volume to final concentrations with 0.1 M NaH_2PO4/Na_2HPO4 buffer pH 7.4).
4. 8 % paraformaldehyde stock solution: 0.8 g paraformaldehyde dissolved in 10 ml water with 70 μl 1 M NaOH, incubate in 37 °C water bath with occasional shaking until just dissolved, prepare fresh or freeze aliquots at −20 °C.
5. Methanol.
6. BT-NGS: PBT with 10 % normal goat serum that had been heat inactivated for 30 min at 56 °C and frozen in aliquots.
7. Antibody stocks (*see* **Note 9**).

2.5 Microscopy and Image Analysis

1. Microscope slides.
2. Coverslips that match the objective to be used.
3. Mounting medium, e.g., Vectashield H-1000 (Vector Labs, Burlingame, CA, USA).
4. Spacers (*see* **Note 10**).
5. Nail polish to seal slides (*see* **Note 10**).
6. Confocal scanning microscope with suitable objectives (*see* **Note 11**).
7. Fiji/ImageJ or other image processing software [48].

3 Methods

3.1 Generation of MARCM Stocks and Crosses

1. MARCM offspring for clone induction is produced from a final cross of two stocks to achieve heterozygosity of the two FRT chromosomes (e.g., *FRT82B*, *tubP-Gal80*× *FRT82B*). Distribute the remaining MARCM elements into the two stocks as convenient (*see* **Note 12**), using standard genetic techniques [49, 50].
2. Grow enough flies of the pair of stocks to be able to collect a sufficient number of eggs within the desired time period (*see* **Note 13**).

3. Collect virgins and males, combine them in bottles or tubes with freshly yeasted food, and keep them at 25 °C for 4–5 days to mate (*see* **Note 14**).

3.2 Egg Collection, Heat-Shock Induction of Clones, Brain Dissection

1. Transfer MARCM crosses into egg collectors, cover with yeasted grape juice plates or fly bottles, and collect at 25 °C in the dark for the desired time period.
2. Keep offspring at 25 °C until the desired stage for heat shock (or dissection, if flp is not induced by heat shock).
3. Immerse sealed grape juice plates or fly bottles in a 37 °C water bath for the desired duration (*see* **Note 15**).
4. Raise larvae at 25 °C on fly food until dissection (*see* **Note 16**). Grape juice plates can be cut into pieces and transferred into fly food bottles.
5. Harvest animals as larvae, pupae, or adults: For pre-wandering larvae, fill bottle with 30 % glycerol, skim larvae off the surface with spoon, wash with tap water, and immobilize on ice for genotyping; take wandering third-instar larvae or pupae off the bottle wall with forceps or a wet brush; Anesthetize adults with CO_2 and keep immobilized on ice.
6. Select offspring of the correct genotype, and discard any offspring with balancer chromosomes.
7. Dissect brains of larvae, pupae, or adults (*see* **Note 17**) in PBS. For immunostaining, keep dissected tissue on ice and fix after 20–30 min of dissection. For subsequent live imaging, dissect at room temperature in live imaging media [14, 51].

3.3 Fixation, Immunostaining, and Microscopy

1. Fix freshly dissected tissue for 30–60 min in 2 or 4 % paraformaldehyde in PBL or PBS (*see* **Note 18**).
2. Wash with PBT four times within 45 min.
3. Preincubate in PBT–NGS solution for 15 min.
4. Incubate in primary antibody in PBT–NGS overnight at 4 °C.
5. Wash with PBT four times within 1 h.
6. Incubate in secondary antibody for 2–3 h at room temperature or, preferably, overnight at 4 °C in the dark.
7. Wash with PBT $\geq$4 times within 1 h.
8. Take off liquid with a pipette, and place a drop of Vectashield onto the brains. Without stirring incubate for several hours to overnight, to slowly diffuse in glycerol without osmotically shocking the tissue. Brains should be imaged soon but can be kept in Vectashield at −20 °C for later processing.
9. Place brains onto the slide, remove Vectashield as much as possible so that surface tension holds the brains in place, and arrange them in rows and columns for microscopy, using an eyelash tool or forceps. Larval brains that are still attached

to discs and cuticle can be dissected on the slide, and clones can be preselected under a stereomicroscope equipped with GFP fluorescence (*see* **Note 19**).

10. Place spacers onto the slide or the coverslip (*see* **Note 10**).
11. Place a cleaned coverslip on top of the brains. Ideally, each brain sits in a drop of Vectashield that just touches the coverslip.
12. Let Vectashield seep in from one corner, slowly to avoid bubbles, until the entire space under the coverslip is just filled.
13. Fix the coverslip to the slide with nail polish (*see* **Note 20**).
14. Image as soon as possible for optimal results. Slides can be stored flat at −20 °C for several months with some deterioration of signal.

4 Notes

1. FRT chromosomes with distally located *tubP-GAL80* transgenes are available for all major chromosome arms [41]. Viability of modular MARCM driver stocks that contain all standard elements in the FRT, *tubP-GAL80* stock (i.e., *hs-flp, enhancer-GAL4*, and *UAS-marker*) is usually excellent for virgin collection (Table 1, and [41]). The males then only provide the wild-type or mutant FRT chromosome, and optional *UAS-gene* constructs.

Table 1
Sample MARCM driver stocks that have been used routinely and remained stable over years (constructed by B. Bello)

FRT chr.	Chr1	Chr2	Chr3	Reference
1	*FRT19A, tubP-GAL80LL1, hsFLP^{1}, w**	*tubP-GAL4, UAS-mCD8::GFP^{LL5}/CyO, ActGFP^{JMR1}*		[19]
2L	*y,w,hsFLP^{122}*	*FRT40A, tubP-GAL80^{LL10}/ CyO,ActGFP^{JMR1}*	*tubP-GAL4^{LL7}, UAS-mCD8::GFP^{LL6}/TM6,Tb,Hu*	[14]
2R	*y,w,hsFLP^{1}*	*FRTG13, tubP-GAL80^{LL2}/ (CyO,actin-GFP^{JMR1})*	*tubP-GAL4^{LL7}, UAS-mCD8::GFP^{LL6}/TM6,Tb,Hu*	B. Bello unpub
2R		*FRTG13, tub-Gal80^{LL2}, hs-Flp^{38}/ CyO, actin-gfp^{JMR1}*	*tubP-GAL4^{LL7}, UAS-mCD8::GFP^{LL6}/TM6,Tb,Hu*	B. Bello unpub
3 L	*y,w,hsFLP^{122}*	*tubP-GAL4, UAS-mCD8::GFP^{LL5}/ CyO,ActGFP^{JMR1}*	*FRT2A, tubP-GAL80^{LL3}/ (TM6,Tb,Hu)*	[82]
3R	*y,w,hsFLP^{122}*	*tubP-GAL4, UAS-mCD8::GFP^{LL5}/ CyO,ActGFP^{JMR1}*	*FRT82B, tubP-GAL80^{LL3}/ (TM6,Tb,Hu)*	[83]

They can be crossed with matching, wild-type FRT chromosomes for labeling, or may include other components for functional analysis. For simple labeling, the FRT82B and FRT40A tubulin drivers yield many eggs and controllable clone frequencies. For mutant analysis, the FRT chromosome used will be dictated by the location of the mutation

2. A ubiquitous driver such as *tubP-GAL4* ([4] or J-P Vincent in [19]) or *Actin-GAL4* [9] will label the whole clone. *elavC155-GAL4* [52] can be useful when studying secondary neuronal projections because it leaves the widely arborized primary neurons [9] or neuroblast-derived glia (own observations) largely unlabeled. Non-ubiquitous GAL4 drivers can be used to label only a subset of neuroblast clones [6], or only certain cell types [11], although most GAL4 drivers are not specific [53]. However, lineage relationships can be either missed if not all cells of a clone are labeled, or wrongly assumed if clones are not as sparse as it seems. In addition, the presence of undetected neighboring clones or clone parts can yield ambiguous results if, for example, non-cell-autonomous effects of mutations are involved. An alternative approach is dual MARCM [27], where in addition to the specific GAL4 driver the GAL80-repressible *tubP-lexA-GAD* simultaneously labels the full clone with a different marker driven by the lexA operator. If the GAL4 driver of choice is weak, it can be useful to recombine the GAL4 driver and/or the UAS-marker transgene onto the free FRT chromosome so that they become homozygous within clones. Drivers homozygous throughout the animal may fail to be repressed by the one copy of *tubP-Gal80*.
3. Any combination of UAS-based expression constructs can be used, many available through Bloomington [44]. The membrane-bound UAS-mCD8GFP reporter [4, 53] labels cell bodies and axonal and dendritic processes, while UAS-mCD4GFP labels finer cell processes but labels cell bodies in the brain less strongly [54]. Pre- and postsynaptic markers can highlight axonal and dendritic regions within clones [55–58]. Adding a nuclear marker such as UAS-nlacZ^{20b} (Y. Hiromi and S. West in [44]) is useful for coexpression studies of nuclear factors [19], and to characterize cells with complicated morphology. UAS-Brainbow [59] and UAS-Flybow [47] can label different clones in different colors. Flybow can also further subdivide clones in later recombination events. In addition to a marker, UAS-driven gene constructs for overexpression, rescue, RNAi knockdown, and modifiers of neural activity can be used.
4. With some hs-flp insertions, clones are rare. The more efficient and less leaky hs-flp^{122} [37, 39], or hs-flp^{38} [60], together with a short heat shock of 6–7 min at 37 °C, yields more brains with sparse clones and fewer empty brains than hs-flp^{1} [61] with a longer heat shock of 30–60 min at 37 °C. Enhancer-driven flp such as ey-flp and repo-flp produces mosaics as well [43, 62, 63].
5. Balancers bearing either *Tb*^{1} (e.g., *TM6B Tb*^{1} *Antp*^{Hu} or [64, 65]) or GFP/RFP-expressing transgenes such as *CyO, actin-gfp*^{JMR1} (Reichart, J.-M. in [44]) can be selected for by eye, or under a stereomicroscope equipped with fluorescence

[44, 50, 66]. Absence of the *FRT, tubP-Gal80* chromosome will cause expression of the GAL4/UAS transgenes used; therefore this chromosome may be selectable under a stereomicroscope without the use of a visible balancer.

6. For fast heat distribution during short heat shocks, grape juice plates or fly bottles with an equally thin layer of fly food work well.
7. Contrast for dissection of brains is best in a transparent dish placed on a black surface and illuminated from the side or the back [66, 67]. Dissection can be performed on ice to arrest development at specific stages, to preserve antigens, and to immobilize and anesthetize animals. A metal computer-cooling element, cut to the size of a flat Styrofoam box and eloxated in black, can be placed into the ice-filled box to provide cold, black surface. The cooling element rests on something other than ice to avoid constant change in focus as the ice melts. A square, flat lid of a glass staining jar or a glass petri plate can hold many larvae, while debris can be pushed away to one side during dissection. The bottom of the dissection dish can optionally be coated with Sylgard 184 silicone polymer (Dow Corning) to aid in delicate dissections. A custom-made hand rest, such as a piece of firm foam that is fit around the ice box, can help to keep hands relaxed and still, and provides a soft surface to place dissection tools. Fixation and staining can be done in clear Eppendorf or PCR tubes; we prefer custom-made multi-well dishes cut from 10 mm thick acrylic glass, covered with a microscope slide. Wells of 15 mm diameter and 5 mm depth allow antibody staining in as little as 70 μl, and visual control of brains under the stereomicroscope when changing solutions.
8. Additional dissection tools: Spring scissors (tip diameter 0.05 mm, Fine Science Tools) for quick dissections of pupae and larvae (*see* **Note 17**); 1 ml syringe with a short, 27 G needle for pupal dissection (*see* **Note 17**); O-rings (Fine Science Tools) or pieces of silicone tubing to adjust the maximal opening angle of forceps and use less force to grab tissue; Eyelash tool (Ted Pella, or custom made; lighter colors are visible under fluorescence illumination) to dissect and orient stained brains on the microscope slide (*see* **Note 19**), and to dislodge brains that adhere to the wall of a pipette; sharpening stone made of True Hard Arkansas rock (Dan's Whetstone Co. or Fine Science Tools) to sharpen damaged tips of forceps, and to some degree scissors, under the stereomicroscope. Tools are most frequently damaged by inadverted contact with parts of the stereomicroscope; thus watch from outside while positioning tools into the field of view, and then locate them through oculars. Minimal force is needed to sharpen or re-bend the tips of tools.

9. Some labs routinely fix in PBL [47], not older than 3 months, and others in PBS or other buffers [46]. PBL is derived from a periodate–lysine solution used to fix glycoproteins [68]; lysine alone may help to preserve certain antigens. Commonly used antibodies for postembryonic brains label either synaptic neuropil or axon tracts, described in [69]: BP106/mouse anti-neurotactin (1:20, DSHB [70]) labels larval and pupal secondary lineages and axon tracts. BP106 requires a 5–10-min methanol incubation after fixation (Kathy Ngo and Volker Hartenstein, pers. comm.) that also makes larval and pupal brains less fragile and does not interfere with many other antibody stainings; BP104/mouse anti-neuroglian (1:20, DSHB; [71]) labels secondary axon tracts in the adult brain; nc82/mouse anti-bruchpilot (1:10; DSHB; [72]) is used for the *Drosophila* standard brain [73] and labels synaptic neuropil. nc82 staining in adult brains greatly improves when incubated over two nights and in 0.5 % Triton X-100 [40, 74]. Chicken anti-GFP (1:1,000, Abcam, Cambridge, UK), rabbit anti-RFP (1:200, Abcam), and rabbit anti-*beta*-Galactosidase (1:1,000, Cappel/MP Biomedicals) can be used for transgenic markers, and Alexa 488/568/647 (1:300, Molecular Probes/Life Technologies) secondary antibodies work well for triple labeling.

10. For third-instar larvae, pupae, and adult brains, a spacer slightly thinner than a coverslip will hold brains in place without squashing them, e.g., two layers of scotch tape. For smaller brains, use one layer of tape, or apply a flexible spacer such as Vaseline to the edges of the coverslip, and gently push down. Minor-quality nail polish will not seal well and may interfere with fluorescence. Hardening mounting media are also in use [9].

11. At 20× magnification, the whole brain can be fully scanned; for single-cell resolution (a neuronal cell body from a postembryonic lineage has a diameter around 4–5 μm), 40× or higher magnification is necessary. At 63× magnification, a fixed brain hemisphere from a wandering third-instar larva just fits into the field of view. For whole-hemisphere scans of brains mounted in glycerol-based Vectashield, a glycerol immersion objective minimizes spherical aberrations [75]. In addition, a long working distance is useful, since working distance of higher magnifying oil objectives can be shorter than the thickness of the brain plus coverslip. Valuable aspects of confocal microscopy and image processing artifacts are discussed in [53].

12. If a transgenic chromosome for a planned stock is difficult to balance for introducing the other elements, consider redistributing the MARCM elements in case the desired stock is not viable. It is not infrequent to obtain around 15 flies or less of the correct genotype from a bottle during the final crossing generations.

13. A bottle with 100–120 virgins and 10–30 males, left to lay eggs within 1–2 h at 25 °C, can give rise to 100–200 live larvae depend-

ing on the fecundity of the stock used, and the time of egg collection, highest during the flies' afternoon/evening [50]. Flies that had been raised under crowded conditions are small and lay few eggs; split up bottles with too many larvae into 2–3 bottles prior to virgin collection. If very large amounts of virgins need to be collected, it can save time to cross a virginator such as *hs-hid (Y)* (M. Van Doren in [76]) into the MARCM driver that can kill off most males in a 1-h-long heat shock at the third instar prior to virgin collection (*see* descriptions of Bloomington stocks 8846, 24638). Each element that is homozygous in the parent stock doubles the amount of larvae of the correct genotype in the offspring; thus, if homozygous viable and fertile, avoid males containing a balancer, and allow a day longer for mating to occur with fewer males.

14. Virgin collections and crosses can be lost instantly when shaken into a bottle with sticky food, and females need to remain well fed with yeast paste. Between egg collections, shake MARCM crosses into fresh bottles with dry surface as soon as the drop of yeast is eaten. If bottles tend to be humid, keep flies less densely, and place a rolled-up paper tissue in the food to absorb moisture when not collecting eggs. Crosses can be kept at 18 °C when not in use to prolong the fertile period. Females lay best during the week following mating, but can be used for 3–4 weeks until the amount of eggs declines.
15. Larvae will crawl rapidly to escape the heat; submerge sealed plates completely, or push down the stopper of the fly food bottle below the surface of the water bath. Larvae crawl onto mite-proof stoppers but not cheap foam ones. The optimal time and duration of heat shock need to be determined experimentally to reach the desired clone density. As an estimate, using hs-flp^{122}, a heat shock of 6 min at 37 °C around hatching most frequently induces 0–2 neuroblast clones per brain, while a heat shock of 8 min most frequently induces 1–5 neuroblast clones per brain (own observations). Background clone induction is usually negligible, but single experiments can show considerable clone frequencies without heat shock (own observations). Certain neuroblast clones are only recovered upon heat shock during certain instars, and in smaller developmental time windows than may be expected from other experiments (B. Bello, personal communication).
16. The density at which larvae are reared greatly affects their developmental rate [38, 50]. In my hands, a maximum of 120 larvae in a bottle of 6 cm diameter will develop rather synchronously, but tends to dry out; add water as needed. If larvae are too dense, split them up into several bottles. Larval and pupal development at 18 °C takes quite exactly half as long as development at 25 °C [50], which can be used to avoid nighttime dissections, unless staging needs to be exact.

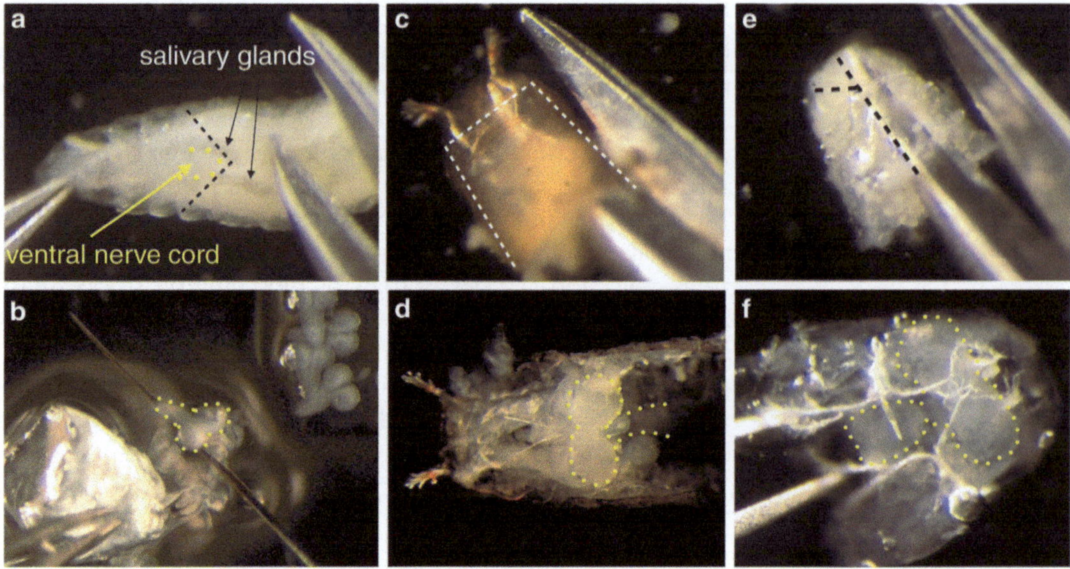

Fig. 2 Dissection of larvae (**a**, **b**), prepupae (**c**, **d**), and pupae (**e**, **f**) using scissors and eyelash tool. Scissors in (**a**) are not positioned at the proposed cut site

17. Dissection is the most time-consuming step in a MARCM experiment; techniques are best collected from several sources and optimized individually [40, 66, 67, 77–79]. If a collection of animals needs to be dissected at a defined developmental stage, some of the intact animals may remain on ice for 2–3 h without greatly affecting the quality of antibody labeling; however, attempt to keep this time as short as possible, for example by preparing a second collection that is 2 h younger than the first.

    Larvae: One quick larval brain dissection technique is to gently tear out the mouth hooks with discs and brain attached (e.g., [78]); the cuticle needs to tear caudal to the anterior spiracles. However, the ventral nerve cord will often break off during the procedure, making later sample orientation more difficult, and the proventriculus can destroy the commissure. Another quick and distortion-free dissection technique is to hold the larva ventral side up at the mouth hooks, locate the ventral nerve cord and neighboring salivary glands through the cuticle, and make an oblique cut with spring scissors at the end of the ventral nerve cord to expose the brain (Fig. 2a). If the cut is placed well, only small amounts of fat body and gut will still be attached. Use the open scissors to invert the anterior end onto the closed forceps, strip the remaining fat body tissue away posteriorly, and remove with one or two cuts. Take care not to cut into the forceps. If the antibody to be used is not sensitive to the presence of fat body, it does not need to be removed completely, which saves time. Young larvae that are too small to be inverted can be opened further with another longitudinal cut, if the brain is not exposed immediately. Still holding the mouth hooks, lift the tissue out of the dissection dish and release into a well filled with PBS and placed on ice. Changing solutions is easier with larger pieces of tissue attached to the brain. Larval brains can be dissected free during mounting in Vectashield, when the tissue is less fragile (Fig. 2b).

Pupae (*see* also [79]): One possibility to dissect pupal brains is to hold the pupa submerged at the anterior spiracles, stab in the middle with one blade of spring scissors to relieve pressure, make one large cut, and then cut into half without squeezing. In prepupae, before head eversion [50, 80], make two long cuts along the sides to the anterior end, lift off the dorsal part like a lid, and remove the dorsal half with an anterior transverse cut between the dorsal and ventral halves (Fig. 2c). An additional cut may be needed to detach the gut. The brain will sit in the ventral half of the puparium case and can be fixed like this (Fig. 2d), since it is fragile especially between 6 and 12 h after puparium formation (APF), and dissected out after fixation and before antibody staining. In older pupae, the anterior half of the pupa can be pulled out from the puparium case with forceps right after cutting the pupa into half. Use forceps to slide a scissor blade underneath the dorsal pupal cuticle all the way to the level of the antennae (Fig. 2e), and make a cut. Open the anterior cuticle further with another mediolateral cut, and free the brain from fat body (Fig. 2f) with a gentle jet of PBS from a 1 ml syringe with short 27 G needle. The brain will adhere to the cuticle only at the retinae. Remove tracheae, and pull the cuticle away from the brain, or if adhesion to the retina is strong, gently tear the cuticle overlying the retina. The water jet can help in dissecting, but if too strong can tear off the optic lobes before the retina is released. Remove the ventral nerve cord if not needed. Transfer pupal brains in a pipette, or in a layer of PBS between almost closed forceps.

Brains from adults, and from older pupae in which the adult cuticle has already formed, can be dissected either from ventral, holding the mouthparts and removing the eye cups one by one [77], or by breaking the cuticle open above the antennae and gently pulling the head capsule apart [67]. If the eyecup is taken off in one piece, the lamina will be removed as well, which greatly improves antibody penetration into the medulla and lobula [78]. If the lamina is needed, the eyecup can be removed in pieces, while removing the red fluorescent pigment as much as possible [81]. The adult brain is covered by large tracheal sacs that need to be removed because they will fill with air during staining and washing. It can be useful to reserve fine forceps for adult brain dissection.

Brightly GFP-labeled clones in freshly dissected or stained tissue are visible under a stereomicroscope equipped with fluorescence; however intensity increases greatly with anti-GFP antibody staining.

18. Avoid removing all liquid when changing solutions to not expose the soft tissue to surface tension, and preserve morphology.
19. For final dissection of larval tissue in Vectashield, line up larval tissue into the tip of a wide-bore Pasteur pipette below the tapered part, place all larvae directly onto the slide, and remove as much Vectashield as possible. Orient a larva so that the brain is visible in side view, place the thin end of an eyelash tool between discs and brain, hold the disc–brain connections against the slide with the eyelash, and pull away the discs with forceps against the elastic resistance of the eyelash (Fig. 2b). Grab the remaining discs with forceps and remove the same way. If the brain is bent over the ventral nerve cord, orient the brain ventral side up, crush the ventral nerve cord against the slide at the site of the bend, and then orient as needed. For best imaging quality, place the most important parts closest to the coverslip.
20. Nail polish will not stick to glass with traces of glycerol; therefore cover all the areas that had Vectashield on them with nail polish. Store slides flat at −20 °C to minimize leaking in case the seal is not perfect.

Acknowledgments

I thank Bruno Bello for constructing and optimizing MARCM stocks, and Bruno Bello, Beate Hartmann, Nadia Riebli, Philipp Kuert, Yanrui Jiang, Susanne Flister, Angela Giangrande, and Heinrich Reichert for sharing material and techniques, and for critical reading of the manuscript. I also thank the imaging facility (Markus Dürrenberger, Oliver Biehlmaier, Alexia Ferrand) and the workshop of the Biozentrum for technical solutions and advice. I acknowledge support via Swiss NSF grant 31003A 140607 to Heinrich Reichert.

References

1. Blair SS (2003) Genetic mosaic techniques for studying Drosophila development. Development 130(21):5065–5072. doi:10.1242/dev.00774 130/21/5065 [pii]
2. Golic KG (1991) Site-specific recombination between homologous chromosomes in Drosophila. Science 252(5008):958–961
3. Xu T, Rubin GM (1993) Analysis of genetic mosaics in developing and adult Drosophila tissues. Development 117(4):1223–1237
4. Lee T, Luo L (1999) Mosaic analysis with a repressible cell marker for studies of gene function in neuronal morphogenesis. Neuron 22(3):451–461. doi:doi:S0896-6273(00)80701-1 [pii]

5. Lee T, Luo L (2001) Mosaic analysis with a repressible cell marker (MARCM) for Drosophila neural development. Trends Neurosci 24(5):251–254. doi:S0166-2236(00)01791-4 [pii]
6. Jefferis GS, Marin EC, Stocker RF, Luo L (2001) Target neuron prespecification in the olfactory map of Drosophila. Nature 414(6860):204–208. doi:10.1038/35102574
7. Lee T, Lee A, Luo L (1999) Development of the Drosophila mushroom bodies: sequential generation of three distinct types of neurons from a neuroblast. Development 126(18): 4065–4076
8. Pereanu W, Hartenstein V (2006) Neural lineages of the Drosophila brain: a three-dimensional digital atlas of the pattern of lineage location and projection at the late larval stage. J Neurosci 26(20):5534–5553. doi:26/20/5534 [pii] 10.1523/JNEUROSCI.4708-05.2006
9. Truman JW, Schuppe H, Shepherd D, Williams DW (2004) Developmental architecture of adult-specific lineages in the ventral CNS of Drosophila. Development 131(20):5167–5184
10. Pereanu W, Shy D, Hartenstein V (2005) Morphogenesis and proliferation of the larval brain glia in Drosophila. Dev Biol 283(1):191–203. doi:S0012-1606(05)00228-9 [pii] 10.1016/j.ydbio.2005.04.024
11. Awasaki T, Lai SL, Ito K, Lee T (2008) Organization and postembryonic development of glial cells in the adult central brain of Drosophila. J Neurosci 28(51):13742–13753. doi:28/51/13742 [pii] 10.1523/JNEUROSCI.4844-08.2008
12. Boone JQ, Doe CQ (2008) Identification of Drosophila type II neuroblast lineages containing transit amplifying ganglion mother cells. Dev Neurobiol 68(9):1185–1195. doi:10.1002/dneu.20648
13. Bowman SK, Rolland V, Betschinger J, Kinsey KA, Emery G, Knoblich JA (2008) The tumor suppressors Brat and Numb regulate transit-amplifying neuroblast lineages in Drosophila. Dev Cell 14(4):535–546. doi:10.1016/j.devcel.2008.03.004; S1534-5807(08)00112-3 [pii]
14. Bello BC, Izergina N, Caussinus E, Reichert H (2008) Amplification of neural stem cell proliferation by intermediate progenitor cells in Drosophila brain development. Neural Dev 3:5. doi:1749-8104-3-5 [pii] 10.1186/1749-8104-3-5
15. Izergina N, Balmer J, Bello B, Reichert H (2009) Postembryonic development of transit amplifying neuroblast lineages in the Drosophila brain. Neural Dev 4:44. doi:1749-8104-4-44 [pii] 10.1186/1749-8104-4-44
16. Viktorin G, Riebli N, Popkova A, Giangrande A, Reichert H (2011) Multipotent neural stem cells generate glial cells of the central complex through transit amplifying intermediate progenitors in Drosophila brain development. Dev Biol 356(2):553–565. doi:S0012-1606(11)01035-9 [pii] 10.1016/j.ydbio.2011.06.013
17. Lin S, Lai SL, Yu HH, Chihara T, Luo L, Lee T (2010) Lineage-specific effects of Notch/Numb signaling in post-embryonic development of the Drosophila brain. Development 137(1):43–51. doi:137/1/43 [pii] 10.1242/dev.041699
18. Truman JW, Moats W, Altman J, Marin EC, Williams DW (2010) Role of Notch signaling in establishing the hemilineages of secondary neurons in Drosophila melanogaster. Development 137(1):53–61. doi:137/1/53 [pii] 10.1242/dev.041749
19. Bello BC, Hirth F, Gould AP (2003) A pulse of the Drosophila Hox protein Abdominal-A schedules the end of neural proliferation via neuroblast apoptosis. Neuron 37(2):209–219. doi:S0896627302011819 [pii]
20. Komiyama T, Johnson WA, Luo L, Jefferis GS (2003) From lineage to wiring specificity POU domain transcription factors control precise connections of Drosophila olfactory projection neurons. Cell 112(2):157–167. doi:S0092867403000308 [pii]
21. Liu Z, Steward R, Luo L (2000) Drosophila Lis1 is required for neuroblast proliferation, dendritic elaboration and axonal transport. Nat Cell Biol 2(11):776–783. doi:10.1038/35041011
22. Kohatsu S, Koganezawa M, Yamamoto D (2011) Female contact activates male-specific interneurons that trigger stereotypic courtship behavior in Drosophila. Neuron 69(3):498–508. doi:S0896-6273(10)01040-8 [pii] 10.1016/j.neuron.2010.12.017
23. Amanai K, Jiang J (2001) Distinct roles of central missing and dispatched in sending the Hedgehog signal. Development 128(24): 5119–5127
24. Wang J, Zugates CT, Liang IH, Lee CH, Lee T (2002) Drosophila Dscam is required for divergent segregation of sister branches and suppresses ectopic bifurcation of axons. Neuron 33(4):559–571. doi:S0896627302005706 [pii]
25. Zheng X, Wang J, Haerry TE, Wu AY, Martin J, O'Connor MB, Lee CH, Lee T (2003) TGF-beta signaling activates steroid hormone receptor expression during neuronal remodeling in the Drosophila brain. Cell 112(3):303–315. doi:S0092867403000722 [pii]

26. Zhu S, Lin S, Kao CF, Awasaki T, Chiang AS, Lee T (2006) Gradients of the Drosophila Chinmo BTB-zinc finger protein govern neuronal temporal identity. Cell 127(2):409–422. doi:S0092-8674(06)01230-X [pii] 10.1016/j.cell.2006.08.045
27. Lai SL, Lee T (2006) Genetic mosaic with dual binary transcriptional systems in Drosophila. Nat Neurosci 9(5):703–709. doi:nn1681 [pii] 10.1038/nn1681
28. Potter CJ, Luo L (2011) Using the Q system in Drosophila melanogaster. Nat Protoc 6(8):1105–1120. doi:nprot.2011.347 [pii] 10.1038/nprot.2011.347
29. Yu HH, Kao CF, He Y, Ding P, Kao JC, Lee T (2010) A complete developmental sequence of a Drosophila neuronal lineage as revealed by twin-spot MARCM. PLoS Biol 8 (8) doi:10.1371/journal.pbio.1000461
30. Kunz T, Kraft KF, Technau, GM, Urbach R (2012) Origin of Drosophila mushroom body neuroblasts and generation of divergent embryonic lineages. Development 139:2510–2522. doi:10.1242/dev.077883
31. Griffin R, Sustar A, Bonvin M, Binari R, del Valle RA, Hohl AM, Bateman JR, Villalta C, Heffern E, Grunwald D, Bakal C, Desplan C, Schubiger G, Wu CT, Perrimon N (2009) The twin spot generator for differential Drosophila lineage analysis. Nat Methods 6(8):600–602. doi:nmeth.1349 [pii] 10.1038/nmeth.1349
32. Wong AM, Wang JW, Axel R (2002) Spatial representation of the glomerular map in the Drosophila protocerebrum. Cell 109(2):229–241. doi:S0092867402007079 [pii]
33. Larsen C, Shy D, Spindler SR, Fung S, Pereanu W, Younossi-Hartenstein A, Hartenstein V (2009) Patterns of growth, axonal extension and axonal arborization of neuronal lineages in the developing Drosophila brain. Dev Biol 335(2):289–304. doi:S0012-1606(09)00926-9 [pii] 10.1016/j.ydbio.2009.06.015
34. Pignoni F, Zipursky SL (1997) Induction of Drosophila eye development by decapentaplegic. Development 124(2):271–278
35. Ito K, Awano W, Suzuki K, Hiromi Y, Yamamoto D (1997) The Drosophila mushroom body is a quadruple structure of clonal units each of which contains a virtually identical set of neurones and glial cells. Development 124(4):761–771
36. de Celis JF, Bray S (1997) Feed-back mechanisms affecting Notch activation at the dorsoventral boundary in the Drosophila wing. Development 124(17):3241–3251
37. Struhl G, Basler K (1993) Organizing activity of wingless protein in Drosophila. Cell 72(4):527–540. doi:0092-8674(93)90072-X [pii]
38. Rooke JE, Theodosiou NA, Xu T (2000) Clonal analysis in the examination of gene function in Drosophila. Methods Mol Biol 137:15–22. doi:1-59259-066-7-15 [pii] 10.1385/1-59259-066-7:15
39. Theodosiou NA, Xu T (1998) Use of FLP/FRT system to study Drosophila development. Methods 14(4):355–365. doi:S1046-2023(98)90591-6 [pii] 10.1006/meth.1998.0591
40. Wu JS, Luo L (2006) A protocol for dissecting Drosophila melanogaster brains for live imaging or immunostaining. Nat Protoc 1(4):2110–2115. doi:nprot.2006.336 [pii] 10.1038/nprot.2006.336
41. Wu JS, Luo L (2006) A protocol for mosaic analysis with a repressible cell marker (MARCM) in Drosophila. Nat Protoc 1(6):2583–2589. doi:nprot.2006.320 [pii] 10.1038/nprot.2006.320
42. Kao CF, Lee T (2010) Genetic mosaic analysis of Drosophila brain by MARCM. In: Zhang B, Freeman MR, Waddell S (eds) Drosphila neurobiology. Cold Spring Harbor Laboratory Press, Cold Spring Harbor, NY, pp 125–140
43. Stork T, Bernardos R, Freeman MR (2012) Analysis of glial cell development and function in Drosophila. Cold Spring Harb Protoc 2012(1):1–17. doi:2012/1/pdb.top067587 [pii] 10.1101/pdb.top067587
44. BDSC http://flystocks.bio.indiana.edu/
45. Ashburner M, Roote J (2000) Laboratory culture of Drosophila. In: Sullivan W, Ashburner M, Hawley S (eds) Drosophila protocols. Cold Spring Harbor Laboratory Press, Cold Spring Harbor, pp 585–599
46. Patel NH (2006) http://www.patellab.org/ > Protocols
47. Hadjieconomou D, Rotkopf S, Alexandre C, Bell DM, Dickson BJ, Salecker I (2011) Flybow: genetic multicolor cell labeling for neural circuit analysis in Drosophila melanogaster. Nat Methods 8(3):260–266. doi:nmeth.1567 [pii] 10.1038/nmeth.1567
48. Eliceiri KW, Berthold MR, Goldberg IG, Ibanez L, Manjunath BS, Martone ME, Murphy RF, Peng H, Plant AL, Roysam B, Stuurmann N, Swedlow JR, Tomancak P, Carpenter AE (2012) Biological imaging software tools. Nat Methods 9(7):697–710. doi:nmeth.2084 [pii] 10.1038/nmeth.2084
49. Greenspan RJ (1997) Fly pushing: the theory and practice of drosophila genetics. Cold Spring Harbor Laboratory Press, Cold Spring Harbor, NY

50. Ashburner M, Golic KG, Hawley RS (eds) (2005) Drosophila: a laboratory manual, 2nd edn. Cold Spring Harbor Laboratory Press, Cold Spring Harbor, NY
51. Cabernard C, Doe CQ (2011) Live imaging of neuroblast lineages within intact larval brains in Drosophila. In: Sharpe J, Wong R, Yuste R (eds) Imaging in developmental biology: a laboratory manual. Cold Spring Harbor Laboratory Press, Cold Sprint Harbor, NY, pp 217–227
52. Lin DM, Fetter RD, Kopczynski C, Grenningloh G, Goodman CS (1994) Genetic analysis of Fasciclin II in Drosophila: defasciculation, refasciculation, and altered fasciculation. Neuron 13(5):1055–1069. doi:doi:0896-6273(94)90045-0 [pii]
53. Ito K, Okada R, Tanaka NK, Awasaki T (2003) Cautionary observations on preparing and interpreting brain images using molecular biology-based staining techniques. Microsc Res Tech 62(2):170–186. doi:10.1002/jemt.10369
54. Han C, Jan LY, Jan YN (2011) Enhancer-driven membrane markers for analysis of nonautonomous mechanisms reveal neuron-glia interactions in Drosophila. Proc Natl Acad Sci U S A 108(23):9673–9678. doi:1106386108 [pii] 10.1073/pnas.1106386108
55. Zhang YQ, Rodesch CK, Broadie K (2002) Living synaptic vesicle marker: synaptotagmin-GFP. Genesis 34(1–2):142–145. doi:10.1002/gene.10144
56. Nicolaï LJ, Ramaekers A, Raemaekers T, Drozdzecki A, Mauss AS, Yan J, Landgraf M, Annaert W, Hassan BA (2011) Genetically encoded dendritic marker sheds light on neuronal connectivity in Drosophila. Proc Natl Acad Sci U S A 107(47):20553–20558. doi:1010198107 [pii] 10.1073/pnas.1010198107
57. Jefferis GS, Potter CJ, Chan AM, Marin EC, Rohlfing T, Maurer CR Jr, Luo L (2007) Comprehensive maps of Drosophila higher olfactory centers: spatially segregated fruit and pheromone representation. Cell 128(6):1187–1203. doi:S0092-8674(07)00204-8 [pii] 10.1016/j.cell.2007.01.040
58. Simpson JH (2009) Mapping and manipulating neural circuits in the fly brain. Adv Genet 65:79–143. doi:S0065-2660(09)65003-3 [pii] 10.1016/S0065-2660(09)65003-3
59. Hampel S, Chung P, McKellar CE, Hall D, Looger LL, Simpson JH (2011) Drosophila brainbow: a recombinase-based fluorescence labeling technique to subdivide neural expression patterns. Nat Methods 8(3):253–259. doi:nmeth.1566 [pii] 10.1038/nmeth.1566
60. Chou TB, Perrimon N (1992) Use of a yeast site-specific recombinase to produce female germline chimeras in Drosophila. Genetics 131(3):643–653
61. Golic KG, Lindquist S (1989) The FLP recombinase of yeast catalyzes site-specific recombination in the Drosophila genome. Cell 59(3):499–509. doi:0092-8674(89)90033-0 [pii]
62. Newsome TP, Asling B, Dickson BJ (2000) Analysis of Drosophila photoreceptor axon guidance in eye-specific mosaics. Development 127(4):851–860
63. Silies M, Yuva Y, Engelen D, Aho A, Stork T, Klambt C (2007) Glial cell migration in the eye disc. J Neurosci 27(48):13130–13139. doi:27/48/13130 [pii] 10.1523/JNEUROSCI.3583-07.2007
64. Lattao R, Bonaccorsi S, Guan X, Wasserman SS, Gatti M (2011) Tubby-tagged balancers for the Drosophila X and second chromosomes. Fly (Austin) 5(4). doi:17283 [pii]
65. Pina C, Pignoni F (2012) Tubby-RFP balancers for developmental analysis: FM7c 2xTb-RFP, CyO 2xTb-RFP, and TM3 2xTb-RFP. Genesis 50(2):119–123. doi:10.1002/dvg.20801
66. Blair SS (2000) Imaginal discs. In: Sullivan W, Ashburner M, Hawley S (eds) Drosophila protocols. Cold Spring Harbor Laboratory Press, Cold Spring Harbor, NY, pp 159–173
67. Ito K (1999) How to dissect Drosophila brains. http://jfly.iam.u-tokyo.ac.jp/html/movie/index.html
68. Van Vactor DL Jr, Cagan RL, Kramer H, Zipursky SL (1991) Induction in the developing compound eye of Drosophila: multiple mechanisms restrict R7 induction to a single retinal precursor cell. Cell 67(6):1145–1155. doi:0092-8674(91)90291-6 [pii]
69. Pereanu W, Kumar A, Jennett A, Reichert H, Hartenstein V (2010) Development-based compartmentalization of the Drosophila central brain. J Comp Neurol 518(15):2996–3023. doi:10.1002/cne.22376
70. Hortsch M, Patel NH, Bieber AJ, Traquina ZR, Goodman CS (1990) Drosophila neurotactin, a surface glycoprotein with homology to serine esterases, is dynamically expressed during embryogenesis. Development 110(4):1327–1340
71. Bieber AJ, Snow PM, Hortsch M, Patel NH, Jacobs JR, Traquina ZR, Schilling J, Goodman CS (1989) Drosophila neuroglian: a member of the immunoglobulin superfamily with extensive homology to the vertebrate neural adhesion molecule L1. Cell 59(3):447–460. doi:0092-8674(89)90029-9 [pii]

72. Wagh DA, Rasse TM, Asan E, Hofbauer A, Schwenkert I, Durrbeck H, Buchner S, Dabauvalle MC, Schmidt M, Qin G, Wichmann C, Kittel R, Sigrist SJ, Buchner E (2006) Bruchpilot, a protein with homology to ELKS/CAST, is required for structural integrity and function of synaptic active zones in Drosophila. Neuron 49(6):833–844. doi:S0896-6273(06)00124-3 [pii] 10.1016/j.neuron.2006.02.008
73. Rein K, Zockler M, Mader MT, Grubel C, Heisenberg M (2002) The Drosophila standard brain. Curr Biol 12(3):227–231. doi:S0960982202006565 [pii]
74. Jiang Y, Reichert H (2011) Programmed cell death in type II neuroblast lineages is required for central complex development in the Drosophila brain. Neural Dev 7:3. doi:1749-8104-7-3 [pii] 10.1186/1749-8104-7-3
75. North AJ (2006) Seeing is believing? A beginners' guide to practical pitfalls in image acquisition. J Cell Biol 172(1):9–18. doi:jcb.200507103 [pii] 10.1083/jcb.200507103
76. Starz-Gaiano M, Cho NK, Forbes A, Lehmann R (2001) Spatially restricted activity of a Drosophila lipid phosphatase guides migrating germ cells. Development 128(6):983–991
77. Yu XM, Wu, J., Sweeney, L., Berdnik, D. (2009) Drosophila adult brain dissection made easy. http://wwwstanfordedu/group/luolab/Linksshtml
78. Morante J, Desplan C (2010) Dissecting and staining Drosophila optic lobes. In: Zhang B, Freeman MR, Waddell S (eds) Drosphila neurobiology. Cold Spring Harbor Laboratory Press, Cold Spring Harbor, NY, pp 157–166
79. Wolff T (2000) Histological techniques for the Drosophila eye part I: larva and pupa. In: Sullivan W, Ashburner M, Hawley S (eds) Drosophila protocols. Cold Spring Harbor Laboratory Press, Cold Spring Harbor, NY, pp 201–227
80. Bainbridge SP, Bownes M (1981) Staging the metamorphosis of Drosophila melanogaster. J Embryol Exp Morphol 66:57–80
81. Williamson WR, Hiesinger PR (2010) Preparation of developing and adult Drosophila brains and retinae for live imaging. J Vis Exp 37:e1936, doi: 1910.3791/1936
82. Kumar A, Bello B, Reichert H (2009) Lineage-specific cell death in postembryonic brain development of Drosophila. Development 136(20):3433–3442. doi:dev.037226 [pii] 10.1242/dev.037226
83. Lichtneckert R, Bello B, Reichert H (2007) Cell lineage-specific expression and function of the empty spiracles gene in adult brain development of Drosophila melanogaster. Development 134(7):1291–1300. doi:dev.02814 [pii] 10.1242/dev.02814

Part II

Other Arthropods

Chapter 7

Dye Coupling and Immunostaining of Astrocyte-Like Glia Following Intracellular Injection of Fluorochromes in Brain Slices of the Grasshopper, *Schistocerca gregaria*

George Boyan and Yu Liu

Abstract

Injection of fluorochromes such as Alexa Fluor® 568 into single cells in brain slices reveals a network of dye-coupled cells to be associated with the central complex. Subsequent immunolabeling shows these cells to be repo positive/glutamine synthetase positive/horseradish peroxidase negative, thus identifying them as astrocyte-like glia. Dye coupling fails in the presence of *n*-heptanol indicating that dye spreads from cell to cell via gap junctions. A cellular network of dye-coupled, astrocyte-like, glia surrounds and infiltrates developing central complex neuropils. Intracellular dye injection techniques complement current molecular approaches in analyzing the functional properties of such networks.

Key words Insect, Brain slices, Dye injection, Immunolabeling, Glia, Gap junctions

1 Introduction

Glia have been shown to play a major regulatory role in establishing neuroarchitecture throughout the insect nervous system [1–4], and those with extensive dendritic projections (or gliopodia) enveloping and infiltrating neuropils such as the central complex are of special interest to studies of brain development [5–9]. Glial/glial and glial/neuronal communication in both invertebrate [10–15] and vertebrate [16–22] nervous systems has been shown to involve gap junctions. In insects such as the grasshopper, gap junctions assemble in glial cells during the latter half of embryogenesis [11] resulting in networks of electrically coupled cells [10].

One accepted indicator of gap junctional communication is the presence of dye coupling as revealed by injected fluorescent dyes [15, 23–25]. In this chapter we describe a combined intracellular dye injection/immunolabeling method for demonstrating dye coupling between cells in unfixed, frozen, brain slices of the embryonic grasshopper *Schistocerca gregaria* and then identifying

Simon G. Sprecher (ed.), *Brain Development: Methods and Protocols*, Methods in Molecular Biology, vol. 1082, DOI 10.1007/978-1-62703-655-9_7, © Springer Science+Business Media, LLC 2014

these cells as being astrocyte-like glia associated with the central complex [26]. Dye coupling is completely blocked in the presence of *n*-heptanol, an acknowledged gap junctional blocker [27–30], in the bathing medium. Intracellular dye injection and immunolabeling are also successful in agarose-embedded brains but the quality of dye coupling is affected by the necessary fixation procedure (*see* ref. 30). Our method using frozen, unfixed, brain slices reveals a network of dye-coupled glia to surround developing central complex modules [26]. Such intracellular studies complement existing molecular approaches for examining glial interactions in brain development [6, 8].

2 Materials

2.1 Frozen Brain Slices

1. Tissue Freezing Medium® (125 mL dripper bottle, Jung).
2. 2-methylbutane maintained at –20 °C for shock freezing tissue.
3. Make a set of glass pipettes, which have been broken back via a diamond knife and the tips flamed blunt in a Bunsen burner. The opening should be large enough to accommodate via suction the heads/brains of all embryo stages being examined. These pipettes will henceforth be referred to as "modified pipettes."
4. Plastic wells (Pelco) filled with tissue freezing medium into the brains are placed for freezing. These wells automatically shape and size the frozen blocks for mounting in the Cryostat (Leica cm3050s).
5. Freeze an appropriate number of Superfrost® Plus (Menzel-Gläser) microscope slides.
6. Pap pen (G. Kisker) to draw a fluid-repelling boundary line around the brain slice on the slide. Store at 4 °C.
7. Indelible blue lab marker (Sigma) for labeling the slide and marking brain slice location (on the glass underside).
8. Prefrozen glass slide container (cuvette) with vertical slots for storing up to ten glass slides.
9. Freezer held at –18 to –20 °C for storing containers with sectioned tissue.
10. Cover slips (17 μm thickness).
11. Fast-drying nail varnish.

2.2 Dissection and Associated Supplies

1. Fine forceps, sizes #3, #5 (both Dumont, Switzerland).
2. Iridectomy scissors (Martin, Germany).
3. Micropipette (Eppendorf) with exchangeable tips.
4. Packages of exchangeable 1 μL tips for antibody delivery, and 50 μL tips for medium exchange (we use Ultratip Greiner Bio-one).

5. Obtain a fine (30 G) needle and matching syringe for injecting the 200 mM KCl into the glass electrode from its blunt end.
6. Binocular microscope (×10 oculars) with adequate zoom capabilities (×10) for dissection of embryos and brains.

2.3 Solutions, Media

1. 0.1 M PBS: 2.6 g/L NaH_2PO_4, 22.5 g/L Na_2HPO_4, 102.2 g/L NaCl, adjust to pH 7.4 with NaOH. Store at 4 °C. Take 100 mL PBS and store over ice on the day of making frozen sections.
2. Prepare an ice-cold 1 % methylene blue solution in 0.1 M PBS and place an aliquot in each of several ice-cold Eppendorf (100 μL) containers. Embryonic grasshopper heads are small (up to 1 mm) so several heads can be placed in one Eppendorf.
3. Physiological saline: 8.8 g/L NaCl, 0.2 g/L KCl, 0.3 g/L $CaCl_2$, 2.5 g/L $MgSO_4$, 0.9 g/L TES buffer. Adjust to pH 7.0 with the buffer. Store at 4 °C. Decant 10 mL for each experimental sitting and allow to reach room temperature prior to the experiment.
4. Fetal Calf Serum (FCS), store frozen (−18 °C). Maintain an aliquot of 10 mL in an Eppendorf at 4 °C. This fluid will be used for dipping the electrode tip into and for internally coating the glass pipettes with which heads/brains are transferred.
5. Aqueous bouin: picric acid 71 %, formaldehyde 24 %, glacial acetic acid 5 %. Fixative for immunolabeling with both anti-glutamine synthetase (for glia) and horseradish peroxidase (for neurons). Store under a fume hood at room temperature.
6. 3.7 % paraformaldehyde (PFA) as the fixative for double glutamine synthetase/repo immunolabeling. Store in a fume hood at room temperature.
7. PBT containing NHS: 1 % normal horse serum (NHS), 0.1 % bovine serum albumin (BSA), PBT (1 % Triton X-100 in 0.1 M PBS). Adjust to pH 7.4 with buffer. Store at 4 °C in the dark.
8. After dye injection and immunolabeling, we use Vectashield® (Vector laboratories) to cover brain slices and minimize bleaching during fluorescence microscopy.
9. The fluorochrome we use here is Alexa Fluor® 568 (Invitrogen, emission maximum 600 nm), but we also used Alexa Fluor® 488 (Invitrogen, emission maximum 517 nm) and Lucifer Yellow CH (Aldrich, emission maximum 528 nm) (*see* ref. 30 and **Note 1**). Alexa Fluor® 568 as a hydrazine has a low molecular weight (790 Da) and so flows readily throughout cells on injection. Store the stock at −18 °C. Make up a 0.1 mL aliquot of 10 mM Alexa Fluor® 568 in 200 mM KCl and store in a glass ampule (3-1801, neoLab, Heidelberg) at 4 °C in the dark. Maintain 10 mL of 200 mM KCl at room temperature for backfilling each micropipette.

10. Make up 50 mL of 1×10^{-5} M *n*-heptanol in TES buffer. This concentration is based on experiments in the embryonic grasshopper brain [26, 30] and in intact, unfixed, *Hydra* [31]. Store at 4 °C.

2.4 Primary Antibodies

1. Anti-glutamine synthetase (BD Transduction Laboratories, mouse anti-6/GS, Nr. 610518). Glutamine synthetase catalyzes the conversion of ammonia and glutamate to glutamine [32]. It is found in glial cells of the vertebrate brain [33, 34] and particularly in astrocytes [35]. The glutamine synthetase antibody recognizes an octamer of identical 45 kDa subunits.
2. Anti-horseradish peroxidase (rabbit, Dianova, 323-005-021). HRP belongs to the cell surface glycoproteins which include cell adhesion and signal molecules [36]. In insects the epitope is neuron-specific [37, 38].
3. Anti-Repo. The expression of the glial-specific homeobox gene *reversed polarity* (*repo*) is revealed by using the anti-Repo primary antibody (Mab 4a3) (*see* ref. 39).

2.5 Secondary Antibodies

1. For anti-glutamine synthetase we used donkey anti-mouse (DAM)-Alexa Fluor® 488 (Invitrogen, emission 519 nm).
2. For anti-HRP and anti-repo, we used either goat anti-rabbit (GAR)-Cy5 (emission 670 nm, Dianova) or goat anti-rabbit (GAR)-Cy3 (emission 570 nm, Dianova). Choose as appropriate to avoid spectral confusion in double-labeling experiments (*see* **Note 3**).

2.6 Electrodes for Dye Injection

1. We used micropipettes made of thin-walled borosilicate glass (GC 100TF-10, Clark Electromedical Instruments), with 1.0 mm outer diameter (O.D.) and 0.78 mm inner diameter (I.D.). An inner filament assists fluid transfer into the tip. Pipettes are packaged into lots of 100 and are precut to a length of 10 cm.
2. To manufacture electrodes you will need an automated electrode puller (e.g., Sutter Instruments, USA) equipped with nitrogen gas cooling of the filament to regulate electrode-tip length. Microelectrodes must suit the immediate purpose and are therefore very individual, but if you have an electrode puller similar to ours, try the following settings: heat 612, pull 186, velocity 100, time 80, nitrogen 1.05, and pressure 5 bar, to make microelectrodes of 5 cm overall length, 7 mm tip length with resistances of 30–40 MΩ when filled with the dye solution (10 mM Alexa Fluor® 568, Invitrogen, in 200 mM KCl), and suitable for filling cell bodies of about 12 μm diameter. The electrode should not be too long otherwise bending and high resistances present problems; if the electrode is too short, it will be blunt and make too large a hole in the cell with

consequent leaking of dye. Our puller makes pairs of electrodes of equal length and quality so make a batch of maximally 20 electrodes at one sitting but use these within 2–3 days. The tips are very fine and since glass is a fluid crystal it will deform with time, making dye injection more difficult.

3. Take a glass or plastic petri dish (10 cm diameter, with cover) and mold fresh malleable plasticine to a flattened block, which is then placed to fit across the diameter of the petri dish. The block should be about 5 cm broad and about 1 cm thick. Take one microelectrode and make repeated impressions in the plasticine block at evenly spaced intervals so that about 6–8 microelectrodes can be stored side by side in the one petri dish. Cover the dish to prevent dust access (glass carries electrostatic charge which attracts dust) and store in a fume hood.
4. A perspex electrode holder for 1.0 mm O.D. glass micropipettes (WPI Instruments, FL, USA). Make sure that the electrode holder has a pressure release opening otherwise the electrode may move during penetration as internal fluid pressures change. The electrode holder contains an internal Ag/AgCl metal site for contact with the electrolyte in the electrode and a pin to which a shielded cable from the head stage (positive pole) of a DC amplifier (see below) is connected.
5. You will need a 4–5 cm length of chlorided silver wire to act as a reference electrode. The simplest method for chloriding is to place about 5 g of silver chloride (AgCl) in a small fireproof container over a Bunsen burner and heat to melting. Repeatedly dip the end (approximately 2 cm) of the silver wire (held by wooden forceps) into the melted AgCl until an even coating has been obtained. This coated length dips into the physiological saline covering the brain slice and the chloride coating will prevent polarization voltages developing between reference and recording electrode. Solder the non-chlorided end of the silver wire to a shielded cable (shielding connects to ground) and attach to the negative pole of the head stage of the DC amplifier. Over time the chloride coating will develop cracks and necessitate repeating the coating procedure.
6. Micromanipulator (e.g., Leitz) equipped with fine adjustment (manual or motorized) of *X*, *Y*, *Z* axes. This guides the electrode holder with its micropipette to the target cell.

2.7 Electronics

1. DC amplifier (we use a Getting 5) equipped with a head stage that is positioned as near to the preparation as possible to reduce electrical interference, and with current passing facility via a virtual ground circuit.
2. A simple 2-channel oscilloscope for monitoring current injection and electrode resistance. Use the electrode resistance-testing capability of the amplifier to check that the electrode has a

resistance of 30–40 MΩ when filled with the fluorochrome used (10 mM Alexa Fluor® 568, Invitrogen, in 200 mM KCl). Resistances can be calculated by injecting repeated 1 nA current pulses (of 100 ms duration) across the bridge circuit of the amplifier and cancelling the evoked voltage shift being viewed on the oscilloscope screen with an internally calibrated current of opposite polarity.

2.8 Fixed-Stage Compound Microscope and Camera

1. In a fixed-stage microscope, the optics not the stage are raised and lowered, allowing focusing of tissues with intracellular electrode in position. You will need a fixed-stage compound microscope (e.g., Zeiss Axioskop 2) equipped with filters for both epifluorescence illumination (tetramethyl rhodamine isothiocyanate, TRITC, emission 580 nm; fluorescein isothiocyanate, FITC, emission 528 nm) and differential interference contrast (DIC)/transmitted light optics. The microscope should stand inside an earthed Faraday cage to minimize electrical interference from lamps and circuits in the building, and on a vibration-free table (preferably supported by compressed air) for stability so as to optimize microelectrode positioning.
2. The microscope should be equipped with a low-power (×10) objective (e.g., Zeiss Plan-Neofluar, 440330, n.A. 0.30) for centering the brain slice and electrode tip and a high-power (×63) water-immersion objective with long working distance (2.1 mm) for monitoring the actual intracellular penetration (e.g., Zeiss Achroplan ceramic-coated ×63 objective, 440067-9901, n.A. 0.95 W). The ceramic renders the objective electrically inert when placed in the saline solution covering the brain slice.
3. A color CCD camera (1.3 MP is adequate) mounted on the microscope, driver software for a computer, a flatscreen monitor (at least 19 in. diagonal and 1,640 × 1,050 pixel resolution is recommended). A micrometer scale etched into a glass slide (Zeiss 473390-9901) is used for checking that the microscope/camera adapter provides a 1:1 image size on the monitor.

2.9 Imaging

1. To reveal fine structures such as gliopodia in 3D, you need to acquire optical sections (a so-called Z stack) of preparations using a confocal laser scanning microscope. We used a Leica TCS SP5 equipped with Leica Hc Pl Apo CS ×20 and Leica Hcx Pl Apo lambda blue ×63 oil immersion objectives (*see* **Note 4**).
2. Fluorochromes were visualized using an argon laser with excitation wavelengths of 488 nm for Alexa Fluor® 488, 578 nm for both Alexa Fluor® 568 and Cy3, and 633 nm for Cy5.
3. Z-stacks of confocal images were processed using public domain software (ImageJ).

3 Methods

3.1 Preparing Frozen Brain Slices

1. Using the #5 forceps and iridectomy scissors, which have previously been dipped in FCS to prevent tissue adhesion, dissect embryos out of the egg into ice-cold 0.1 M PBS and free embryos from surrounding membranes. For grasshoppers, embryos are staged according to percentage of development (*see* ref. 40).
2. For younger embryos (<60 %), decapitate embryos with a single snip of the iridectomy scissors. For older embryos (>70 %), the cuticle becomes an obstacle during cryosection. Therefore open the head capsule by cutting away the front with the iridectomy scissors, cut through the left and right optic stalks, and extract the brain carefully from the head capsule.
3. Take a modified glass pipette and suck up FCS repeatedly to coat the inside. This prevents tissue adhering to the glass. Use the modified pipette to now suck up single heads/brains and transfer to an ice-cold 1 % methylene blue solution in 0.1 M PBS for 5 min. to stain the tissue surface and thus facilitate visual orientation of tissue within the frozen blocks for sectioning.
4. Transfer the heads/brains, unfixed, into the wells containing Tissue Freezing Medium®. Shock freeze the wells by placing them in a cuvette containing 2-methylbutane maintained at −20 °C. Do not add sucrose as this adversely affects the quality of later dye coupling.
5. Orient the frozen block containing the head/brain in the Cryostat according to the desired plane of sectioning. We generally sectioned horizontally with respect to the body axis and at 16 μm thickness using a Leica low profile 819 microtome blade. Replace this blade as required. This slice thickness allowed individual cells surrounding the central body of the brain to be easily targeted for intracellular dye injection under DIC optics and without damaging primary glial processes.
6. Each brain slice (still frozen and attached to the blade) is transferred by direct application onto a separate frozen Superfrost® Plus (Menzel-Gläser) microscope slide. For older embryos the depth of the brain could mean that 20 or so slides are necessary. This 1:1 procedure is important because it means that only one brain slice is being investigated (at room temperature) at any given time. Any additional brain slices on the slide would dry out during the experiment.
7. Draw a complete ring around the section with the Pap pen for frozen slides. This ring will limit the spread of the saline solution covering the tissue slice and allow maximum fluid contact between the slide and the water-immersion objective.

We also marked the location of the brain slice itself by placing a dot on the underside of the glass slide with the indelible marker.

8. Transfer all the frozen slides from one brain to the same prefrozen glass slide container and store containers in a freezer at −18 °C. We stored sectioned tissue for up to 5 days without noticeable changes in tissue preservation or dye coupling.

3.2 Intracellular Dye Injection

1. In preparation for dye injection, transfer the frozen slide with its single brain slice to the fixed-stage microscope. Immediately cover with physiological saline at room temperature to a depth of at least 5 mm to allow good contact with the immersion objective. Too little fluid and the surface tension forces generated by the objective might tear the slice from the slide. Allow 1–2 min for temperature acclimatization. Confirm using DIC optics at ×10 magnification that the preparation is centered within the optical field of the microscope. This is critical because once the ×63 objective is in use, repositioning of the slice is limited by the restricted visual field offered by the objective.
2. Take the Eppendorf containing FCS from the refrigerator and allow to come to room temperature.
3. Remove an electrode carefully from the array within the petri dish (too sudden a movement may cause the tip to snap off!). Fill the electrode with the fluorochrome solution by placing its blunt end vertically into the solution and allowing it to travel via capillary action along the internal filament into the tip. This will take about 1–2 min. The entire electrode should then be filled with 200 mM KCl from its blunt end via the fine syringe needle to enable a good electrical contact with the Ag/AgCl electrode within the electrode holder. Dip the electrode tip several times into the FCS to act as a lubricant facilitating penetration and subsequent removal of the electrode from the target cell.
4. Slide the electrode into the perspex microelectrode holder (1.0 mm diameter, WPI Inc., FL, USA). Ensure there are no salt residues on the electrode holder by wiping down with a tissue wetted with aqua dest.
5. Position the electrode holder into an appropriate micromanipulator equipped with adjustment (manual or motorized) of *X*, *Y*, *Z* axes. The shielded cable from the microelectrode holder is connected to the positive pole on the head stage of a DC amplifier (in our case a Getting 5) equipped with current passing facility via a virtual ground circuit.
6. Place the tip of the chlorided silver wire acting as a reference electrode into the physiological solution covering the preparation and connect to the negative pole of the amplifier head stage.

7. Using transmitted light, position the microelectrode tip so that it is centered in the field of view of the ×10 objective on the compound microscope. Lower the electrode via the micromanipulator so that the tip dips into the saline covering the preparation, and continue lowering until the tip is located just above the surface of the brain slice.
8. Now switch on your preamplifier and use the electrode resistance-testing capability of the amplifier to check that the electrode has a resistance of 30–40 MΩ when filled with the fluorochrome. Resistances can be calculated by injecting a 1 nA current pulse (of 50 ms duration) across the bridge circuit of the amplifier and cancelling the evoked voltage shift being viewed on the oscilloscope screen with an internally calibrated current of opposite polarity. If the resistance exceeds 50–60 MΩ, reject the electrode as it will probably block on dye injection. Also *see* **Note 2** for checking your system for electrical neutrality.
9. Switch to the high-power (×63) water-immersion objective. Lower the objective into the saline and continue lowering until the microelectrode tip is in view. Center the microelectrode tip and lower to the tissue surface while constantly changing the focus so that the target cell also comes into view. The angle of deflection of the microelectrode holder will have to be adjusted so that the microelectrode tip does not travel through too much nontarget tissue (we use a 15° depression angle). The steeper the angle, the closer the tip will have to be positioned to the cell before penetration. Microelectrode tip and target cell are both kept in focus and the electrode tip is propelled into the cell via the micromanipulator (Fig. 1a).
10. Intracellular penetration of the target cell is monitored using the DIC optics (transmitted light) of the Zeiss Axioskop 2 microscope. We performed penetrations exclusively of the cell soma to avoid damage to, and inadvertent staining of, projections from neighboring cells. On penetration, switch from transmitted light to epifluorescence (for Alexa Fluor® 568, use rhodamine or TRITC filters; *see* ref. 30 for settings).
11. Commence dye injection by applying constant hyperpolarizing current via the amplifier and observe the initial passing of dye into the cell (Fig. 1b). The fluorochrome carries a charge so use negative current to inject dye into the cell. Do not exceed 5 nA injected current as monitored on an oscilloscope screen. Use the microscope shutter to block the fluorescence excitation and continue injection in the dark. Unblock occasionally to check injection progress on the computer screen. Make photographs with the camera software for documentation purposes (*see* **Note 3**). In our case dye injection proceeded for

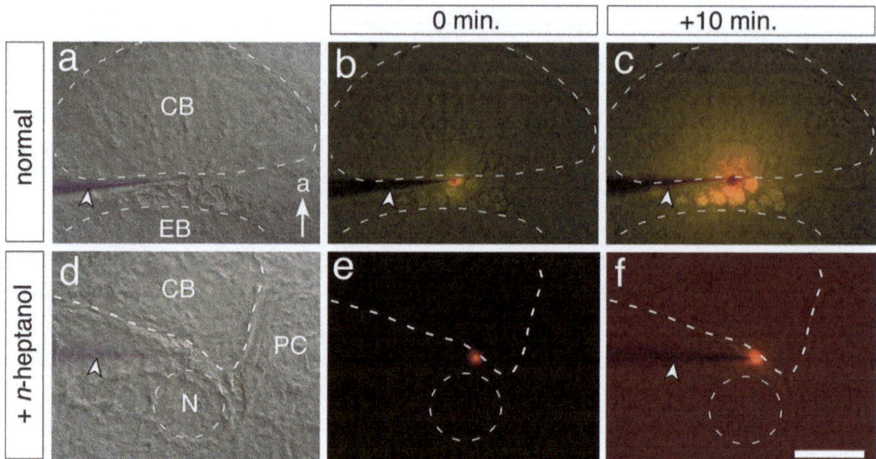

Fig. 1 Intracellular injection of Alexa Fluor® 568 reveals dye-coupled glia cells associated with the central complex of the embryonic grasshopper brain. Photomicrographs of horizontal embryonic brain slices: (**a–c**) normal, (**d–f**) after addition of bathing medium containing 1×10^{-5} M *n*-heptanol (a gap junctional blocker). (**a**, **d**) DIC images show dye-filled (*purple*) electrode (*white arrowheads*) on penetration of a single target cell. Central complex neuropils (*CB* central body, *EB* ellipsoid body, *N* nodulus) are outlined *dashed white*. *Arrow* indicates anterior (**a**). Combined transmitted light/fluorescence images show the brain slices on commencement (**b**, **e**: 0 min) and after 10 min. (**c**, **f**) Dye injection. Note the large collection of dye-coupled cells normally (**c**) and their complete absence when *n*-heptanol is present (**f**). Scale bar represents 70 μm. Panels modified from ref. 26

10–15 min beyond which there was no change in the quality of staining or dye coupling (Fig. 1c). Constantly monitor electrode resistance on the oscilloscope (*see* **Note 4**).

12. On completion of staining, switch off the amplifier, revert to transmitted light, and use the micromanipulator to carefully withdraw the microelectrode tip from the cell. Remove the reference electrode from the solution. Clean the ×63 objective with distilled water to remove salt residues.

3.3 Gap Junctional Blocker

1. Alexa Fluor® 568 has a small molecular weight [41] so that its passage between glia cells might be mediated by gap junctions. We tested this possibility by applying TES buffer containing 1×10^{-5} M *n*-heptanol—a proven gap junctional blocker (*see* refs. 28, 29)—and then repeating the dye injection experiment as described above (Fig. 1d).
2. In each experiment involving *n*-heptanol we found that only the penetrated cell was stained after 10 min of injection with Alexa Fluor® 568 (Fig. 1e). This shows that the injection of the dye itself was not affected by the presence of *n*-heptanol. Complete absence of dye coupling (Fig. 1f) supports the presence of gap junctions between the labeled cells in our brain slices.

3.4 Immunolabeling

1. Immunolabeling is performed directly on the brain slice still attached to the glass slide. Remove the slide from the microscope and immediately apply fixative (aqueous Bouin or paraformaldehyde as appropriate) to the saline covering the slice. Drain the fixative/saline mixture from the slice after 1–2 min and replace with fresh pure fixative. Keep slides overnight at 4 °C in the dark.
2. After fixation drain the fixative using a micropipette with an exchangeable 50 μL tip and replace with 0.1 M PBS as a washing solution. Exchange this solution at least 6 times over the course of 1 h. Bring a 5 mL aliquot of preincubation medium to room temperature. Then exchange the washing solution with preincubation medium for 2 h at room temperature to block unspecific binding sites.
3. The following primary antibodies were then applied to the brain slice, either singly or in combination depending on the data required: (a) anti-glutamine synthetase, dilute 1:200 in the preincubation medium; (b) anti-horseradish peroxidase, dilute 1:150 in preincubating medium; and (c) anti-repo, dilute 1:500 in preincubation medium. Drain the preincubation medium currently covering the slice and replace with fresh medium containing the primary antibody(s). In each case, slices remain exposed to primary antibodies for 3 days at 4 °C in the dark.
4. After exposure to the primary antibody(s), drain medium from the brain slice with a micropipette and wash sections by repeated exchanges of 0.1 M PBS using a fresh pipette tip. Then replace with the same preincubation medium as above to which the relevant secondary antibody has been added as follows: for anti-glutamine synthetase, DAM-Alexa Fluor® 488:1:300 dilution; for anti-HRP, GAR-Cy5 or GAR-Cy3:1:150 dilution; and for anti-Repo, GAR-Cy5 or GAR-Cy3:1:150 dilution. *See* **Note 5** on avoiding spectral confusion. Exposure of the brain slice to the secondary antibody(s) was for 24 h at 4 °C in the dark.
5. Incubation medium was then removed and replaced by several changes of 0.1 M PBS. The brain slices remained in PBS overnight at 4 °C in the dark and were subsequently covered in Vectashield® (Vector laboratories) to minimize photobleaching. Cover slips were applied and sealed to the glass slide with fast-drying nail varnish to prevent movement under the oil immersion objectives used in confocal microscopy.

3.5 Imaging

1. We acquired optical sections of preparations with a Leica TCS SP5 confocal laser scanning microscope equipped with ×10 oculars and ×20 and ×63 oil immersion objectives (*see* **Note 6**).
2. Confocal images following double immunolabeling carried out on brain slices prepared as above or via agarose embedding

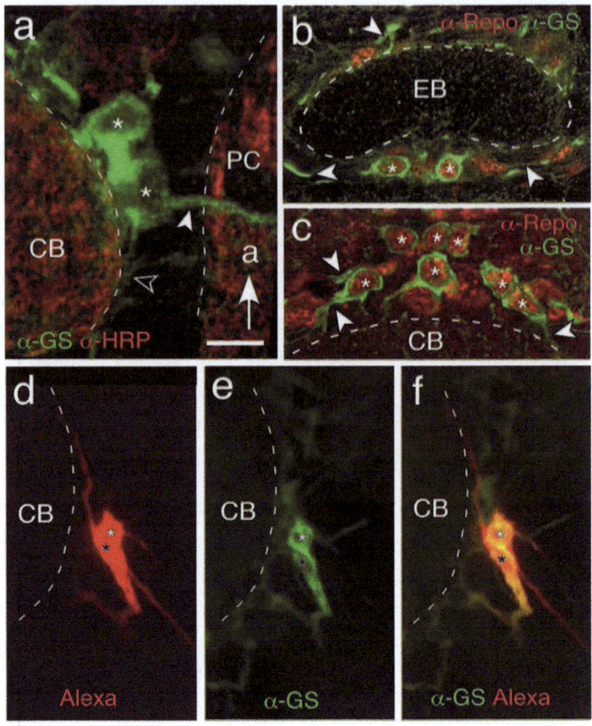

Fig. 2 Immunolabeling confirms glial identity of cells. (**a**) Confocal image of a brain slice following agarose embedding and double immunolabeling shows cells associated with the central body (CB) to be glutamine synthetase (GS) positive (*green*)/HRP negative (*red*). Two cells (*white stars*) direct glia podia around the CB (*open white arrowhead*) as well as (*white arrowhead*) into neighboring neuropil of the protocerebrum (PC). *Arrow* indicates anterior (*a*). (**b, c**) All GS-positive cells (*green*) associated with the ellipsoid body (*EB*) (**b**), and central body (*CB*) (**c**), co-express the glial-specific nuclear antigen Repo (*red, white stars*) confirming their identity as astrocyte-like glia. Note the extensive GS-positive gliopodia (*white arrowheads*) surrounding these neuropils. (**d–f**) Confocal images confirm glial identity of a pair of dye-coupled cells (*black, white stars*) associated with the central body (*CB*) following Alexa Fluor® 568 injection in a brain slice. (**d**) *Red channel* shows dye-injected (*black star*) and dye-coupled (*white star*) cells following dye injection. (**e**) *Green channel* shows both cells to be glutamine synthetase positive. (**f**) Combined channels confirm co-labeling (*yellow*) of anti-GS and Alexa and therefore glial identity of both cells. Scale bar in (**a**) represents 12 μm, 25 μm in (**b**), 20 μm in (**c–f**). Panel (**a**) modified from ref. 26; panels (**b, c**) modified from ref. 9

reveal that the cells surrounding central complex neuropils are both glutamine synthetase positive/HRP negative (Fig. 2a) and glutamine synthetase positive/repo positive (Fig. 2b, c), confirming their identity as astrocyte-like glia. Confocal examination of a brain slice following injection of Alexa Fluor® 568 into a single cell associated with the central body reveals dye-coupled cells (Fig. 2d) which on immunolabeling prove to be glutamine synthetase positive (Fig. 2e, f), consistent with their being astrocyte-like glia (Fig. 2b, c).

4 Notes

1. The Alexa Fluor® 568 fluorochrome was preferred over Alexa Fluor® 488 and Lucifer Yellow (which also revealed dye coupling) because of its low molecular weight, staining intensity, and because its emission wavelength is sufficiently distant so as to prevent spectral overlap with those of the secondary antibodies used in the subsequent immunohistochemistry. For a comparison of results using these various fluorochromes, *see* [30].
2. You will need to check the electrical neutrality of your recording/injection system with an artificial electrode (a 10 MΩ resistor in parallel with a 4.7 pF ceramic capacitor) connected to your preamplifier and oscilloscope to ensure that no 60-cycle interference is present. Once the real recording and reference electrodes are in place, you will need to check that no voltage displacement is present on the oscilloscope. If there is, it is a sign that current is flowing across the electrode resistance and, depending on its polarity, this could lead to uncontrolled dye release into the tissue prior to penetration of a cell. Salt bridges caused by saline residues on microscope or electrode holder are a common source of polarizations. Wipe these down regularly using ion-free water.
3. To allow better definition of stained cells and considerably reduce background when the brain slice was simultaneously illuminated with transmitted light on the Zeiss Axioskop 2 fluorescence microscope, the injected Alexa® Fluor 568 was also excited at violet excitation wavelengths (430–440 nm) and, using the camera-specific software (Visicapture™), captured at yellow/orange emission wavelengths (535 nm).
4. If resistance rises sharply under current injection, it indicates that the electrode is blocking and dye is no longer flowing freely into the cell. If this happens try to relieve the block with repeated pulses of positive current. If this does not work, break off injection, turn on fluorescence, and examine the quality of staining. If adequate, apply fixative to the preparation. Otherwise start again with a fresh electrode and another target cell elsewhere in the slice. Never reuse an electrode as tissue invariably adheres to the tip causing blockage.
5. In experiments that involved combinations of antibodies with Alexa Fluor® 568, the fluorochrome conjugated with each secondary antibody needs to emit at a distinctly different wavelength to avoid spectral confusion. The advantage of the Cy5

fluorochrome in this regard is that it signals in infrared (670 nm) and so can be allocated any color by imaging software after confocal scanning.

6. Tip: Use the ×63 objective, rather than digitally zooming the ×20 objective, to capture high-resolution morphological details of the stained cells.

Acknowledgements

This work was supported by DFG grant BO 1434/3–5 and the Graduate School of Systemic Neuroscience, University of Munich. We thank S. Götz for assistance with dye injection and immunolabeling of glia (Fig. 2d–f).

References

1. Jones BW, Fetter RD, Tear G, Goodman CS (1995) Glial cells missing: a genetic switch that controls glial versus neuronal fate. Cell 82: 1013–1023
2. Jones BW (2001) Glial cell development in the *Drosophila* embryo. Bioessays 23:877–887
3. Hidalgo A (2003) Neuron-glia interactions during axon guidance in *Drosophila*. Biochem Soc Trans 31:50–55
4. Klämbt C (2009) Modes and regulation of glial migration in vertebrates and invertebrates. Nat Rev Neurosci 10:769–779
5. Vanhems E (1995) Insect glial cells and their relationships with neurons. In: Vernadakis A, Roots B (eds) Neuron-Glia interrelations during phylogeny: II. Plasticity and regeneration. Humana, Totowa, NJ, pp 49–77
6. Awasaki T, Lai S-L, Ito K, Lee T (2008) Organization and postembryonic development of glial cells in the adult central brain of *Drosophila*. J Neurosci 28:13742–13753
7. Edwards TN, Meinertzhagen IA (2010) The functional organisation of glia in the adult brain of *Drosophila* and other insects. Prog Neurobiol 90:471–497
8. Awasaki T, Lee T (2011) New tools for the analysis of glial cell biology in *Drosophila*. Glia 59:1377–1386
9. Boyan G, Loser M, Williams L, Liu Y (2011) Astrocyte-like glia associated with the embryonic development of the central complex in the grasshopper *Schistocerca gregaria*. Dev Genes Evol 221:141–155
10. Schofield PK, Swales LS, Trehern JE (1984) Potentials associated with the blood-brain barrier of an insect: recordings from identified neuroglia. J Exp Biol 109:307–318
11. Swales LS, Lane NJ (1985) Embyronic development of glial cells and their junctions in the locust central nervous system. J Neurosci 5:117–127
12. Schmidt J, Deitmar JW (1996) Photoinactivation of the giant neuropil glial cells in the leech *Hirudo medicinalis*: effects on neuronal activity and synaptic transmission. J Neurophysiol 76:2861–2871
13. Alexopoulos H, Böttger A, Fischer S et al (2004) Evolution of gap junctions: the missing link? Curr Biol 14:879–880
14. Phelan P, Goulding LA, Tam JLY et al (2008) Molecular mechanism of rectification at an identified electrical synapse in the *Drosophila* giant fibre system. Curr Biol 18:1955–1960
15. Koussa MA, Tolbert LP, Oland LA (2011) Development of a glial network in the olfactory nerve: role of calcium and neuronal activity. Neuron Glia Biol 6(4):245–261
16. Zahs KR, Newman E (1997) Asymmetric gap junctional coupling between glial cells in the rat retina. Glia 20:10–22
17. Newman EA (2001) Propagation of intercellular calcium waves in retinal astrocytes and Müller cells. J Neurosci 21:2215–2223
18. Theiss C, Meller K (2002) Aluminum impairs gap junctional intercellular communication between astroglial cells in vitro. Cell Tissue Res 310:143–154
19. Houades V, Koulakoff A, Ezan P et al (2008) Gap junction-mediated astrocytic networks in the mouse barrel cortex. J Neurosci 28: 5207–5217

20. Tanaka M, Yamaguchi K, Tatsukawa T et al (2008) Connexin43 and bergmann glial gap junctions in cerebellar function. Front Neurosci 2:225–233
21. Parys B, Cote A, Gallo V et al (2010) Intercellular calcium signaling between astrocytes and oligodendrocytes via gap junctions in culture. Neuroscience 167:1032–1043
22. Rela L, Bordey A, Greer CA (2010) Olfactory ensheathing cell membrane properties are shaped by connectivity. Glia 58:665–678
23. Hossain MZ, Ernst LA, Nagy JI (1995) Utility of intensely fluorescent cyanine dyes (Cy3) for assay of gap junctional communication by dye-transfer. Neurosci Lett 184:71–74
24. Ball KK, Gandhi GK, Thrash J et al (2007) Astrocytic connexin distributions and rapid, extensive dye transfer via gap junctions in the inferior colliculus: implications for [^{14}C] Glucose metabolite trafficking. J Neurosci Res 85:3267–3283
25. Lanosa XA, Reisin HD, Santacroce I et al (2008) Astroglial dye-coupling: an in vitro analysis of regional and interspecies differences in rodents and primates. Brain Res 1240:82–86
26. Boyan GS, Liu Y, Loser M (2012) A cellular network of dye-coupled glia associated with the embryonic central complex in the grasshopper Schistocerca gregaria. Dev Genes Evol 222:125–138
27. Burt JM, Spray DC (1980) Single-channel events and gating behavior of the cardiac gap junction channel. Proc Natl Acad Sci U S A 85:3431–3434
28. Weingart R, Bukauskas FF (1998) Long-chain n-alkanols and arachidonic acid interfere with the Vm-sensitive gating mechanisms of gap junction channels. Eur J Physiol 435:310–319
29. Juszczak GR, Swiergiel AH (2009) Properties of gap junction blockers and their behavioural, cognitive and electrophysiological effects: animal and human studies. Prog Neuropsychopharmacol Biol Psychiatry 33:181–198
30. Boyan GS, Niederleitner B (2011) Patterns of dye coupling involving serotonergic neurons provide insights into the cellular organization of a central complex lineage of the embryonic grasshopper *Schistocerca gregaria*. Dev Genes Evol 220:297–313
31. Chapman JA, Kirkness EF, Simakov O et al (2010) The dynamic genome of *Hydra*. Nature 464:592–596
32. Wedler FC, Horn BR (1976) Catalytic mechanisms of glutamine synthetase enzymes. J Biol Chem 251:7530–7538
33. Martinez-Hernandez A, Bell KP, Norenberg MD (1977) Glutamine synthetase: glial localization in brain. Science 195:1356–1358
34. van der Hel WS, Notenboom RGE, Bos IWM et al (2005) Reduced glutamine synthetase in hippocampal areas with neuron loss in temporal lobe epilepsy. Neurology 64:326–333
35. Ward M, Jobling A, Puthussery T et al (2004) Localization and expression of the glutamate transporter, excitatory amino acid transporter 4, within astrocytes of the rat retina. Cell Tissue Res 315:305–310
36. Snow PM, Patel NH, Harrelson AL et al (1987) Neural-specific carbohydrate moiety shared by many surface glycoproteins in *Drosophila* and grasshopper embryos. J Neurosci 7:4137–4144
37. Jan LY, Jan YN (1982) Antibodies to horseradish-peroxidase as specific neuronal markers in *Drosophila* and grasshopper embryos. Proc Natl Acad Sci U S A 79:2700–2704
38. Haase A, Stern M, Wächtler K et al (2001) A tissue-specific marker of Ecdysozoa. Dev Genes Evol 211:428–433
39. Halter DA, Urban J, Rickert C et al (1995) The homeobox gene repo is required for the differentiation and maintenance of glial function in the embryonic nervous system of *Drosophila melanogaster*. Development 121: 317–322
40. Bentley D, Keshishian H, Shankland M et al (1979) Quantitative staging of embryonic development of the grasshopper, *Schistocerca nitens*. J Embryol Exp Morphol 54:47–74
41. Weber PA, Chang HC, Spaeth KE et al (2004) The permeability of gap junction channels to probes of different size is dependent on connexin composition and permeant-pore affinities. Biophys J 87:958–973

Part III

Molluscs

Chapter 8

Methods in Brain Development of Molluscs

Andreas Wanninger and Tim Wollesen

Abstract

Representatives of the phylum Mollusca have long been important models in neurobiological research. Recently, the routine application of immunocytochemistry in combination with confocal laser scanning microscopy has allowed fast generation of highly detailed reconstructions of neural structures of even the smallest multicellular animals, including early developmental stages. As a consequence, large-scale comparative analyses of neurogenesis—an important prerequisite for inferences concerning the evolution of animal nervous systems—are now possible in a reasonable amount of time. Herein, we describe immunocytochemical staining protocols for both whole-mount preparations of developmental stages—usually 70–300 μm in size—as well as for vibratome sections of complex brains. Although our procedures have been optimized for marine molluscs, they may easily be adapted for other (marine) organisms by the creative neurobiologist.

Key words Immunocytochemistry, Fluorescence, Antibody staining, Whole-mount, Neurogenesis, Free-floating vibratome sections, Complexity, Cephalopods, Brains, Lophotrochozoa

1 Introduction

The Mollusca constitutes one of the most diverse animal phyla, comprising, to name but a few, small worm-shaped groups in the millimeter range, valve-bearing polyplacophorans, bivalves, and snails, as well as the cephalopods with their highly complex behavior [1]. The wide variety in molluscan gross morphology is also represented by the neuroanatomical features of its representatives. As such, while basal taxa as well as small-sized individuals often exhibit a weakly ganglionated central nervous system (CNS) that mostly consists of interconnected longitudinal nerve cords, "higher" gastropods and all cephalopods have sophisticated brains that result from fusion of various ganglia [2–4]. Despite the long history of gastropods and cephalopods as model systems in neurophysiological and neuroanatomical research [5, 6], surprisingly little is known concerning the ontogeny of the nervous system on the one hand and precise neurotransmitter and neuropeptide distribution within the adult molluscan CNS on the other.

Simon G. Sprecher (ed.), *Brain Development: Methods and Protocols*, Methods in Molecular Biology, vol. 1082, DOI 10.1007/978-1-62703-655-9_8, © Springer Science+Business Media, LLC 2014

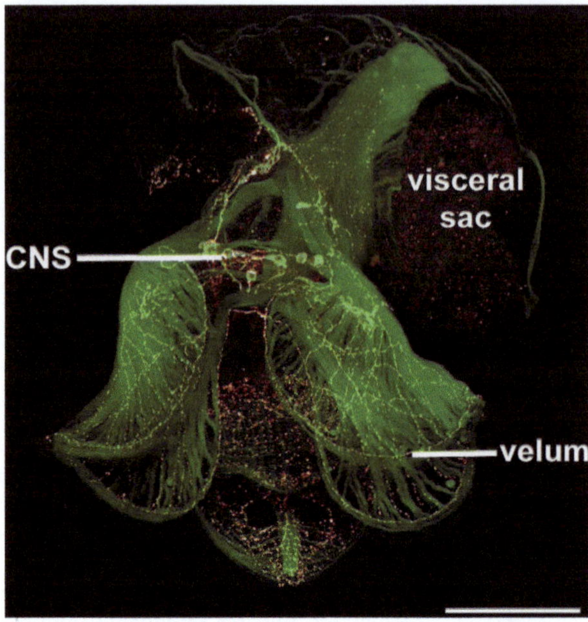

Fig. 1 Maximum intensity projection of a whole-mount staining of a veliger larva of the marine gastropod *Aplysia californica* (dorsal view, anterior faces downwards). The serotonergic nervous system is labeled in *red* and the musculature (by F-actin staining using phalloidin) in *green*. Scale bar: 100 μm

Since immunocytochemical staining in combination with confocal microscopy and 3D reconstruction techniques, which allow detailed high-throughput analyses, have now become routine lab procedures, this picture is likely to change rapidly. Indeed, a number of comparative studies on invertebrate—and hence also molluscan—neurogenesis have become available in the past few years and have injected important novel data into the discussion on functional as well as evolutionary aspects concerning molluscan nervous systems [7–15]. Herein, we present an easy-to-do and reliable protocol for whole-mount studies of the molluscan nervous system of minute specimens, including embryonic and larval stages (*see* Fig. 1), as well as a convenient procedure for nervous tissue staining of free-floating vibratome sections (*see* Fig. 2) that allows for precise determination of immunoreactive substances in complex brains of gastropods and cephalopods.

2 Materials

Use deionized H_2O for the preparation of all solutions. Buffer solutions may be stored at room temperature. Sera and antibodies are stored undiluted at –20 °C and may be kept in the fridge for a few days after reconstitution. Avoid repeated freezing and thawing.

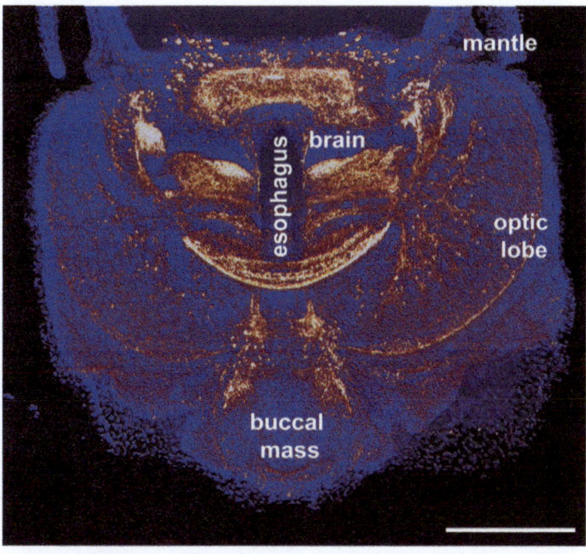

Fig. 2 Maximum intensity projection of a 100 μm thick free-floating vibratome section of the CNS of the squid *Loligo vulgaris* (dorsal view, anterior faces downwards). FMRF amide-like immunoreactivity is labeled in *red* and perikarya are *blue* (DAPI). Scale bar: 300 μm

2.1 Whole-Mount Staining of Neural Tissue

1. Make solution for relaxation of live embryos and larvae. 7.14 % stock solution: 38.1 g $MgCl_2 \times 6H_2O$, 100 mL water. Use parts of it to make a 3.5–3.6 % $MgCl_2$ solution.
2. 4 % Paraformaldehyde (PFA) in 0.1 M phosphate buffer (PB) for fixation: 33.48 g of $Na_2HPO_4 \times 2H_2O$, 7.93 g $NaH_2PO_4 \times H_2O$, and 100 mL H_2O. Alternatively: 26.7 g Na_2HPO_4, 6.9 g NaH_2PO_4, and 100 mL H_2O. Adjust pH to 7.3. Dissolve 4 g of PFA powder in 100 mL PB or make a 1:10 dilution of custom-made 4 % PFA solution in PB (*see* **Notes 1** and **2**).
3. Washing solution: 0.1 % NaN_3 to PB is recommended to avoid bacterial or fungal growth (not necessary for the fixative).
4. Many molluscan larvae bear shells, which are best decalcified after fixation in 0.05 M EGTA (dissolve 1.9 g EGTA powder in 100 mL H_2O) (*see* **Note 3**).
5. PBT: 0.2–10 % Triton X-100 (concentration depending largely on specimen size and tissue), 0.1 M PB + 0.1 % NaN_3.
6. Block-PBT solution: 6 % normal goat serum (NGS) in PBT.
7. Antibody solution: dilute all primary and secondary antibodies at the respective working concentration in Block-PBT (*see* **Note 4**).
8. Fluorescent counterstains: e.g., DAPI (for nuclei), phalloidin or phallacidin (for F-actin including musculature).

9. Clearing medium: Murray's Clear ("BBA"), a 2:1 mixture of benzyl benzoate:benzyl alcohol; prepare also a 1:1 dilution of 100 % EtOH and BBA as intermediate.
10. 75 % EtOH.
11. Mounting medium (e.g., Vectashield, Fluoromount G).

2.2 Staining of Vibratome Sections

1. Vibratome.
2. Dissection forceps.
3. Paper towels.
4. Rubber mold.
5. Razor blade.
6. Instant glue.
7. Coated objective slides.
8. Cover slip.
9. Vibratome embedding medium: 24.2 g ovalbumin in 66 mL H_2O. Adjust the stirrer to minimize agglutination of the powder and skimming of proteins. Close beaker with lid to avoid desiccation. Heat up 25 mL H_2O to 50 °C in a water bath. Add 5.5 g of gelatin from porcine skin in little portions to avoid agglutination of the powder. Close beaker with lid to avoid desiccation. Slowly combine both solutions as soon as gelatin and albumin are dissolved. Add 1.2 g NaN_3 to avoid fungal and/or bacterial growth. Embedding medium may be stored up to ½ year at 4 °C. The entire process takes approximately 5 h.
10. Elvanol mounting medium [16]: 5 g Mowiol, 20 mL PB, pH 7.3. Prepare in a covered Erlenmeyer flask and stir for 16 h at room temperature (RT). Add 10 mL of glycerol and stir again for 16 h at RT. Centrifuge solution at either 4,000 rpm for 15 min, 3,000 rpm for 20 min, or 2,000 rpm for ½ h at RT. Aliquot supernatant into 1.5 mL portions in Eppendorf tubes and store at −20 °C.
11. Coating medium for objective slides: Thoroughly rinse objective slides in vial containing a mix of 50 % acetone and 50 % pure EtOH. Dry objective slides with paper towels and incubate at 60 °C for 24 h. Heat up 100 mL H_2O to 50 °C in a covered Erlenmeyer flask in a water bath.

    Add 1 g gelatin and 0.1 g chrome alum (potassium chromium sulfate) when temperature reaches 50 °C and stir for 15 min. Filter solution and transfer into a vial. Submerse objective slides thrice in coating medium for 3 s per step. Dry coated objective slides at 40 °C in oven overnight.

3 Methods

All procedures are carried out at RT unless stated otherwise.

3.1 Fixation

Identical for whole-mount and vibratome-sectioned samples:

1. Animals are carefully anesthetized by adding 7.14 % $MgCl_2$ to the seawater or by cooling the samples on ice prior to fixation, especially important for larvae that may retract into their shells. For marine species, eventually replace all seawater by 3.5 % $MgCl_2$ to avoid precipitation of the PB salts once you add the fixative.
2. Specimens are fixed in 4 % PFA in 0.1 M PB, for 1–3 h at RT (or overnight at 4 °C for larger specimens).
3. Samples are rinsed in 0.1 M PB thrice for 10 min and twice for 1 h at RT. They are stored in 0.1 M PB + 0.1 % NaN_3 at 4 °C. Alternatively, samples may be stepped into 75 % EtOH and stored at −20 °C.

3.2 Whole-Mount Immunocytochemistry

1. Decalcify in 0.05 M EGTA if necessary (samples are best watched under polarized light to check progress of decalcification).
2. Transfer specimens into Block-PBT. Incubate for 2–24 h at 4 °C.
3. Incubate specimens overnight (18–24 h) in primary antibody(ies) at 4 °C (*see* **Notes 5** and **6**).
4. Wash specimens for 2–24 h in Block-PBT at 4 °C with a minimum of four changes.
5. Incubate specimens overnight (18–24 h) in secondary antibody(ies) at 4 °C in the dark (*see* **Note 7**).
6. If nucleic acid counterstain is desired, add drops of DAPI (working concentration: 1 %) at least 1 h prior to the subsequent washes.
7. If phalloidin staining is desired (e.g., for neuropil staining), add 2.5 % phalloidin or phallacidin at least 1 h prior to the subsequent washes.
8. Wash specimens for 2–24 h in PB without NaN_3 at 4 °C with a minimum of four changes in the dark (*see* **Note 8**).
9. If no clearing is required, mount specimens on objective slides, preferably in a medium with antifade reagent (e.g., Vectashield, Fluoromount G).
10. Larger and nontransparent specimens (e.g., yolky embryos and larvae) may require clearing. Step samples into H_2O and an ascending EtOH series (5–10 min per step with three changes in 100 % EtOH) after **step 6**. Then transfer them into a 1:1 dilution of EtOH:BBA (10 min) and finally into 100 % BBA. Mount specimens on objective slides in this solution. OBS: EtOH will abolish phalloidin and phallacidin staining. If any of

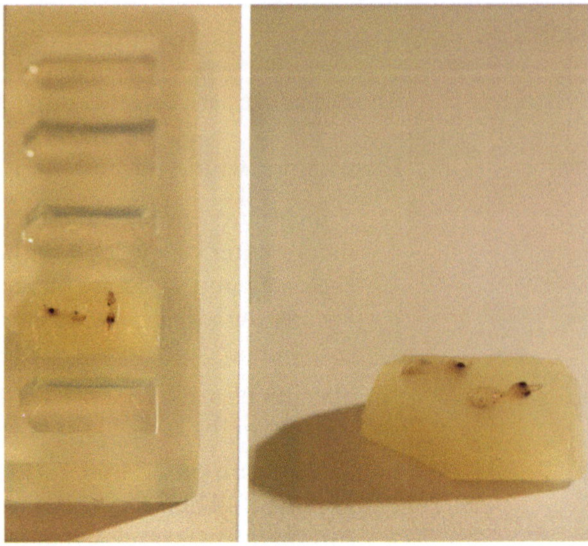

Fig. 3 Hatchlings of the squid *Loligo vulgaris* embedded in gelatin-albumin medium in a rubber mold for vibratome sectioning (*left*). Trimmed block in pyramidal shape ready for sectioning (*right*)

these stains are used, dehydration should be carried out in an isopropanol series with very short steps (30 s each), followed by 1:1 isopropanol:BBA and 100 % BBA application.

3.3 Immunocytochemistry on Vibratome Sections

1. Step specimens into 0.1 M PB if stored in 75 % EtOH.
2. Remove all hard parts from sample (e.g., beak, radula, eye lens, or shell) manually or by decalcification (*see* Subheading 3.2, **step 1**) since they might interfere with sectioning.
3. Excise portion of embedding medium, transfer to small beaker, and incubate at 60 °C in oven for 15 min.
4. Dry sample gently with paper towels (do not overdry).
5. Pour embedding medium into rubber mold and quickly submerse specimen (*see* **Note 9**).
6. Center specimen in rubber mold with forceps and ensure its proper orientation for the sectioning process (*see* **Note 10**).
7. Store rubber mold at 4 °C for 15 min to solidify embedding medium.
8. Store rubber mold in a 10 mL 37 % formaldehyde and 90 mL 0.1 M PB solution at 4 °C overnight. This step will interlink proteins of the embedding medium and the specimen.
9. Transfer specimen in rubber mold into 0.1 M PB and store at 4 °C until further processing (*see* **Note 11**).
10. Carefully remove embedded sample block from rubber mold (*see* Fig. 3).

11. Trim block with razor blade to pyramidal shape and cut off a single corner. This will facilitate orientation of the sections (*see* Fig. 3).
12. Fill tray with chilled 0.1 M PB.
13. Glue trimmed sample with instant glue on stage and let dry for 10 s.
14. Submerse sample in 0.1 M PB and adjust thickness of sections (*see* **Note 12**).
15. Adjust begin and end, frequency, and speed of sectioning process.
16. Section trimmed block with a vibratome and collect all sections of interest with forceps. Ensure that problematic hard parts of the specimen are located at the distal end of each section. By doing so, only a little portion of the specimen will rupture in case that these parts are impossible to section.
17. Apply immunocytochemistry protocol to the sections of interest as described above (*see* Subheading 3.2, **steps 2–6**).
18. Mount vibratome sections on coated objective slides.
19. Orient and dry sections with forceps and paper towels (do not overdry).
20. Apply approximately 100 μl of mounting medium alongside the objective slide. Gently cover sections with a cover slip and avoid air bubbles.
21. Let sample harden in the dark at 4 °C for a few hours to several days.

4 Notes

1. Either 5× or 10× PB stock solutions may be made. Against common belief, it is not necessary to store these at 4 °C. If so, the PB salts may form crystals, which have to be dissolved by heating.
2. As with PB, the PFA fixative may be stored at RT. If stored at 4 °C, PFA may form a cloudy precipitation at the bottom of the container. On the other hand, formaldehyde molecules may cross-link after a while (days to weeks) and may hamper rapid tissue penetration which may result in poor tissue preservation. It is always best to use freshly made fixative.
3. PFA and EGTA powders dissolve poorly. Heat the solution but avoid boiling. Use fume hood. Let solution cool down and adjust pH to 7.3.
4. Concentrations of antibodies vary depending on the species investigated. As a rule of thumb, one may start with a 1:400

dilution for commercially available primary antibodies (e.g., anti-serotonin) and 1:100–1:300 for commercial secondaries, but often concentrations as low as 1:2,000 may be feasible.

5. For double or multi-labeling with antibodies, the primary antibodies must be raised in different hosts. Antibodies may be applied as "cocktail" of the respective primaries and secondaries in their respective working concentrations.
6. Perform negative controls by omitting the primary and secondary antibodies (only one component at a time), respectively, in independent experiments. This should prevent labeling of neuronal elements and ensures that fluorescence signal due to autofluorescence or unspecific binding is not misinterpreted as positive signal. In addition, a primary antibody can be incubated in its respective antigen with conjugated bovine serum albumin. This preabsorption experiment should not reveal any immunoreactivity, suggesting that the given antibody specifically binds to the respective antigen.
7. Since the secondary antibodies bear the signal-producing fluorochrome, all steps from secondary antibody incubation onwards should be carried out in the dark. Subsequent washes and mounting is best done by using a stereomicroscope with illumination from below, whereby the white light is shielded by a red filter (simple plastic sheet is sufficient).
8. NaN_3 quenches fluorescence. Therefore, it is vital to omit NaN_3 in the last washes prior to mounting.
9. Compared to protocols that use, e.g., resins, embedding specimens with gelatin-albumin medium is quick and straightforward. However, since this medium does not penetrate the tissue, delicate internal structures may be damaged during the sectioning or subsequent mounting process. Hence, it is important to ensure proper exposure of the entire specimen to the embedding medium.
10. The orientation of the block and the embedded specimen should always be in accordance with its stability during the sectioning process, i.e., one large side will be cut horizontally (*see* Fig. 3).
11. Blocks should not be stored for longer than 2 days since the embedding medium will soften.
12. The section thickness should be determined experimentally when employing different antibodies or direct stains (e.g., phalloidin) on different species. The working distance and other parameters of the microscope objectives should also be taken into account. As a rough estimate, physical sections with a thickness of 50–100 μm have been proven well-suited for cephalopod and gastropod brains.

Acknowledgments

We thank Rudi Loesel (RWTH Aachen University) for advice concerning vibratome sectioning. This work was supported by the FWF (Austrian Science Fund) grant P24276-B22 to A.W.

References

1. Haszprunar G, Wanninger A (2012) Molluscs. Curr Biol 13:R510–R514
2. Bullock TH, Horridge GA (1965) Structure and function in the nervous systems of invertebrates. Freeman and Company, San Francisco, p 1611
3. Lin MF, Leise EM (1996) Gangliogenesis in the prosobranch gastropod *Ilyanassa obsoleta*. J Comp Neurol 374:180–193
4. Shigeno S, Tsuchiya K, Segawa S (2001) Embryonic and paralarval development of the central nervous system of the loliginid squid *Sepioteuthis lessoniana*. J Comp Neurol 437: 449–475
5. Chase R (2002) Behavior & its neural control in gastropod molluscs. Oxford University Press, Oxford, p 314
6. Nixon M, Young JZ (2003) The brains and lives of cephalopods. Oxford University Press, Oxford, p 392
7. Marois R, Carew TJ (1997) Projection patterns and target tissues of the serotonergic cells in larval *Aplysia californica*. J Comp Neurol 386:491–506
8. Friedrich S, Wanninger A, Brückner M, Haszprunar G (2002) Neurogenesis in the mossy chiton, *Mopalia muscosa* (Gould) (Polyplacophora): evidence against molluscan metamerism. J Morphol 253:109–117
9. Wanninger A, Haszprunar G (2003) The development of the serotonergic and FMRF-amidergic nervous system in *Antalis entalis* (Mollusca, Scaphopoda). Zoomorphology 122:77–85
10. Braubach OR, Dickinson AJG, Evans CCE, Croll RP (2006) Neural control of the velum in larvae of the gastropod, *Ilyanassa obsoleta*. J Exp Biol 209:4676–4689
11. Voronezhskaya EE, Nezlin LP, Odintsova NA, Plummer JT, Croll RP (2008) Neuronal development in larval mussel *Mytilus trossulus* (Mollusca: Bivalvia). Zoomorphology 127: 97–110
12. Todt C, Wanninger A (2010) Of tests, trochs, shells, and spicules: development of the basal mollusk *Wirenia argentea* (Solenogastres) and its bearing on the evolution of trochozoan larval key features. Front Zool 7:6
13. Wollesen T, Loesel R, Wanninger A (2009) Pygmy squids and giant brains: mapping the complex cephalopod CNS by phalloidin staining of vibratome sections and whole-mount preparations. J Neurosci Methods 179:63–67
14. Wollesen T, Cummins SF, Degnan BM, Wanninger A (2010) FMRFamide gene and peptide expression during central nervous system development of the cephalopod mollusk, *Idiosepius notoides*. Evol Dev 12:113–130
15. Wollesen T, Nishiguchi MK, Seixas P, Degnan BM, Wanninger A (2012) The VD1/RPD2 α1-neuropeptide is highly expressed in the brain of cephalopod mollusks. Cell Tissue Res 348:439–452
16. Rodriguez J, Deinhardt F (1960) Preparation of a semipermanent mounting medium for fluorescent antibody studies. Virology 12: 316–317

Part IV

Xenopus Protocols

Chapter 9

In Situ Hybridization and Immunostaining of Xenopus Brain

Kai-li Liu, Xiu-mei Wang, Zi-long Li, Rong-qiao He, and Ying Liu

Abstract

The dynamic expression pattern analysis provides the primary information of gene function. Differences of the RNA and/or protein location will provide valuable information for gene expression regulation. Generally, in situ hybridization (ISH) and immunohistochemistry (IHC) are two main techniques to visualize the locations of gene transcripts and protein products in situ, respectively. Here we describe the protocol for the whole brain dissection, the in situ hybridization and immunostaining of the developing *Xenopus* brain sections. Additionally, we point out the modification of in situ hybridization for microRNA expression detection.

Key words Expression pattern, Brain dissection, Section, Immunostaining, In situ hybridization, *Xenopus*, microRNA

1 Introduction

In situ hybridization and immunostaining are widely used practical techniques to detect the locations of gene transcript and protein product in situ, respectively. Moreover, the co-application of in situ hybridization and immunostaining in one section is increasingly employed for simultaneously observing location of the gene transcripts of interest as well as the spatial distribution of another gene product at the biochemical level.

For the gene expression analysis in early developing *Xenopus* brain, usually earlier than stage 35/36 (st.35/36), in situ hybridization and immunostaining are often conducted with fixed whole-mount embryo followed by sectioning for following analysis. Whole-mount in situ hybridization (WISH) was adapted according to Harland [1], with modifications [2–5]. Whole-mount immunohistochemistry (WIHC) was carried out according to the described [3, 4]. Whole-mount in situ hybridization for detecting microRNA (miRNA) expression was performed according to our previous work [6–9].

Simon G. Sprecher (ed.), *Brain Development: Methods and Protocols*, Methods in Molecular Biology, vol. 1082,
DOI 10.1007/978-1-62703-655-9_9,

However, as embryonic development proceeds, it gradually encounters difficulty for probe and antibody to penetrate into the brain and other tissues completely if the embryos or tissues are too large. This low permeability can cause poor signal-to-background ratio in ISH and IHC. Therefore, it is necessary to dissect brain and other tissues at later embryonic (st.37/38 and later) and adult stages. Here, we describe the method for dissecting and preparing the brain for in situ hybridization and immunostaining. Particularly, we will present the updated approach for examining miRNA expression.

2 Materials

2.1 Tools for Xenopus Embryo Manipulation, Brain Dissection

1. A stereotype microscope is required for observation during embryo or tissue dissection. If the embryos were injected with *gfp* mRNA, a stereotype fluorescent microscope is needed.
2. Forceps with blunt or sharp tips (e.g., Sigma Tweezers style #5) are used to hold and dissect embryos or tissues, which will be placed on agarose plate for operation.
3. Small sharp surgical scissors for eye operation are used to cut skin and nerve.
4. Stainless-steel bone clamp is adequate for opening adult head skull to obtain the intact brain and separate nerve fiber.
5. Superfrost® Plus slides or other Poly-L-Lysine coated slides.
6. Hairloop is used for moving embryos gently.
7. Whatman No. 2 filter paper.

2.2 Chemical Reagents

Representative chemical reagents for ISH and IHC are listed as follows:

1. HEPES, tricaine methanesulfonate (Sigma).
2. Formaldehyde.
3. Paraformaldehyde (PFA).
4. Tween-20.
5. Methanol.
6. Ethanol.
7. Sucrose.
8. OCT.
9. Paraffin.
10. Xylene.
11. Proteinase K.
12. BSA (BSA V; Sigma).

13. Polyvinylpyrrolidone (PVP-40).
14. Ficoll 400 (phamacia).
15. Formamide (redistilled).
16. Torula RNA (Type IX, Sigma).
17. Heparin.
18. CHAPS.
19. Blocking reagent (Roche).
20. Lamb serum.
21. Anti-digoxigenin–alkaline-phosphatase antibody (Roche cat. No. 11093274910).
22. Tetramisole (Sigma cat. No. L9756-5G).
23. NBT/BCIP (Roche cat. No. 11681451001).
24. BM purple (Roche cat. No. 1144207400).
25. Fast Red Tablets (Roche cat. No. 11496549001).
26. DAB substrate kit (zsbio).
27. AEC substrate kit (zsbio).
28. Hoechst 33258 (Sigma).

2.3 Solutions for In Situ Hybridization and Immunostaining

Solutions for in situ hybridization should be prepared using ultrapure water (18.2 MΩ at 25 °C). For IHC, double distilled water (ddH_2O) is feasible. Chemical reagents are the analysis pure:

1. MMR: 0.1 M NaCl, 2.0 mM KCl, 1 mM $MgSO_4$, 2 mM $CaCl_2$, 5 mM HEPES (pH 7.8), 0.1 mM EDTA. For 1 L 10× MMR, add NaCl 58.44 g, KCl 1.49 g, $MgCl_2 \cdot 6H_2O$ 2.03 g, $CaCl_2$(anhyd) 2.22 g, and HEPES 11.92 g into 800 mL water, mix dissolve and adjust pH to 7.5 with NaOH, and bring to 1 L with water. Autoclave and store at room temperature (RT).
2. MS222 anesthetic solution: MS222 0.2 mg/mL in 0.1× MMR, pH 7.5–7.8.
3. 1× PBS: For 1 L 10× PBS, dissolve $NaH_2PO_4 \cdot 2H_2O$ 2.89 g, $Na_2HPO_4 \cdot 12H_2O$ 26.73 g, and NaCl 102.2 g in 800 mL ultrapure water, and then adjust pH to 7.4 with NaOH. Bring volume to 1 L and autoclave. Store this solution at RT. Dilute the stock solution with water before use and adjust pH if necessary.
4. MEMFA: 0.1 M MOPS (pH 7.4), 2 mM EGTA, 1 mM $MgSO_4$, 3.7 % formaldehyde (*see* **Note 1**).
5. 4 % PFA/1× PBS: For 50 mL, dilute 10 mL 20% PFA with 5 mL 10× PBS with water prior to use (*see* **Note 2**).
6. 20 % Sucrose/1× PBS: For 50 mL, dissolve 10 g RNase-free sucrose with 1× PBS, sterilize the solution through 0.45 μM filter, and store at 4 °C.

7. 50× Denhardt's solution: 1 % BSA, 1 % PVP-40, 1 % Ficoll 400. For 50 mL, dissolve 0.5 g BSA, 0.5 g PVP-40, and 0.5 g Ficoll 400 in water, sterilize the solution through 0.22 μM filter, and store at −20 °C.
8. 20× SSC: Dissolve 175.3 g NaCl and 88.2 g sodium citrate in 800 mL of water, adjust to pH 7.0 with 1 N HCl. Make up to 1,000 mL of water. Autoclave and store the solution at RT.
9. Hybridization buffer (for ordinary RNA probe): 50 % formamide, 5× SSC, 0.1 % Tween-20, 100 μg/mL Heparin, 1 mg/mL torula RNA, 1× Denhardt's solution, 0.1 % Tween-20, 10 mM EDTA, 0.1 % CHAPS. Sterilize the solution with 0.22 μM filter and store it at −20 °C.
10. Hybridization buffer (for LNA probe for miRNA): 50 % Formamide, 5× SSC, 0.1 % Tween-20, 50 μg/mL Heparin, 500 μg/mL yeast tRNA or torula RNA, 10 mM Citric acid (pH 6.0). Filter the solution with 0.22 μM filter and store it at −20 °C.
11. Washing solution (post hybridization): 50 % formamide, 2× SSC, 0.1 % Tween-20 for ordinary probe; 2× SSC for LNA probe.
12. PBT: 1× PBS, 0.1 % Tween-20.
13. MABT: 100 mM maleic acid, 150 mM NaCl, 0.1 % Tween-20, pH 7.5 (*see* **Note 3**).
14. Blocking buffer: 2 % blocking reagent, 20 % heat inactivated sheep serum in MABT for ordinary probe; 1 % blocking reagent, 1 % heat inactivated sheep serum in MABT for LNA probe (*see* **Note 4**).
15. Alkaline-phosphatase buffer: 100 mM NaCl, 100 mM Tris–HCl pH 9.5, 50 mM $MgCl_2$, 0.1 % Tween-20, 2 mM tetramisole (*see* **Note 5**).
16. NBT/BCIP staining solution: For every mL of alkaline-phosphatase buffer, add 1 μL of NBT and 3.5 μL of BCIP (*see* **Note 6**).
17. Fast Red staining solution: Add one Roche tablet in 2 mL 0.1 M Tris–HCl, pH 8.2, vortex 2 min and centrifuge 30 s to discard the red precipitates.

3 Methods

3.1 Brain Dissection and Fixation

Induction of ovulation in females, in vitro fertilization, embryo culture, and staging are carried out as described [10, 11]:

1. For tadpoles (st.37/38 and later), transfer the embryos in a small dish (diameter 35 mm), and make them anesthetized by replacing the culture solution with MS222 solution (*see* **Note 7**).

2. When the embryos stop moving, place them into agarose plate with precooling 1× PBS. Use the forceps with blunt tips to hold the embryo at the trunk; use another forceps with sharp tips to make a cut through the spinal cord at the dorsal truck just behind the hindbrain. Tip up the skin of the front cut with forceps and carefully peel off the skin of the dorsal head to expose the brain. Then insert the tips of the forceps from the cut to the bottom of the brain, and lift up the brain by a careful forward movement of the forceps and shear the nerve bundles with forceps tips or scissors. Transfer the dissected brain immediately into fixation solution with a pipette or forceps.
3. For the brain dissection of developing frogs to adult, make the frog anesthetized in the MS222 for around 20 min and then place it on the ice. Cut off the spine and tear the skin from foramen magnum toward the head with a sharp scissors and then use bone clamp to open the skull and expose the brain (*see* **Note 8**). The procedure of dissecting brain follows as above (**step 2**). If this step takes long time, to prevent the decay of brain tissue, add several drops of fixation solution onto the brain to fix it in situ for 1 h before taking it out (*see* **Note 9**).
4. Fix the dissected brain in MEMFA or 4 % PFA at 4 °C overnight (*see* **Note 10**).
5. For whole-mount in situ hybridization or paraffin section preparation. Then the brain could be dehydrated gradually by replacing the fixation solution with 25 % ethanol/PBS, 50 % ethanol/50% PBS, 75% ethanol, 100% ethanol sequentially, 5–10 min each; and stored in 100% ethanol at −20 °C (*see* **Note 11**). For whole-mount immunohistochemistry, ethanol should be replaced with methanol. For cryosection section preparation, go on Subheading 3.2 directly without dehydration.

3.2 Cryosection Preparation and Pretreatment Prior to Hybridization

1. Pipette off the fixation solution and add precooling 1× PBS to wash for 3 times.
2. Pipette off the PBS, and add 20 % sucrose/1× PBS to cryoprotect the brains at 4 °C for 4 h to overnight until the brains sink to the bottom.
3. Label molds the name and the direction of specimens. Transfer the brain(s) into the mold and remove the sucrose solution as much as possible. Add appropriate volume (enough to immerse the whole specimen) of OCT in the molds. Orientate the brains in right directions with a long syringe needle, and then carefully move the mold on the iron plate upon dry ice or into the 70 % ethanol precooling at −80 °C. After the OCT has been completely frozen, store the block at −80 °C before sectioning.
4. Slightly tickle or press the mold to take out the frozen block, mount the block on the specimen stand of cryostat at right direction, and cut the edges to modify it into right shape before

sectioning. Cut 10–15-μm sections of the embedded brain at −24 °C.

5. Collect sections one by one on ready-to-use Superfrost® Plus slides or other Poly-L-Lysine coated slides, air-dry the sections, and store the slides at −80 °C.
6. The stored sections should be defrosted at RT for at least 30 min prior to use for hybridization or immunostaining.

3.3 Paraffin Section Preparation and Pretreatment Prior to Hybridization

1. Dissolve paraffin at 60 °C before the day of embedding, and keep it at 60 °C.
2. Wash the stored dehydrated brains with 100 % ethanol, and then transfer specimens to xylene by gradually replacing the solution with 75 % ethanol/25 % xylene, 50 % ethanol/50 % xylene, 25 % ethanol/75 % xylene, and 100 % xylene, and shake at RT for 5–10 min each time.
3. Wash the brains with new 100 % xylene, pour the specimens with xylene into a small beaker (specimens should be immersed in xylene), and incubate at 60 °C for 20 min (*see* **Note 12**).
4. Embedding: Add the same volume of prewarmed paraffin, incubate at 60 °C for 45 min, and then replace with 100 % paraffin, incubated at 60 °C for 20 min. Replace with 100 % new paraffin, incubated at 60 °C for 3 h; wash specimens with 100 % new paraffin, incubate at 60 °C for 20 min; replace with new paraffin, pour specimens with paraffin into prewarmed mold or plastic disc, and set the specimens at the right orientation and positions with prewarmed needle. Turn off the oven, and leave the paraffin block with specimens to be solidified slowly in the oven overnight (*see* **Note 13**).
5. Modify the embedded block and cut 5–10 μm sections in a microtome. Transfer sections on the surface of 0.2 % ethanol on slides prewarmed at 37 °C (*see* **Note 14**). Discard the solution when sections have been completely extended, dry the sections at 37 °C overnight, and then store the slides at 4 °C.
6. For hybridization or immunostaining. Dewax the section in 100 % xylene by three washes, 10 min each time, 100 % ethanol wash twice, and then rehydrate the section by washes with graded alcohol sequentially reverse to the dehydration, 2 min each wash.

3.4 In Situ Hybridization of Sections

3.4.1 Pretreatment of Sections and Hybridization

1. Digoxigenin (DIG)- or fluorescein-labeled antisense RNA probes should be generated prior to hybridization step. The labeled LNA probe for miRNA is available from EXIQON Company.
2. Wash the sections with PBT twice at RT, 2–5 min each wash.
3. For ordinary RNA probe, the sections could be applied for hybridization (**step 4**) directly. For the LNA probe for miRNA,

rinse sections in PBT with 1 μg/mL proteinase K at RT for 10 min, followed with two washes in PBT with 2 mg/mL glycine, 5 min each wash, and three washes in PBT, 3 min each wash. Then refix the sections in the 4 % PFA solution for 15 min, followed with three washes in PBT, 3 min each time (*see* **Note 15**).

4. Place slides horizontally in humid hybridization chamber bottomed with filter paper immersed in 1× SSC/50 % formamide. Prewarm hybridization buffer with 1 μg/mL probe at 60–65 °C for ordinary probe or at the temperature suggested by the manufacturer for LNA probe for 5–10 min. Add 120–150 μL probe mix per slide, and carefully cover slide with a coverslip. Seal the box with plastic film and transfer the box into oven prewarmed at the right temperature to hybridize overnight (*see* **Note 16**).

3.4.2 Washing Steps and Antibody Visualization

1. Set a glass trough with washing solution stand in water bath. Warm up the solution and the glass trough to the hybridization temperature. Add 100 mL washing solution to a glass trough for ten slides.
2. Transfer slides to the glass trough with prewarmed washing solution (*see* **Note 17**), and take away the coverslips after they detach off the slides (usually in 1–2 min). Then, for ordinary probe, make two washes at 65 °C with stirring or shaking, 30 min each; for LNA probe, make two washes in 2× SSC at RT, 10 min each, followed with another wash in 0.2× SSC at RT for 10 min.
3. Wash the slides 2 times in MABT at RT, 30 min each for ordinary probe, or five times in PBT, 3 min each, for LNA probe.
4. Take the slide out one by one with forceps, wipe away the excessive solution with filter paper, and then lay the slides on the holder of the humid chamber with filter paper soaked in 1× PBS or water. Add around 1mL blocking buffer per slide to cover all the sections (without coverslip), and block for a half to 1 h at RT.
5. Discard the blocking buffer, add 150 μL antibody mix (antibody diluted 1:2,000 to 1:8,000 in blocking buffer) per slide, cover with a piece of parafilm cut as the same size of coverslip, and incubate the slide in a humid chamber at RT overnight.
6. Discard the antibody solution, transfer the slides in a glass trough with MABT, and make five washes on a shaker, 30 min each wash at RT (*see* **Note 18**), followed with another five washes in PBT, 5 min each at RT.
7. Staining: Wash the slides three times in alkaline-phosphatase buffer for 5 min each at RT. Take out the slides and set them in a humid chamber, add 150 μL NBT/BCIP or Fast Red staining buffer per slide, cover with parafilm, and then develop

in the dark for usually 1 h to overnight with NBT/BCIP or 20 min to 6 h with Fast Red, depending on the abundance of the RNA.

8. Stop the staining reaction by two washes in PBT, 1× PBS sequentially. Then the sections could be mounted for observation (go **step 13**) and go on with subsequent immunostaining (*see* Subheading 3.5) or the following counterstaining (e.g., Hoechst nuclear staining) for histological examination.
9. Hoechst staining for counterstaining of cell nuclei: Add several drops of freshly prepared 1 μg/mL Hoechst 33258/1× PBS to the slides, cover with a piece of parafilm to spread the staining solution to all section, incubate for around 10 min at RT to stop the staining reaction, and wash sections with 1× PBS twice, 3 min each.
10. Mounting: For fluorescent staining, mount in a water-soluble, non-fluorescing mounting medium, e.g., Aqua-Poly/Mount (Polysciences) for observation. For chemical staining, dehydrate and clear sections by graded alcohols two washes in 100 % ethanol and two washes in xylene, 2–3 min each, and then mount in Histomount or Canadian gel solution (*see* Fig. 1).

3.5 Immunohistochemistry of Sections

1. For paraffin sections, antigen should be retrieved in 0.01 M sodium citrate pH 6.0 at 80 °C for 20 min. For cryosections, this step can be omitted. When HRP-conjugated second antibody is applied, 0.3 % H_2O_2 in 1× PBS treatment for 10–30 min would be needed to inactivate the endogenous peroxidase.
2. Rinse sections with PBS for three times, then PBT for three times, 5 min for each time.
3. Add around 1 mL 10 % goat serum/2 % BSA/PBT per slide to cover all the sections, and block for 30 min.
4. Discard the blocking buffer, and overlay with 150 μL primary antibody solution (antibody diluted in blocking solution as recommended by the manufacturer), covered with coverslips and incubated for 6 h at RT or overnight at 4 °C in a humidified chamber.
5. Rinse the sections with PBT for three times, 5 min each.
6. Re-block sections with 10 % goat serum/2 % BSA/PBT for 30 min. Then incubate sections with second antibody (dilute antibody in the blocking solution as recommended by the manufacturer) in the wet chamber for 2 h at RT or overnight at 4 °C.
7. Wash sections with PBT for 3 times, with PBS for 3 times, 5 min.
8. If the second antibody is conjugated with fluorescent dye like FITC and TRITC, sections can be directly mounted and observed under fluorescent microscope. If the second antibody is AP or HRP conjugated, the substrate such as BM purple/

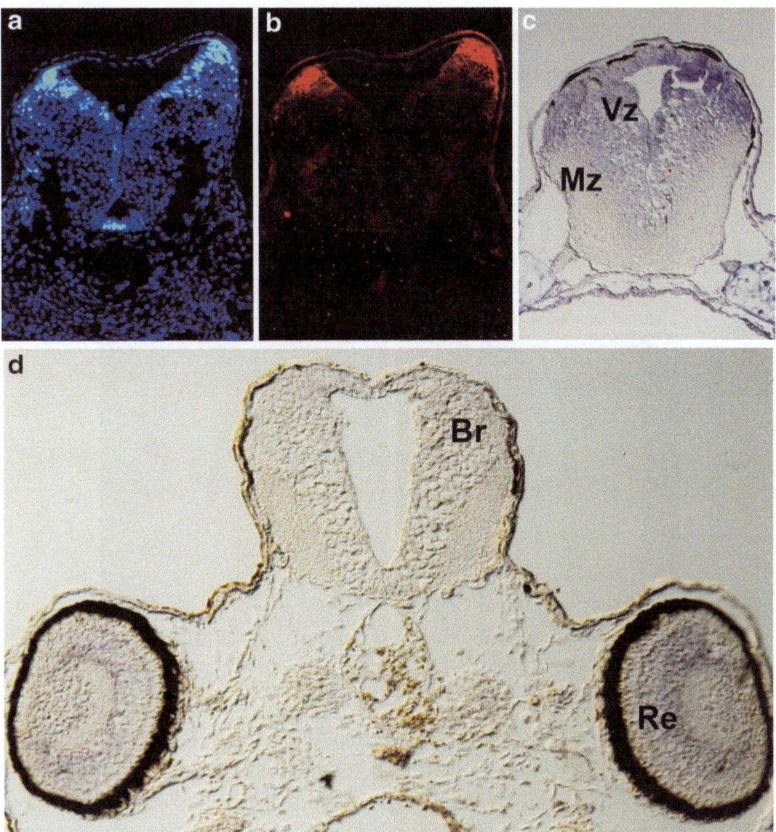

Fig. 1 In situ hybridization (ISH) of st.46 *Xenopus* brain sections for detecting transcripts of *zic2* (**a**, **b**), *ath3*/*neurod4* (**c**), microRNA-181 (**d**). ISH for *Zic2* was carried out in cryosection, stained by using AP substrate Fast Red (*red*), which shows expressed in the dorsal region of the developing brain (**b**), the nuclei of which were counterstained with Hoechst (**a**). ISH for *ath3*/*neurod4* and microRNA-181 were conducted in paraffin sections and stained by applying AP substrate BM *purple* (**c**, **d**). *Ath3*/*neurod4* was expressed higher in ventricular zone (proximal) while lower in marginal zone (distal) of developing brain/neural tube (**c**). microRNA-181 is rarely detectable in developing brain while relatively highly expressed in the retina inner nuclear layer (**d**). *Br* brain, *Re* retina, *Vz* ventricular zone, *Mz* marginal zone

FastRed for AP and DAB or AEC for HRP should be added on the slides for staining as described in Subheading 3.4 or the kit instruction (*see* Fig. 2). The post staining washes and mount are as described above.

4 Notes

Generally, the volume of solution for fixation and washing should be no less than 10 volumes of the specimens unless indicated. During the process of ISH or IHC, tissue sections should not be dry after initiating PBS washing, especially when the antibody

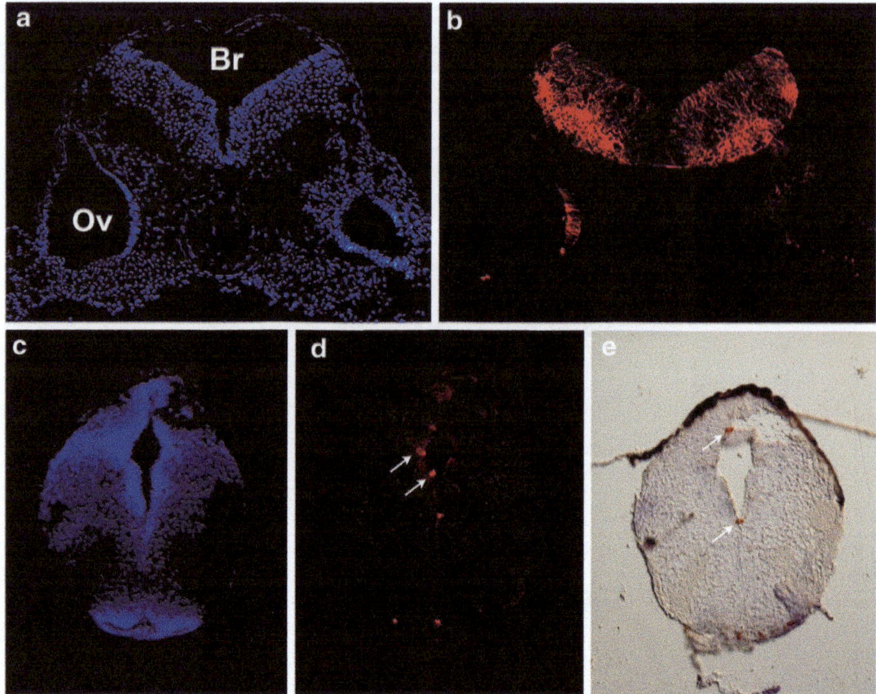

Fig. 2 Immunostaining of *Xenopus* brain sections for detecting protein locations of N-tubulin (**a**, **b**) and phosphohistone-H3 (pH3) (**c**–**e**). IHC for n-tubulin was carried out in cryosection in (**a**, **b**), stained by TRITC-conjugated secondary antibody showing red fluorescence and expressed in the whole developing brain/neural tube and otic vesicle (**b**), the nuclei of which were counterstained with Hoechst (**a**). IHC for pH3 was conducted in paraffin sections and stained with TRITC-conjugated secondary antibody (**c**, **d**) and HRP substrate AEC (**e**). pH3 is specifically expressed in the proliferating cells in the midline of the developing brain as shown by the *arrows*. *Br* brain, *Ov* otic vesicle

is incubated. Otherwise, the background will appear to be too high. Sense probe control and blank control should be set in in situ hybridization. For immunohistochemistry, the blank control without adding the primary antibody is also required. The signal-to-background ratio of in situ hybridization or immunohistochemistry depends on the abundance of transcript/antigen of interest and also the quality of the antibody. Therefore, the final quality of the transcript/protein detection may be optimized in modulating the corresponding step when necessary:

1. The solution with formaldehyde is all right when stored at 4 °C and used within a week at most of the cases. But it's better to make the 10-times solution without formaldehyde (10× MEM) as stock solution and add fresh formaldehyde prior to use. For 100 mL 10× MEM, add MOPS 20.93 g, 0.2 M EGTA 10 mL, and 1 M $MgSO_4$ 1 mL, adjust pH to 7.4 with NaOH, and filter sterilized and aliquoted with 50 mL plastic tube, stored at −20 °C.

2. 20 % PFA is commercially available. For self-preparation of 100 mL solution, heat 80 mL distilled water at 60–65 °C. Add 20 g paraformaldehyde and slowly add drops (<100 μL) of 10 N NaOH, with stirring on hot plate in fume hood until solution becomes clear; this should take no longer than 30–60 min! Add distilled water to 100 mL, and then filter with Whatman No. 2 filter paper. Aliquot with 50 mL plastic tube, stored at −20 °C.
3. 5× MAB stock solution could be prepared without Tween-20, and NaOH pellets could be applied for the pH adjustment. Autoclave and store the solution at RT. Dilute the stock and add Tween-20 prior to use. It'll be easier to get the right volume of the dense Tween-20 by cutting the sharp pipette tip a little.
4. The blocking solution could be prepared prior to use by diluting 5× MAB, 10 % blocking reagent, 100 % serum (inactivated by incubated at 55 °C for 30 min), and Tween-20 with water. The 10 % blocking reagent and inactivated serum should be aliquoted and stored at −20 °C.
5. The solution should be freshly prepared prior to use by diluting the 5 M NaCl, 1 M Tris–HCl pH 9.5, and 1 M $MgCl_2$ with water. Add Tween-20 and tetramisole and mix well prior to use. The solution works well without tetramisole for in situ hybridization for ordinary probe in our experiences.
6. The ready-to-use mix is also available, as Sigma cat. No. B1911; BM purple could be apply for ordinary RNA probe staining.
7. The brain should be dissected within 10 min after the tadpoles fall asleep, or the neurons/embryos would die.
8. Pay attention to avoid damaging the brain and bleeding. Be patient and operate gently and carefully.
9. The in situ fixation should be conduit in hood. Add more fixation solution from time to time to keep the brain immersed in the fix.
10. For immunohistochemistry, Dent's fixative (80 % methanol/20 %DMSO) is another optional for some antibodies.
11. For small embryonic brain, the dehydration step can be simplified by directly replacing the fix with 100 % alcohol.
12. Extension of incubation time in xylene will lead to damage of the tissue.
13. The whole embedding process should be conduit in oven. The solid paraffin appeared should be remelted before going to the next step.
14. Avoid the attachment of sections with the slide before the sections have been completely extended.

15. The proteinase K and glycine solution should be freshly prepared with 10 or 20 mg/mL stock; gentle mixing is required to avoid denaturing the proteinase. Avoid shaking the slides during the proteinase and postfixation treatment.
16. For LNA probe, hybridization temperature of Tm-22 °C is suggested as described [12], but lower temperature could be applied in our experience, especially for non-abundant miRNAs. In addition, a 30 min prehybridization step is optional by adding 200 μL hybridization buffer without probe for each 60 mm slides; a piece of parafilm cut as the size of coverslip could be applied to spread the hybridization buffer and cover the section. But the parafilm "coverslip" could not be applied at hybridization temperature higher than 38 °C. In addition, the parafilm and hybridization buffer should be removed carefully to avoid the damage of sections.
17. This step needs to be fast. Never let the slides cool down.
18. This step is optional for probes give low background or faint signal.

Acknowledgments

We're grateful to Federico Cremisi for sharing the protocol of in situ hybridization with cryosections. This work was supported by the NSFC No. 31271387.

References

1. Harland RM (1991) In situ hybridization: an improved whole-mount method for Xenopus embryos. Methods Cell Biol 36:685–695
2. Lupo G, Liu Y, Qiu R, Chandraratna RA, Barsacchi G, He RQ, Harris WA (2005) Dorsoventral patterning of the Xenopus eye: a collaboration of Retinoid, Hedgehog and FGF receptor signaling. Development 132: 1737–1748
3. Liu Y, Lupo G, Marchitiello A, Gestri G, He RQ, Banfi S, Barsacchi G (2001) Expression of the Xvax2 gene demarcates presumptive ventral telencephalon and specific visual structures in Xenopus laevis. Mech Dev 100:115–118
4. Lan L, Liu M, Liu Y, He R (2007) Expression and antibody preparation of POU transcription factor qBrn-1. Protein Pept Lett 14:7–11
5. Lan L, Liu M, Liu Y, Zhang W, Xue J, Xue Z, He R (2006) Expression of qBrn-1, a new member of the POU gene family, in the early developing nervous system and embryonic kidney. Dev Dyn 235:1107–1114
6. Decembrini S, Bressan D, Vignali R, Pitto L, Mariotti S, Rainaldi G, Wang X, Evangelista M, Barsacchi G, Cremisi F (2009) MicroRNAs couple cell fate and developmental timing in retina. Proc Natl Acad Sci USA 106(50):21179–21184
7. Liu K, Liu Y, Mo W, Qiu R, Wang X, Wu JY, He R (2011) MiR-124 regulates early neurogenesis in the optic vesicle and forebrain, targeting NeuroD1. Nucleic Acids Res 39: 2869–2879
8. Qiu R, Liu Y, Wu JY, Liu K, Mo W, He R (2009) Misexpression of miR-196a induces eye anomaly in Xenopus laevis. Brain Res Bull 79:26–31
9. Qiu R, Liu K, Liu Y, Mo W, Flynt AS, Patton JG, Kar A, Wu JY, He R (2009) The role of miR-124a in early development of the Xenopus eye. Mech Dev 126:804–816
10. Sive HL, Grainger RM, Richard RM (2002) Early development of Xenopus laevis: a laboratory manual. Cold Spring Harbor Laboratory Press, Cold Spring Harbor, NY

11. Nieuwkoop PD, Faber J (1994) Normal table of Xenopus laevis (Daudin): a systematical and chronological survey of the development from the fertilized egg till the end of metamorphosis. Garland, New York

12. Kloosterman WP, Wienholds E, de Bruijn E, Kauppinen S, Plasterk RH (2006) In situ detection of miRNAs in animal embryos using LNA-modified oligonucleotide probes. Nat Methods 3:27–29

Chapter 10

Microinjection Manipulations in the Elucidation of Xenopus Brain Development

Cristine Smoczer, Lara Hooker, Saqib S. Sachani, and Michael J. Crawford

Abstract

Microinjection has a long and distinguished history in *Xenopus* and has been used to introduce a surprisingly diverse array of agents into embryos by both intra- and intercellular means. In addition to nuclei, investigators have variously injected peptides, antibodies, biologically active chemicals, lineage markers, mRNA, DNA, morpholinos, and enzymes. While enumerating many of the different microinjection approaches that can be taken, we will focus upon the mechanical operations and options available to introduce mRNA, DNA, and morpholinos intracellularly into early stage embryos for the study of neurogenesis.

Key words Microinjection, Morpholino, mRNA, Lineage marker, *Xenopus*, Over-expression, Knockdown, Transgenic, Mutant

1 Introduction

Microinjection approaches have facilitated numerous studies on neurogenesis; however, fine injection needles drawn from heated glass capillaries have been used for many other purposes, especially in studies employing amphibian eggs and embryos. For example, the first nuclear transfer and cloning exercises were accomplished in frog [1, 2]. Injection rigs can be used to introduce materials both inter- and intracellularly, and the gamut of injectables runs from nuclei, to peptides, antibodies, biologically active chemicals, lineage markers, mRNA, DNA constructs, morpholinos, and enzymes. In *Xenopus laevis*, ovulation and insemination are easily induced, and the large size and accessibility of eggs and embryos makes the species ideal for many studies on early neural patterning. Moreover, cheap and effective transgenic approaches have been developed; however, the lack of suitable stem cell lines still precludes knockouts by homologous recombination. The pseudotetraploid nature of the *X. laevis* genome also imposes limits upon the scope experiments.

Simon G. Sprecher (ed.), *Brain Development: Methods and Protocols*, Methods in Molecular Biology, vol. 1082, DOI 10.1007/978-1-62703-655-9_10,

1.1 Intercellular Injection

The object of intercellular injection has been to introduce agents either between germinal layers or into the blastocoel.

1.1.1 Localized Agents

In studies of neural activation and posterior transformation, retinoic acid was co-injected with the lipophilic marker DiI (1,1′-Dioctadecyl-3,3,3′,3′-tetramethylindocarbocyanine iodide) between deep and surface ectoderm in early gastrula [3]. The approach permits the introduction of a teratogen that remains relatively localized over the course of neurogenesis. Moreover, the introduction of many varieties of marker (such as Invitrogen's CellTracker and MitoTracker or BrdU; *see* ref. [4] for review) offers a plethora of opportunities for consideration in lineage marker studies. Finally, while there are not many examples of the virally mediated introduction of genes to amphibians, *vaccinia* was successfully injected and used to ectopically introduce *Sonic hedgehog* into axolotl tissues, and this approach could be considered in *Xenopus* [5].

1.1.2 Blastocoel

Peptides have been injected into the blastocoel either to activate [6–8] or to competitively block cell surface or matrix-mediated pathways [8].

A similar approach that used antibodies as a blocker has also been successfully used [9]. Other, more complex mixtures containing biological agents have been tested in this manner, for example, the contents of germinal vesicles [10].

1.2 Intracellular Injection

Intracellular injection, particularly if accompanied by inclusion of a fluorescent lineage marker, provides a useful entrée to study gene perturbation effects and cell lineages. Considerable versatility is introduced by the availability of different developmental stages for injection. For example, nuclear and DNA injections typically take place at the one-cell stage; however, individual blastomeres can be injected when effects upon restricted lineages are desired [11]. Injection at the 2-cell stage permits gene perturbation on one side of the future embryo—the other side serves as contralateral control [12].

Intracellular approaches can facilitate subsequent analysis through the grafting of heterologously treated tissues or through the production of embryos that serve as donors for animal cap explants [13].

1.2.1 Exogenous DNA

Several decades ago, linearized DNA was microinjected into fertilized *Xenopus* eggs to study tissue-specific expression of the introduced gene [14]. Although a majority of the transgenic embryos were highly mosaic, a very small percentage survived to adulthood and transmitted the transgene to the next generations [15, 16].

1.2.2 REMI (Restriction Enzyme-Mediated Integration)

Many years passed following pioneering nuclear transfer studies [1, 17] before nuclear transplantation coincident with the introduction of plasmid evolved to facilitate the generation of transgenic frogs [18].

The latter technique was developed to address the high mosaicism that accompanies microinjection solely of linearized DNA. In order to control mosaicism, an in vitro fertilization procedure facilitates transgene incorporation into decondensing pronuclei. Sperm nuclei, restriction endonuclease, and linearized DNA were co-injected into eggs by means of an infusion pump (sperm nuclei are too big to pass through the narrow needle used in most microinjections) [18, 19]. Despite the reduced incidence of mosaicism, there are major drawbacks to this approach: the large bore needle that is required to introduce nuclei damages and impairs the viability of injected embryos; the concomitant restriction enzyme-mediated integration (REMI) does not always occur immediately—mosaic embryos still result; concentrations of restriction endonuclease have to be carefully titered to avoid genome fragmentation and aneuploidy; and finally, integration copy number and variable penetrance/expressivity can complicate phenotype interpretation [18, 20, 21]. The long generation time of *X. laevis* renders its use as a genetic model problematic, and typically, numerous transgenics must be assessed in an experiment to control for integration site-specific variations. With a shorter generation time and diploid genome, *Xenopus* (*Silurana*) *tropicalis* is better suited for experiments requiring founder lines and genetic crosses.

1.2.3 *I-Sce I* Meganuclease

An approach utilizing a rare-cutting meganuclease is based on the presence of 18 bp restriction sites for *I-Sce I* within the *Xenopus* genome. The transgene construct is flanked by *I-Sce I* restriction sites. Following digestion with the meganuclease, the mixture is microinjected close to the forming nucleus of fertilized eggs. This procedure is simple and effective and yields high survival rates among transgenic embryos that can later transmit the transgene to their offspring [22–24]. The disadvantages of the method reflects the tendency for random and multiple insertions of the transgene both singly and as concatemers, as well as a degree of mosaicism [24].

1.2.4 Phi-C31 Integrase

Integrase-mediated transgene insertion relies on *Phi-C31*-mediated recombination between *attB* sites flanking the transgene construct and pseudo-*attP* sites present in the amphibian genome. Exogenous DNA and the mRNA encoding *Phi-C31* integrase are co-injected into fertilized eggs. This procedure results in relatively efficient single-copy transgene incorporation at specific sites [25, 26]. Efficacy can be improved by deploying a vector that brackets the transgene with chicken β-globin 5′-HS4 insulators [25, 26]. The same insulators have been used to flank *I-Sce I* sites and thereby improve transgene activity in that system as well [27]. The disadvantage of the approach is that the vector becomes somewhat large, and this, in conjunction with the repetitive nature of the insulators, renders the construct prone to recombination during amplification.

1.2.5 Lineage Markers

Several strategies are available to deploy fluoresceinated lineage markers. Intracellular injection of fluorescently labelled hydrophilic dextrans has a well-tried history [28, 29]. An advantage that dextrans confer is that they tend to diffuse relatively well throughout the injected cell, and yet they remain bounded by their large molecular mass to that cell and its progeny. Tissues derived from injected embryos can be combined with wild type or differently injected embryos via heterotypic grafting [12, 30]. Alternatively, the mRNA for GFP [31] or fluorescent fusion markers [32] can be deployed either by co-injection or as bi-cistronic transcripts derived from IRES vectors.

1.2.6 Biologically Active In Vitro Synthesized mRNAs

Transcripts are routinely injected into *Xenopus* embryos to perturb neurological development. Ectopic gain-of-function experiments are straightforward [33]. Alternatively, chimeric constructs that deploy fused activator (*VP16*) or repressor (*engrailed*) domains are used to deliver phenotypes that can reveal whether a candidate gene exerts an activating or repressive effect [34]. Mutations can be introduced to alter the behavior or to define the role of encoded sub-domains. For example, a point mutation introduced to *Mix.1* impaired the DNA binding functions of its encoded homeodomain [34]. Finally, mRNA "rescue" experiments can be mounted to assess hierarchical relationships between genes. If morpholino-mediated knockdowns or chimeric activator/repressor transcripts are introduced, then the capacity of potential downstream signalling partners can be assessed by co-injecting them in complementation studies. All of these approaches suffer some disadvantages: transcripts will be nonspecifically distributed throughout all the progeny of an injected blastomere; ectopic mRNAs have a limited lifespan—their sensitivity to degradation limits their utility to early developmental stages; and different species of transcript might vary with regard to longevity and distribution pattern in a context-specific manner.

1.2.7 Morpholino Antisense Oligonucleotides

Morpholinos can be injected to abrogate transcript translation [35] or, at least in zebrafish, to specifically inhibit the generation of a particular alternatively spliced isoform [36]. Their use is described elsewhere in this volume.

2 Materials

A good source for general recipes and handling instructions for *Xenopus* and related reagents is available and recommended [37].

2.1 Frogs and Embryos

1. *Xenopus laevis* males and females (*Xenopus* I, Michigan, USA). Wild caught specimens preferred.
2. Human chorionic gonadotropin (HCG) (AkzoNobel or Sigma). Prepare a 1,000 U/ml solution in the manufacturer provided buffer. Unused remainder can be stored at –20 °C for several weeks.

3. Single-use tuberculin syringe, 1 cc, 26-gauge needle.
4. Tricaine methanesulfonate MS-222 (Sigma). Prepare 1 l of a 0.2 % solution buffered to pH 7.2 with sodium bicarbonate fresh the morning of use in water and store at 4 °C.
5. Scissors, forceps, small finger bowl or 80 ml beaker, and Petri dishes.
6. A means to incubate embryos at 12–13 °C.
7. Petri dishes, capillary tubes (*not* heparin coated or chemically treated), and water-resistant tape (masking tape or the colored paper tape normally used to label lab ware is fine).
8. Pasteur pipettes with small rubber bulb for the transfer of embryos. These can be cut and flame polished to leave an opening of 1.5–2 mm.

2.2 Solutions

2.2.1 Steinberg's Solution

Stock Part A (20×): 1.16 M NaCl, 13.4 mM KCl, 16.6 mM $MgSO_4$, 6.7 mM $Ca(NO_3)_2$, pH to 7.4.

Stock Part B (20×): 92.5 mM Tris base, 0.08 N HCl.

Prepare both parts using distilled water, autoclave, and store at room temperature.

Dilutions of stock solution for typically used Steinberg's solutions:

	200 %	100 %	80 %	20 %
Part A per liter	100 ml	50 ml	40 ml	10 ml
Part B per liter	100 ml	50 ml	40 ml	10 ml

2.2.2 MBS (Modified Barth's Saline, Stored in Two Parts)

Part A. 10× MBS: 880 mM NaCl, 25 mM $NaHCO_3$, 10 mM KCl, 50 mM HEPES of pH 7.8, 10 mM $MgSO_4$. Adjust pH to 7.8 with NaOH, top up with distilled water, and then autoclave. Store at room temperature and use within the month.

Part B. 0.1 M $CaCl_2$—into 7 ml aliquots. Store frozen.

On the day of experiments, prepare 1× MBS by adding 100 ml of 10× MBS and 7 ml 0.1 M $CaCl_2$ to distilled water—volume adjusted to 1 l.

2.2.3 L-Cysteine

2 % L-cysteine hydrochloride (Sigma or ICN) is dissolved in H_2O and adjusted to pH 8.0 with NaOH. Prepare fresh on the morning of use.

2.2.4 Ficoll

Ficoll 400 is dissolved in 0.3× MBS in a ratio 2–5 % weight to volume, depending upon the needle size. Typically, RNA and morpholino injections deploy very fine needles and embryos only require 2–3 % Ficoll for structural support during injection and recovery. Larger needles required for nuclear transfer will require 5 % Ficoll. Prepare the morning of use and refrigerate to 12–13 °C.

2.3 Glass Needles

1. Microinjection needles. Drawn from borosilicate glass capillaries (1 mm outside diameter and 0.58 mm inside diameter—Harvard Apparatus Ltd., Edenbridge, UK, or Drummond 3-000-203-G/X) using a microneedle puller (such as model PC-10, Narishige, Tokyo, Japan) (*see* **Note 1**).
2. 1 cc inoculation (tuberculin) syringe with a 26-gauge needle.
3. Mineral oil (Eppendorf).
4. Micromanipulator.
5. Positive displacement injector (e.g., Drummond Nanoject II—*see* **Note 2**).
6. Stereomicroscope.
7. Paraplast film or glass slide.

3 Methods

3.1 Preparation of Frogs and Fertilization of Eggs [38]

1. Select two females exhibiting prominent red cloacae. Inject each with 0.75 cc HCG and incubate overnight (if frogs are to be kept at 18 °C, inject at 6:00 pm in the afternoon the day before needed or at 11:00 if females are to be held at room temperature 22–23 °C).
2. Next morning, check to see if the females are shedding eggs. If they are, then anesthetize the male in tricaine methanesulfonate (MS-222) and sacrifice by cervical section or decapitation. Remove the testes to 200 % Steinberg's solution and keep cool either on ice or in a refrigerator.
3. Macerate one quarter testes in 2 ml of 80 % Steinberg's solution and draw up and down in 1 cc syringe (no needle), to break the tissue up further and to evenly distribute the sperm.
4. Strip eggs from the female into a Petri dish and add 0.5 cc sperm suspension. Leave at room temperature for 10 min.
5. After 10 min has elapsed, flood the fertilized embryos with 0.1× MBS.
6. In the meantime, prepare fresh 2 % L-cysteine solution.
7. Successfully fertilized embryos rotate so that their dark animal pole faces up, and the cortex visibly contracts.
8. Once rotation has occurred, remove the embryos to a vessel for de-jellying (finger bowl or 80 ml beaker). Remove as much of the 0.1× MBS as easily possible, and flood with L-cysteine solution. Swirl them *very gently* periodically. After 2–4 min, the jelly should have dissolved, and the eggs will pack close together in the bottom of the beaker when it is tilted.
9. Immediately decant the bulk of the solution being careful never to fully expose the embryos to air, and gently replace

with 0.1× MBS to wash off the L-cysteine. It is important to remove all jelly, but to minimize the duration of contact with L-cysteine.

10. Repeat this 9 more times taking great care not to violently agitate the embryos.
11. Remove to a Petri dish of 0.1× MBS, and cool the embryos to 14 °C.
12. For subsequent manipulation/transfers of embryos, we use a Pasteur pipette that has been cut and flame polished at the end leaving an opening approximately 1.5–2 mm wide (wider that an embryo by a generous margin).

Good quality and undamaged embryos will look round and firm and will not adhere to the plastic (*see* **Note 3**).

3.2 Preparation of the Injection Apparatus

1. We recommend prepared needles be placed in a large Petri dish and baked to degrade any RNases. Only experience will help you to determine what parameters of forge or flame will work best to produce an optimal needle. You need needles that can easily discharge a fixed volume of injectable fluid, but that will not leave a wound large enough to cause cytoplasmic leaking from punctured embryos.
2. Under a stereomicroscope, break off the tip of the needle to open it. Prepared needles will be sharp enough that the ends will be discernable only under a stereomicroscope.
3. Using a 1 cc syringe equipped with a 26-gauge needle, backfill the glass microinjection needle with mineral oil. A tiny droplet of oil should become barely discernable at the tip. If air bubbles are introduced, begin again with a fresh needle.
4. Fix the microinjector into an orientation that is angled approximately 45° downward. It is good practice to clean the plunger of the microinjector with 70 % ethanol prior to use.
5. Taking great care to avoid the introduction of air bubbles, load the needle onto the microinjector.
6. Empty the microinjector of oil to the extent that the positive displacement plunger has gone as far as it can without reaching the constricted end of the glass needle.
7. Under the stereomicroscope, fill the needle from a droplet of injectable substance that is placed center field upon paraplast or a glass slide. Exercise caution to avoid sucking up air bubbles.
8. For mRNA injections, mRNA is diluted in RNase-free water which is drawn up into the needle from glass slides which have been washed with 70 % ethanol, baked (65–95 °C for 2 h), cooled to room temperature, and wiped with RNase AWAY® solution (Life Technologies).

3.3 Preparation of Injectables Morpholinos

Morpholinos can be obtained through Gene Tools LLC, and their administration and qualities are described elsewhere in this volume.

3.4 Preparation of Injectables Morpholinos mRNA

1. Prepare a workspace that is essentially RNase-free. This can be done cheaply by wiping surfaces and pipettors with a solution of 0.1 N NaOH and 0.1 % SDS.
2. mRNA is prepared by in vitro transcription from linearized template. Template must be of high quality and RNase-free. We have found Ambion's mMESSAGE mMACHINE (Life Technologies–Invitrogen) to provide good quality transcript. Capped transcript is diluted to the desired concentration in freshly distilled or nano-pure RNase-free water (older aliquots of water absorb CO_2 and acidify with time. *See* **Note 4**).
3. The quality and size of transcript should be checked on a fresh gel (1 % agarose in 1× TAE) that is prepared using RNase-free gel casting trays, gel boxes, loading tips, running and loading buffers, etc.
4. We recommend aliquoting transcript in known quantities suitable for one-time use—repeated freeze–thaw cycles damage transcript integrity, and repeated handling increases the risk of RNase contamination. Store the transcripts in aliquots at −80 °C.
5. Fluoresceinated dextran is a poor choice of marker to monitor the efficacy of mRNA injection since it is hard to acquire the fluor free of RNase activity. We typically co-inject with transcript for GFP and use this component both as a vital marker (under UV illumination) and to balance total mRNA concentrations to match control injected quantities.
6. Normally, a dilution series is prepared in order to develop a dose–response curve.

3.5 Injection of Embryos

One of the challenges in later assessment is to separate un-injected or poorly injected embryos from viable experimental specimens. Vital markers facilitate assessment of injection efficacy at relatively early stages (e.g., during early neurulation), when the label remains intense. It is advantageous to ensure that a fluorescent marker (fluoresceinated dextran), tag (e.g., fluorescein- or lissamine-tagged morpholino), or mRNA for GFP (green fluorescent protein) is part of the injection strategy. A directly tagged morpholino or bi-cistronic marker transcript permits more direct estimations of intensity and distribution of product within an embryo. Co-injected markers like dextran or mRNA for GFP help by inference only. Experiments usually start with a characterization of the effects of an injected substance following introduction at the one-cell stage. Once a dose–response curve has been established, injection into specific blastomeres can ensue. Generally speaking, when injected embryos are going to be used to produce explants for culture or grafting, they are injected at the one-cell stage:

1. Prepare a few Petri dishes as platforms for the injection procedure by placing a capillary tube (*not* heparin coated or chemically treated) so that it aligns down the center of the plate. Fix it in place at both ends using small strips of tape.
2. Place a prepared Petri injection platform under the stereomicroscope and extrude a few milliliters of Ficoll solution down one side of the centered capillary tube.
3. Gently drop a line of de-jellied embryos into the Ficoll. Let them sit for a minute or 2 (*see* **Note 5**).
4. Gently remove as much of the Ficoll solution as possible without disrupting the aligned embryos. Use a tissue on the opposite side of the capillary to wick away much of the remaining Ficoll leaving the embryos damp, but not submerged. Surface tension will keep them in place for injection.
5. Lower the injection needle and pierce each embryo at a relatively steep angle near, but not into, the area likely to house the nucleus. The sperm entry site is usually visible as a slight difference in pigmentation: an ideal injection latitude is between the animal pole and this discoloration point. Keep the injection movement fluid and smooth. Piercing can be quick, but withdrawal should be a little slower, and the whole process for each injection will lower to just over a second with practice.
6. When all embryos in the line have been injected, raise the glass injection needle, and reserve the injection rig somewhere safe and convenient while the Petri dish is prepared for removal.
7. Gently flood the injected embryos with Ficoll solution by gently adding the Ficoll from the capillary side of the line or from either end. The object is to submerge them without floating too many since this would introduce stress via surface tension. Floaters can usually be sunk by gently adding Ficoll solution dropwise overtop of them.
8. Remove the embryos to a cool incubator (12–13 °C) for at least 1 h.
9. If you plan to inject different concentrations of substance in a single sitting, start with the lowest concentration and then reuse the needle. This can be done between specimen lots by expelling all aqueous fluid and refilling with a higher concentration of the same reagent.
10. Preferably after an hour or two, but certainly before gastrulation (long-term storage in Ficoll inhibits gastrulation movements), the Ficoll solution should be replaced with 0.1× MBS. Replace the solutions by serially and gently pouring 0.1× MBS onto the

plate and then decanting the solution until the Ficoll has been largely replaced. Embryos may float, but can be immersed again by dripping fluid gently overtop.

11. At this point survey the embryos under a stereomicroscope and remove all dead or nondividing embryos. Remove the surviving embryos to a fresh dish of 0.1× MBS. Continue to incubate at 12–13 °C.
12. Each day, be sure to change the 0.1× MBS and to remove dead embryos before they render the media toxic.
13. If a fluorescent marker has been employed, use a fluorescence stereomicroscope to remove unsuccessfully injected embryos from the plates. Incubate the remainder to the developmental stage desired.

4 Notes

1. Needles can also be pulled manually over a small flame, but this takes practice and skill. Affix a modified large gauge needle to a stand, and use it as a miniature gas burner to heat a small region of a glass capillary. Be careful to keep flammables away from the metal that can become very hot. Roll the capillary between your fingers above the flame until the glass become soft and ductile, and then pull the two halves apart while lifting the glass away from the flame. This takes caution, care, and practice.
2. We have previously employed a homemade air pressure displacement rig to inject embryos. This is accomplished by attaching a "y" of tubing to a source of pressurized air. One arm of the "y" is partially clamped, and the other goes to the glass needle held in a micromanipulator. A glass needle is back-filled with injectable fluid and then plugged into one end of the air delivery tube. The clamp on the other arm is adjusted to allow the bulk of air to escape while permitting the fluid to remain in the needle. When delivery of injection fluid is desired, the partially clamped tube is pinched by folding to prevent any air from bleeding out, and this in turn raises pressure sufficient to expel injection fluid. Injection volumes are calibrated by time and diameter, and recalibration has to be repeated periodically. The process is slow and cumbersome, but surprisingly accurate and reproducible.
3. Flattened or floppy eggs/embryos or sticking to the bottom of the Petri dish reflect damaged materials. Possible causes include poor fertilization (check sperm motility and Steinberg's solution pH), poor egg quality from the female, too much time in L-cysteine, incomplete washing in 0.1× MBS after de-jellying,

or rough handling via too much agitation or sharp transfer pipettes—this damages the fertilization membrane and permits direct contact of the egg/embryo surface with the Petri dish.

4. The use of DEPC (diethylpyrocarbonate) to inactivate RNase is common in many RNA-based experiments. Unfortunately, it may be toxic to embryos, and we have found that it is unnecessary to use DEPC-treated water as a carrier for injections.
5. If embryos leak substantially after injection, there may be several causes: the eggs were not exposed to Ficoll long enough to lend support; the needle is too thick; or finally, the needle injection path was not stable and tore the membrane too much either during penetration or, more likely, during withdrawal.

Acknowledgment

M.J.C. is supported by the Natural Sciences and Engineering Research Council (NSERC) of Canada Grant #203549.

References

1. Briggs R, King TJ (1952) Transplantation of living nuclei from blastula cells into enucleated frogs' eggs. Proc Natl Acad Sci USA 38(5): 455–463
2. Gurdon JB (2006) Nuclear transplantation in Xenopus. Methods Mol Biol 325:1–9
3. Drysdale TA, Crawford MJ (1994) Effects of localized application of retinoic acid on Xenopus laevis development. Dev Biol 162(2): 394–401
4. Dolbeare F (1995) Bromodeoxyuridine: a diagnostic tool in biology and medicine, part II: oncology, chemotherapy and carcinogenesis. Histochem J 27(12):923–964
5. Roy S, Gardiner DM, Bryant SV (2000) Vaccinia as a tool for functional analysis in regenerating limbs: ectopic expression of Shh. Dev Biol 218(2):199–205
6. Otto C, Schutz G, Niehrs C, Glinka A (2000) Dissecting GHRH- and pituitary adenylate cyclase activating polypeptide-mediated signalling in Xenopus. Mech Dev 94(1–2): 111–116
7. Cooke J, Smith JC (1989) Gastrulation and larval pattern in Xenopus after blastocoelic injection of a Xenopus-derived inducing factor: experiments testing models for the normal organization of mesoderm. Dev Biol 131(2): 383–400
8. Saint-Jeannet JP, Dawid IB (1994) Vertical versus planar neural induction in Rana pipiens embryos. Proc Natl Acad Sci USA 91(8): 3049–3053
9. Purcell L, Gruia-Gray J, Scanga S, Ringuette M (1993) Developmental anomalies of Xenopus embryos following microinjection of SPARC antibodies. J Exp Zool 265(2): 153–164
10. Kao KR, Elinson RP (1985) Alteration of the anterior-posterior embryonic axis: the pattern of gastrulation in macrocephalic frog embryos. Dev Biol 107(1):239–251
11. Shi J, Severson C, Yang J, Wedlich D, Klymkowsky MW (2011) Snail2 controls mesodermal BMP/Wnt induction of neural crest. Development 138(15):3135–3145
12. Khosrowshahian F, Wolanski M, Chang WY, Fujiki K, Jacobs L, Crawford MJ (2005) Lens and retina formation require expression of Pitx3 in Xenopus pre-lens ectoderm. Dev Dyn 234(3):577–589
13. Zuber ME, Gestri G, Viczian AS, Barsacchi G, Harris WA (2003) Specification of the vertebrate eye by a network of eye field transcription factors. Development 130(21):5155–5167
14. Rusconi S, Schaffner W (1981) Transformation of frog embryos with a rabbit beta-globin gene. Proc Natl Acad Sci USA 78(8):5051–5055
15. Etkin LD, Pearman B, Ansah-Yiadom R (1987) Replication of injected DNA templates in Xenopus embryos. Exp Cell Res 169(2): 468–477

16. Etkin LD, Pearman B (1987) Distribution, expression and germ line transmission of exogenous DNA sequences following microinjection into Xenopus laevis eggs. Development 99(1):15–23
17. Gurdon JB (1962) The transplantation of nuclei between two species of Xenopus. Dev Biol 5:68–83
18. Kroll KL, Amaya E (1996) Transgenic *Xenopus* embryos from sperm nuclear transplantations reveal FGF signaling requirements during gastrulation. Development 122(10): 3173–3183
19. Amaya E, Kroll KL (1999) A method for generating transgenic frog embryos. Methods Mol Biol 97:393–414
20. Bronchain OJ, Hartley KO, Amaya E (1999) A gene trap approach in Xenopus. Curr Biol 9(20):1195–1198
21. Sparrow DB, Latinkic B, Mohun TJ (2000) A simplified method of generating transgenic Xenopus. Nucleic Acids Res 28(4): E12
22. Loeber J, Pan FC, Pieler T (2009) Generation of transgenic frogs. Methods Mol Biol 561: 65–72
23. Ogino H, McConnell WB, Grainger RM (2006) High-throughput transgenesis in Xenopus using I-SceI meganuclease. Nat Protoc 1(4): 1703–1710
24. Pan FC, Chen Y, Loeber J, Henningfeld K, Pieler T (2006) I-SceI meganuclease-mediated transgenesis in Xenopus. Dev Dyn 235(1): 247–252
25. Allen BG, Weeks DL (2005) Transgenic Xenopus laevis embryos can be generated using phiC31 integrase. Nat Methods 2(12): 975–979
26. Allen BG, Weeks DL (2006) Using phiC31 integrase to make transgenic Xenopus laevis embryos. Nat Protoc 1(3):1248–1257
27. Sekkali B, Tran HT, Crabbe E, De Beule C, Van Roy F, Vleminckx K (2008) Chicken beta-globin insulator overcomes variegation of transgenes in Xenopus embryos. FASEB J 22(7): 2534–2540
28. Wetts R, Fraser SE (1991) Microinjection of fluorescent tracers to study neural cell lineages. Development Suppl 2:1–8
29. Gimlich RL, Braun J (1985) Improved fluorescent compounds for tracing cell lineage. Dev Biol 109(2):509–514
30. Keller R, Tibbetts P (1989) Mediolateral cell intercalation in the dorsal, axial mesoderm of Xenopus laevis. Dev Biol 131(2):539–549
31. Zernicka-Goetz M, Pines J, Ryan K, Siemering KR, Haseloff J, Evans MJ, Gurdon JB (1996) An indelible lineage marker for Xenopus using a mutated green fluorescent protein. Development 122(12):3719–3724
32. Itoh K, Sokol SY (2011) Polarized translocation of fluorescent proteins in Xenopus ectoderm in response to Wnt signaling. J Vis Exp (51). doi:2700 [pii] 10.3791/2700
33. Wolanski M, Khosrowshahian F, Kelly LE, El-Hodiri HM, Crawford MJ (2009) xArx2: an aristaless homolog that regulates brain regionalization during development in Xenopus laevis. Genesis 47(1):19–31
34. Lemaire P, Darras S, Caillol D, Kodjabachian L (1998) A role for the vegetally expressed Xenopus gene Mix.1 in endoderm formation and in the restriction of mesoderm to the marginal zone. Development 125(13):2371–2380
35. Bayramov AV, Martynova NY, Eroshkin FM, Ermakova GV, Zaraisky AG (2004) The homeodomain-containing transcription factor X-nkx-5.1 inhibits expression of the homeobox gene Xanf-1 during the Xenopus laevis forebrain development. Mech Dev 121(12):1425–1441
36. Pei W, Noushmehr H, Costa J, Ouspenskaia MV, Elkahloun AG, Feldman B (2007) An early requirement for maternal FoxH1 during zebrafish gastrulation. Dev Biol 310(1):10–22
37. Sive H, Grainger RM, Harland RM (2000) Early development of Xenopus laevis. A laboratory outline, 1st edn. Cold Spring Harbor Laboratory Press, Cold Spring Harbor
38. Drysdale TA, Elinson RP (1991) Development of the *Xenopus laevis* hatching gland and its relationship to surface ectoderm patterning. Development 111:469–478

Chapter 11

Morpholino Studies in Xenopus Brain Development

Jennifer E. Bestman and Hollis T. Cline

Abstract

Antisense morpholino oligonucleotides (MOs) have become a valuable method to knock down protein levels, to block mRNA splicing, and to interfere with miRNA function. MOs are widely used to alter gene expression during development of Xenopus and zebra fish, where they are typically injected into the fertilized egg or blastomeres. Here, we present methods to use electroporation to target delivery of MOs to the central nervous system of *Xenopus laevis* or *Xenopus tropicalis* tadpoles. Briefly, MO electroporation is accomplished by injecting MO solution into the brain ventricle and driving the MOs into cells in the brain with current passing between two platinum plate electrodes, positioned on either side of the target brain area. The method is straightforward and uses standard equipment found in many neuroscience labs. A major advantage of electroporation is that it allows spatial and temporal control of MO delivery and therefore knockdown. Co-electroporation of MOs with cell-type specific fluorescent protein expression plasmids allows morphological analysis of cellular phenotypes. Furthermore, co-electroporation of MOs with rescuing plasmids allows assessment of specificity of the knockdown and phenotypic outcome. By combining MO-mediated manipulations with sophisticated assays of neuronal function, such as electrophysiological recording, behavioral assays, or in vivo time-lapse imaging of neuronal development, the functions of specific proteins and miRNAs within the developing nervous system can be elucidated. These methods can be adapted to apply antisense morpholinos to study protein and RNA function in a variety of complex tissues.

Key words *Xenopus*, Electroporation, Morpholino oligonucleotide, Antisense, Knockdown, In vivo imaging, Neuron, Brain, CNS phenotype

1 Introduction

Antisense morpholino oligonucleotides (MOs) are DNA analogs that are widely used interfere with and knock down gene expression. Through the substitution of the morpholine ring for DNA's riboside moiety, MOs were conceived and designed to overcome the instability of other DNA antisense oligonucleotides. By virtue of their chemistry, MOs and their targets are not recognized and degraded by the RNAse H and therefore offer a stable and specific steric-blocking method to interfere with RNA processing. When injected into embryos, antisense MO technology has been shown

Simon G. Sprecher (ed.), *Brain Development: Methods and Protocols*, Methods in Molecular Biology, vol. 1082, DOI 10.1007/978-1-62703-655-9_11,

to suffer from off-target effects [1], as do other antisense technologies (i.e., RNA interference (RNAi; [2])), and so a series of controls have been recommended when MOs are used [3]. The length of MOs (25 bases) affords MO technology a higher degree of target sequence specificity than other knockdown technologies such as short interfering RNAs (siRNAs), for example, where in addition to the target gene, siRNAs have been shown to repress the translation of other genes with partial complementarity [2, 4]. It is possible to make siRNAs work in *Xenopus* by supplementing RISC (RNA-induced silencing complex) components with exogenous expression [5, 6], thereby permitting genetically regulated cell-type specific or spatial and temporal control of knockdown. Nevertheless, the ease of use, faster onset of knockdown, and fewer problems with target specificity are among the advantages of MO technology [3].

The major application of MOs is to interfere with translation of specific gene targets. When MOs are designed to bind to the sequence immediately surrounding the start codon, they inhibit the 40S ribosomal subunit from scanning the 5′UTR and finding the start codon (*see* **Note 1**). Therefore, MOs prevent translation in a highly effective and gene-specific manner. MOs have also been used to interfere with RNA splicing [7, 8], with microRNAs and their targets [9, 10], and other RNA-protein interactions [11].

The *Xenopus* tadpole model system has many advantages for CNS research. The tadpoles develop externally and are transparent when they are young. The CNS is positioned directly under the skin making it accessible for experimental manipulation and direct observation. Furthermore, barriers to MO use in other model systems, namely, MO delivery and cell uptake, are relatively insignificant.

In *Xenopus,* MOs have been administered by a variety of methods. A common method is to inject MOs into one cell of *Xenopus* embryos during the two-cell stage. The developing frog will then have one affected side and a non-affected control side [12, 13]. Other methods that allow more tissue and temporally specific MO administration include lipofection [14, 15] and electroporation [16–21]. MOs that are tagged with fluorophores (lissamine or carboxyfluorescein) are charged molecules, which allow them to be efficiently delivered to cells through electroporation, which we will detail below.

Here, we describe the methods we use to make observations of neurodevelopmental events in the developing optic tectum, the area of the *Xenopus* tadpole brain that receives direct input from the retinal ganglion cell axons coming from the eye and that controls visually guided behaviors. We have also used electroporation to target other areas of the CNS and spinal cord, and methods we describe here could easily be adapted for those tissues [20, 22, 23]. We co-electroporate fluorophore-tagged MOs and fluorescent protein reporter plasmids (e.g., GFP, CFP, YFP, tdTomato) and use confocal or 2-photon microscopy to take 3D stacks of the transfected cells to measure whether interfering with genes of interest with MO knockdown effects the development of the CNS.

2 Materials

1. Borosilicate glass capillary pipettes. We suggest 1.0 mm outer × 0.75 mm inner diameters and single barrel with filament (e.g., World Precision Instruments).
2. Micropipette puller (P-97 Flaming/Brown, Sutter Instruments). Tapered micropipettes for injecting reagents through the tadpole skin and into the ventricle of the brain must be crafted. Ideal micropipettes are shaped so that they are wispy and sealed at the tip. Before use, the tip is broken with forceps to create a ~20 μm tip. The pipette tip must be rigid enough to penetrate the tadpole's skin without producing an excessively large hole, but still pass MO solution into the brain ventricle without clogging. Instructions for injection micropipettes, which are specific to the puller's filament type (box or trough or differing diameters) and gauge of glass, can be found in Chapter 2 of the micropipette fabrication manual, Pipette Cookbook [24].
3. Microinjector (e.g., Picospritzer II, General Valve Corporation).
4. Dissection microscope (e.g., Wild M8, Leica).
5. Two coarse micromanipulators to hold the micropipette injection needle and the platinum electrode (e.g., MM33, Märzhäuser Wetzlar).
6. A square pulse generator (e.g., Grass SD9 Stimulator). Settings: 1.6 ms pulse duration, 30–70 V, and 1–5 pulses. Equipment that provides precise interpulse intervals is not necessary.
7. 3 μF capacitor (e.g., Part Number, 97F5437, Genteq Capacitors). The capacitor is wired between the stimulator output and the platinum electrode to add an exponential decay waveform to the square pulse.
8. Insulated wires with alligator and/or banana clips to connect the output of the Grass Stimulator to the capacitor and from the capacitor to the platinum electrode leads.
9. Micropipette holder. Must have an opening to fit the outer diameter of the micropipette glass (recommended 1.0 mm), a port for the input tubing of the picospritzer, and a mount onto a rod that fits the micromanipulator. As shown in the photo in Fig. 1a and in the diagram in Fig. 1b. For example, MP series holders from Warner Instruments, Hamden, CT.

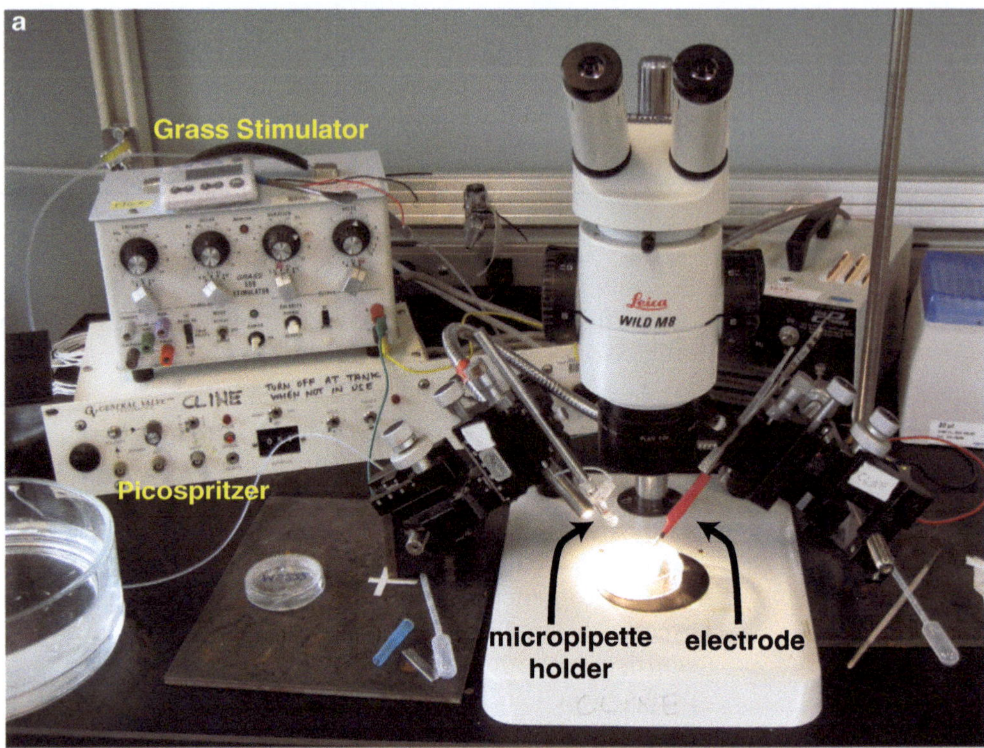

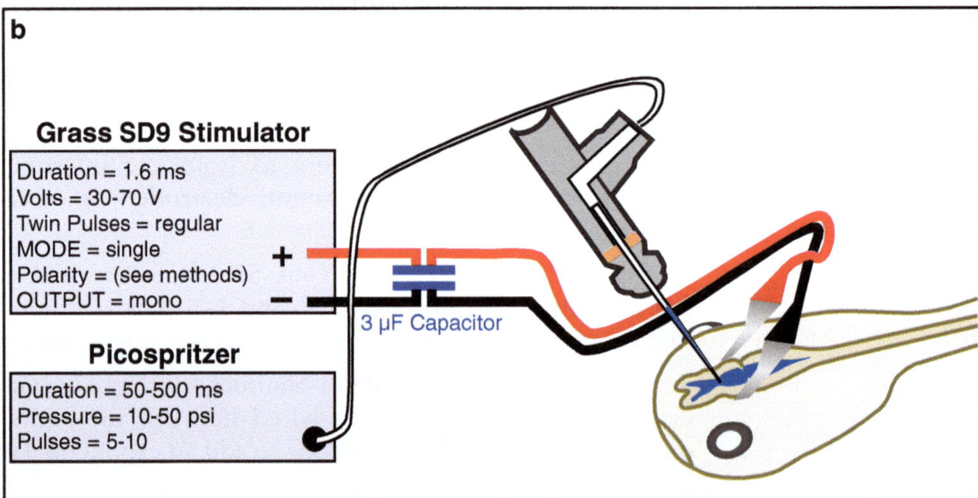

Fig. 1 Configuration of electroporation equipment. (**a**) A typical setup of the apparatus needed to electroporate includes a Grass Stimulator square pulse generator, a Picospritzer pressure injector, dissecting microscope, and micromanipulators to hold the electrode and micropipette for injection. (**b**) Simple wiring diagram and settings for the Grass Stimulator and Picospritzer and diagram to indicate the position of the injection micropipette and the plates of the electrode

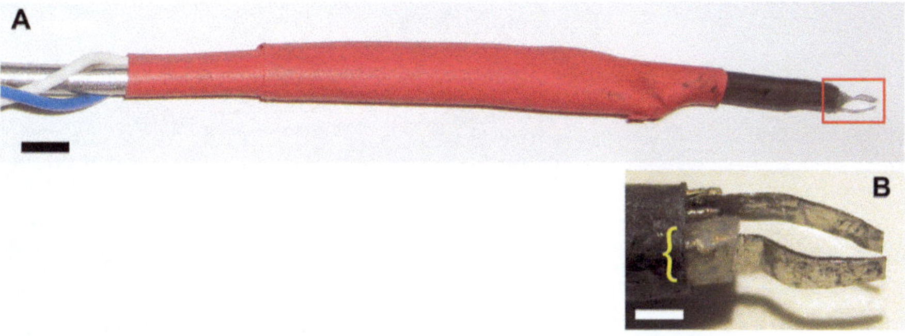

Fig. 2 Custom platinum electrode. (**a**) The plates of the electrode are constructed out of platinum foil that is soldered to insulated wire leads. The plates are isolated from one another with shrink tubing. A larger piece of shrink tubing holds the aligned plates in place, and a still larger piece of shrink tubing holds the electrode onto the rod that fits in the micromanipulator. Scale bar = 5 mm. (**b**) A magnified view of the tip of the electrode showing the folded platinum foil that is folded and cut into ~1 mm wide plates. The *bracket* indicates the shrink tubing that insulates electrodes from one another. Scale bar = 2 mm

10. Custom platinum electrode. Description of the fabrication of the electrode (Figs. 1b and 2) is given in Subheading 3.2, **step 2**.
11. Apparatus for housing/moving/positioning tadpoles. Tadpoles can be reared in large (200–500 ml) bowls (lower left, Fig. 1a). After imaging experiments begin in which individual animals are followed over time, we house tadpoles in plastic 6-well tissue culture plates with lids. We move individual tadpoles around by sucking them into disposable plastic transfer pipettes with openings cut large enough (with a razor blade) to encompass the tadpoles without injury. For electroporation under the microscope, we transfer the tadpoles to an inverted petri dish with wet tissue on top of it. The petri dish then can be slid around and the tadpole positioned under the microscope. A fine paintbrush can be used to adjust the position of the tadpoles.

    We make custom imaging chambers. To do this, we make thin (5–10 mm) sheets of Sylgard elastomer (Dow Corning), cut a square piece of the Sylgard that fits on top of a microscope slide, and then carve tadpole-shaped recesses in the Sylgard so that the top of the tadpoles' heads are at the same level as the surface of the Sylgard sheet (Fig. 3). Tadpoles are placed in the recesses with a bit of MS-222 solution, and glass coverslips are placed directly on their heads/on the Sylgard surface. If the Sylgard is clean and the surface is smooth and flat, the glass coverslip will make a tight seal. Check to make sure the head and brain are not squished. Brisk blood flow through the brain must be maintained throughout the imaging protocol.

Fig. 3 Tadpole imaging chamber. We carve tadpole-shaped recesses in blocks of Sylgard elastomer. The chambers are placed on a microscope slide so that they can be locked onto the stage of the microscope. The recesses must be large enough to hold the tadpole so that the top of their heads is just at the surface of the Sylgard and the tadpole is not compressed and does not shift during image acquisition. To hold the tadpoles in place under a water immersion lens, we place a slide coverslip over the tadpoles on the surface of the Sylgard

12. 100× Steinberg's stock: 0.5 g KCl, 0.8 g $Ca(NO_3)2{\cdot}4H_2O$, 2.1 g $MgSO_4{\cdot}7H_2O$, 34 g NaCl, and 119 g HEPES to 1 l dH_2O, pH to 7.4, and keep at 4 °C.
13. 1× Steinberg's working solution: 100 ml 100× Steinberg's stock to 10 l dH_2O. Add 500 μl penicillin/streptomycin (5,000 U/ml/5,000 mg/ml, Invitrogen), pH to7.4. Keep and use at room temperature.
14. Anesthesia: 0.02 % MS-222/Tricaine (3-Aminobenzoic acid ethyl ester, Sigma-Aldrich) dissolved in 1× Steinberg's working solution, pH 7.0.
15. Endotoxin-free plasmid DNA encoding gene of interest/fluorescent protein. Plasmids should be prepared with Qiagen EndoFree Plasmid Maxi Kit, for example. Vectors with CMV promoters (e.g., Clontech pEGFP and similar) are used at 1–5 μg/μl. Vectors with Gal4/UAS repeats (e.g., [25]) are used at 0.1–1 μg/μl.
16. Lissamine- or carboxyfluorescein-tagged morpholino (GeneTools), diluted in endotoxin-free dH_2O; working concentrations for electroporation are 0.1–0.5 mM.
17. ~1 % Fast Green stock solution, prepared with endotoxin-free molecular grade water and filtered through a syringe filter. Fast Green is added to the DNA/morpholino solution to equal ~0.01–0.1 % so that the injected solution can be seen in the ventricle of the brain.

18. 1 ml disposable plastic syringe for loading micropipettes with plasmid/morpholino reagents. By melting the center of the syringe with the plunger pulled out over a flame and pulling the tip (or carefully letting the end fall to the floor or countertop), it will form a wispy capillary that is >1 mm in diameter. Cut the tip of the pulled syringe until you find it is not sealed and is sufficiently small to fit inside the back of the micropipettes (*see* Fig. 3 in ref. 18). This syringe can be kept and reused to load micropipettes with plasmid and MO reagents.
19. *Xenopus laevis* tadpoles. Fertilized eggs are acquired from hormone-induced matings of albino *Xenopus laevis* frogs in our colony. Tadpoles raised at 23 °C with a 12 h light/12 h dark diurnal cycle until used for experiments. Alternatively, tadpoles can also be purchased as fertilized eggs/young embryos from commercial sources (e.g., Nasco, Xenopus Express, or Xenopus One).
20. Aquarium air pump. Not necessary, but recommended to add to the bowl to speed recovery from anesthesia. For example, Tetra 77851"Whisper" Air Pumps for 10 gallon tank.
21. Compound microscope equipped with epifluorescence and appropriate filter cubes and a low-power (20×) air objective, optional (e.g., Nikon Optiphot-2). This microscope is used to screen quickly tadpoles in order to determine whether sufficient cells are transfected and MOs are taken up into cells.
22. A multiphoton or confocal microscope system equipped with filters and/or lasers appropriate for the fluorescent molecules in the experiment (e.g., PerkinElmer UltraVIEW spinning disc confocal system using a Nikon Eclipse FN1 with a 25× 1.1NA water immersion objective).
23. Incubator for tadpole housing. Set to 23 °C with a 12 h light/12 h dark diurnal cycle.

3 Methods

3.1 Animal Husbandry

Tadpoles are reared as groups of ~100 in large bowls at 23 °C with a 12 h light/12 h dark diurnal cycle. For our experiments, we typically use animals that are between Nieuwkoop and Faber stage 46 (5 days postfertilization, dpf) and stage 49 (~12 dpf) [26], though electroporation is effective on younger and old animals. Once experiments begin, single tadpoles are kept in 6-well tissue culture plates so that time-lapse images from identified individual tadpoles can be acquired. All procedures are conducted with *Xenopus* tadpoles anesthetized with 0.02 % MS-222 solution. Before electroporation or imaging protocols, the tadpoles are transferred into the MS-222 solution, and within minutes, they are immobile, unresponsive to touch, and ready for the procedure. Afterward, they are revived within minutes of placing them back to their rearing containers with Steinberg's solution.

3.2 Preparing the Electroporation Equipment

1. Configuring the electroporation equipment. We use an equipment configuration shown in Fig. 1a, so that the Grass Stimulator and Picospritzer are within arm's reach while looking through the microscope. The wiring diagram is shown in Fig. 1b. What is not visible in the photo in Fig. 1a is the 3 μF capacitor shown in the wiring diagram in Fig. 1b. Before you begin the electroporation procedure, practice adjusting the micropipette and electrode so that they both can reach where the stage/tadpole in the field of view is. Minimizing the time that the tadpoles are anesthetized and on the stage of the microscope by preparing the equipment will help their recovery from the procedure.
2. Fabricating the electrode. Two platinum foil plates are fabricated from new or discarded filaments from a Sutter puller that are folded and cut to a shape that fits the contour of the tadpole head/brain (Fig. 2). Typical platinum electrodes are approximately 0.5–1 mm wide and 5–10 mm long and are soldered to wire leads. These plates must remain electrically isolated but secured so that the tips of the plates are ~0.5 mm apart and can straddle the area of the brain to be electroporated. We use heat shrink tubing around the soldered connection of one electrode (bracket, Fig. 2b) and then shave down the tubing so the second platinum plate can be positioned very close to it. Next, position the plates so they are parallel and use heat shrink tubing to secure them together. Lastly, secure the plates with shrink-wrap along a rod that fits in the micromanipulator. Connect the wire leads as shown in Fig. 1b.

3.3 In Vivo Tectal Cell Transfection of Plasmid Constructs and MOs with Electroporation

1. Preparation of transfer materials for electroporation. Using molecular biology grade, endotoxin-free water, prepare 10–20 μl working solutions of plasmid, MO, and Fast Green in microcentrifuge tube. Recommended solution concentrations are 1–5 μg/μl for CMV-promoter plasmids, 0.1–1 μg/μl for Gal4-UAS plasmids, 100–500 μM MO (*see* **Note 2**), and 0.01–0.1 % Fast Green. Though there are reports that some MOs become unstable over time [27], we have had success keeping these working solutions at 4 °C for weeks and repeatedly drawn from them.
2. Loading micropipettes. Using the disposable 1 ml syringe that has been melted and pulled to a <0.5 mm capillary, carefully draw 1–2 μl of plasmid into the capillary part of the syringe. Be careful not to let the vacuum pressure pull the solution into the body of the syringe because it will be difficult to expel it. Insert the syringe tip into the micropipette and expel the solution.

3. Insert the pipette into the micropipette holder. It must be secured well or the pressure of the Picospritzer will eject the micropipette from the holder.
4. Break back the tip of the glass micropipette by brushing it with a tissue or breaking it with forceps until it has a tip size of ~20 μm.
5. Check that the micropipette is open by using low pressure (10 psi) and short durations (10 ms) to eject the plasmid/MO. It should form a small droplet at the tip of the micropipette.
6. Check that the electrode is approximately in the correct position and can reach the stage under the microscope where the tadpole will be placed.
7. Anesthetize tadpoles. Initially anesthetize only the tadpole that will be used so that there is no chance of overexposing the tadpole to the anesthetics. As you become more familiar and faster with the procedure, five to ten tadpoles can be anesthetized simultaneously. To do this, suck up the tadpole into the transfer pipette and squirt it into the MS-222 solution. Depending upon how much Steinberg's is moved with the tadpole, the MS-222 will become dilute and will need to be replaced after multiple tadpoles are transferred.
8. Place an anesthetized tadpole along with a little MS-222 solution on the tissue-covered petri dish using the disposable plastic transfer pipette. Gently arrange the tadpole with the paintbrush and by moving the petri dish platform so that the brain is aligned with the glass micropipette as in Fig. 1b. Younger animals (stage 42 and younger) that still have enlarged yolky guts tend to tip onto their sides, and using the paintbrush to curl their tails up along the side of the tadpoles will help orient them so their dorsal side is up. Alternately, the tadpoles can be propped up against ridges formed in the tissue to orient them. Keep the tadpole moist with the MS-222 solution.
9. Insert tip of the micropipette through the skin, into the ventricle of the brain. The angle of the micropipette must not be too shallow or it will slide/glance off the skin. The angle of the pipette is limited by working distance of the microscope. The tadpole's head has a slight downward angle between the midbrain and its mouth, which can help the pipette penetrate through the skin when the tadpole is oriented as in Fig. 1b.
10. Eject the plasmid/MO/Fast Green solution from the pipette. The approximate amount of plasmid/MO solution is ~10 nl, but because the pipette may break as it penetrates the skin, the blue color of the Fast Green filling the ventricle is a better indicator of the volume. The *Xenopus* tectum has two lobes that are separated by a large ventricle. When we wish to electroporate a large number of tectal cells, we fill the ventricle until the

caudal/lateral corner of the midbrain ventricle is blue from the Fast Green Dye and the triangular shape of the ventricle is clear. Overfilling the ventricle will put pressure on the tissue and may cause damage. Ejecting less plasmid/MO will result in fewer electroporated cells. Depending on the bore of the micropipette and the pressure of the Picospritzer, filling the ventricle may take 10 or more pulses, but 5–10 pulses of 30 ms duration at 20 psi are a good starting point. Ejecting more, shorter duration and lower pressure pulses is preferred because it causes less damage to the brain. As the micropipette is reused on multiple tadpoles, its tip may break and the pulse duration and/or pressure should be reduced so that single Picospritzer ejections do not harm the tadpole.

11. You may choose to leave the micropipette in place or back it up and out of the way.
12. Positioning the electrode. Lower down the platinum electrode so that it just spans the target area of the brain. The edges of the electrode tips should be depressed firmly down on the skin of the tadpole. The shape of the electrode and the orientation of how it is placed on the tissue can both be used to target areas of the brain for transfection. Placement of the electrodes is important to consider because positively charged molecules (e.g., lissamine-conjugated MOs) will move toward the negative pole, while negatively charged molecules (e.g., DNA and carboxyfluorescein-conjugated MOs) will move toward the positive pole of the electrode during electroporation. Electroporation of untagged MOs, however, has also been shown in some tissues [28]. By orienting the electrode and electroporating with a single polarity, it is possible to drive reagents in a single direction in order to target areas of the tissue for transfection. The Grass Stimulator is capable of reversing the polarity of the voltage pulses, which can be used to electroporate reagents that have opposite charges. To target both lobes of the tectum, we place the electrode so that the tips of the electrode plates press down just beyond the lateral edge of each side of the tectum.
13. Electroporation. Apply voltage pulses by tapping the toggle switch on the Grass Stimulator. The interpulse interval has not been found to affect electroporation efficiency. Keep in mind when co-electroporating plasmid and MOs, that MOs are more easily/efficiently electroporated than plasmid vectors (*see* **Note 3**). As a rule, increasing the voltage and the number of pulses will increase the number of cells that are transfected and the expression levels of individual cells. Efficiency of plasmid transfection is also affected by the expression vector used (as discussed in Subheading 3.2). Regardless of the voltage level used, delivering more than 5 pulses (with one polarity)

is not advised because higher number of pulses may damage the tissue. For an application where we want to transfect ~25 cells in each lobe of the tectum with a Gal4-UAS plasmid (i.e., a high efficiency expression vector) and a lissamine-tagged MO, we typically apply 2–3, 1.6 ms duration, and 30 V pulses of each polarity. With a CMV-promoter-driven fluorescent reporter combined with a lissamine-tagged MO, the setting would change slightly to 4–5, 1.6 ms duration, and 40 V pulses of each polarity, because the expression vector drives weaker expression. Signs to look for when electroporating are discussed in **Note 4**.

14. Retract electrode and micropipette. Often, the electrode plates may need to be cleaned (dabbing them with a folded tissue).
15. Place tadpole into recovery chamber. This can be done by moving the tadpole while it is on the petri dish cover over a larger bowl and then gently squirting it off the tissue with a plume of Steinberg's from the transfer pipette. We return tadpoles to a bowl of Steinberg's solution and add the output of an aquarium air pump to the bowl until the tadpoles are revived and swimming.
16. Tadpoles are returned to 23 °C incubator and 12 h light/12 h dark diurnal cycle.

3.4 In Vivo Imaging of Xenopus Tadpoles to Screen for CNS Phenotypes

1. We wait about 24 h after electroporation before the first image is acquired to allow sufficient fluorescent protein to be expressed. Another factor to consider in deciding when to collect images is the rate of protein turnover of the gene targeted by the MO. Images should be acquired over an interval when the MO is capable of knocking down protein levels.
2. Individual tadpoles should be transferred into MS-222 and anesthetized.
3. Anesthetized tadpoles in the MS-222 solution are transferred to the Sylgard imaging chamber, which is mounted on a microscope slide (Fig. 3). Under a dissecting microscope, tadpoles are positioned within the carved indentation of the Sylgard so that the dorsal side of the tadpole is up. Place a coverslip over the tadpole/Sylgard and press it down so that it attaches to the wet surface of the Sylgard and presses gently on the surface of the tadpole's head. A bit of liquid from the tadpole chamber on the Sylgard will ensure that the coverslip sticks to the Sylgard. If tadpoles are selected for confocal/multiphoton image acquisition, care must be taken so that the tadpole does not shift midway through the acquisition of the z-stack. *See* **Note 5** on positioning tadpoles.
4. (Optional) Prescreen tadpoles for imaging experiments. Tadpoles may be viewed quickly under the screening microscope

with an air objective in order to prioritize them based on the number of reporter-expressing cells and the quality of plasmid and MO transfection. The tadpoles and CNS can be inspected for damage at this point (*see* **Note 6**). Simply secure the microscope slide with the Sylgard chamber on the stage of the microscope. Similar to Subheading 3.3, **step 15**, after prescreening, individual tadpoles are returned to a larger bowl with an aquarium pump until they are revived. Once swimming, each identified tadpole is transferred to a 6-well plastic tissue culture dish so that individuals can be identified and imaged.

5. Acquire confocal/multiphoton z-stack image. Sylgard chamber holding the tadpole is placed under the microscope and positioned so that the field of view is centered on the region of interest. Top and bottom levels of the stack are set, filters and laser(s) are selected, settings are modified to maximize signal and minimize saturation, and images are acquired.
6. Tadpoles are given time to revive in a bowl with the aquarium pump input before transferred back to their well of the 6-well cell culture dish.
7. Return tadpoles to 23 °C incubator and 12 h light/12 h dark diurnal cycle.

3.5 MO Uptake and Fluorescent Protein Expression in the Tectum

MOs are more efficiently electroporated into tissue than are DNA plasmids. MOs are taken up by more cells over a physically larger area of the brain than seen by GFP expression from co-electroporated plasmids. Figure 4 shows two typical examples of right tectal lobes ~24 h after electroporation of a GFP plasmid vector and lissamine-tagged control MO. The animals were injected with 0.1 mM lissamine-tagged control MO and 0.5 μg/μl turboGFP plasmid under the control of the Gal4/UAS expression system described in Bestman et al. [29]; the electrode spanned the tectum as shown in Fig. 1b and electroporated with 3, 1.6 ms, and 40 V pulses with each polarity.

The confocal stacks in Fig. 4 show that the majority of the tectal cells have taken up the MO (magenta signal), and relatively few cells express the GFP (green). The right side of the CNS of the tadpole is lightly outlined in Fig. 4a. The right tectal lobe is shown in the bright-field image (Fig. 4b) and in the flattened confocal image of a 50 μm stack taken ~20 μm below the surface of the skin (Fig. 4c). The image in Fig. 4d is a flattened 180 μm stack through the right tectal lobe, with magnified views of cell bodies (Fig. 4d (i), (ii)) and distal processes (Fig. 4d (iii–vi)). These images show that the MO signal in the tectum is extensive and that the GFP-expressing cells also contain MO. The differences in the number of cells expressing GFP and the MO may be due to the different size and charge of MOs and plasmid DNA. Adjustment of the DNA to MO ratio may increase the correlation between distribution of

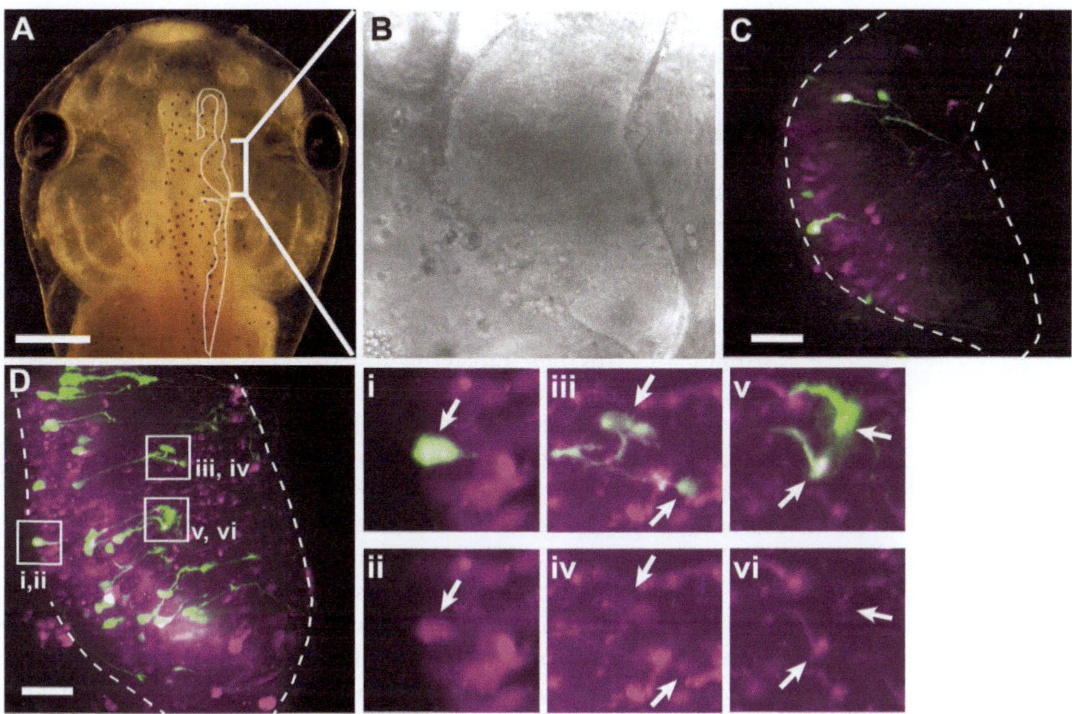

Fig. 4 Examples of MO uptake and fluorescent protein expression in two electroporated tectal lobes. The tadpoles were electroporated 24 h prior to image acquisition with 0.1 mM lissamine-tagged MO (*magenta*) and 0.5 μg/μl turboGFP (*green*). (**a**) A tadpole head. The right side of the brain is outlined and the *bracket* indicates the tectum, which is magnified in (**b**). (**b**) A bright-field image of the right tectal lobe with the plane of focus about 60 μm below the surface of the skin. (**c**) An image of a flattened 50 μm confocal stack taken about 20 μm below the surface of the skin showing the distribution of the lissamine-tagged control MO and turboGFP-expressing cells in the right tectal lobe. Scale bar = 50 μm. (**d**) An image of a flattened 180 μm confocal stack taken through the right tectal lobe of a stage 46 tadpole. The *boxes* in (**d**) are magnified to show the distribution of MO in the cells. Scale bar = 50 μm. (*i, ii*) MO fills the soma of the turboGFP-expressing cell, as well as many of its neighbors. The *arrows* indicate the position of the turboGFP-expressing cell. (*iii–vi*) We see that many cells also show evidence of the MO signal extending >100 μm into the distal processes of the cells. *Arrows* indicate the distal tips of tectal cells expressing turboGFP and show the lissamine MO signal. Panels (*i*)–(*vi*) are 50 μm square

MOs and expression of reporter from DNA plasmids [30]. If it is critical for the experimental design to have precise co-delivery of MOs and plasmid, single-cell electroporation, where transfer material is delivered to a single cell by delivery of voltage pulse through a modified patch pipette, may be a better approach for combined delivery of MO and plasmid reporter [18].

3.6 Verification of Target Knockdown and Analysis of Morpholino Efficacy and Specificity

Control and knockdown verification experiments for MO use have been outlined by Eisen and Smith [3] and Bedell et al. [27] and also can be found on the GeneTools website (http://www.gene-tools.com). We note here that the methods suggested in these resources have been largely developed for experiments where MOs are administered by the injection into fertilized eggs or blastomeres. With embryo injection methods, it is possible to calculate precisely the concentration of MO that the cell receives. A recommended strategy to investigate specificity of knockdown is to test whether the knockdown phenotype decreases severity with decreased concentrations of MO. With electroporation, the MO concentration injected into the ventricle is known, but the diluted MO concentration within the ventricle is hard to estimate because it is not a closed system. More importantly, the efficiency of MO transfer from the ventricle into cells by electroporation is difficult to estimate and whether transfer is linearly related to ventricular MO concentration is unknown. Therefore, electroporation may effectively concentrate MOs in target cells, so injecting a range of MO concentrations into the ventricle may not result in a corresponding range of MO concentrations in target cells in a complex tissue like the brain. Figure 4 gives an indication of the heterogeneity of MO distribution in electroporated optic tecta and suggests that MO distribution is uneven across the tissue.

The specificity of an MO on the knockdown of a target gene can be analyzed through a variety of experiments. First, the results of experimental MO electroporation should be compared to the results of electroporation of control MOs. MO knockdown can also be compared to other results of knockdown methods, such as the electroporation of dominant-negative plasmid DNA constructs or other interfering reagents [31–33]. Electroporation also works well to coadminister a "rescue" construct to replace the gene targeted by MO knockdown along with the MO [34].

Several authorities recommend using multiple MOs against the same target transcript to validate specificity of knockdown. For translation-blocking MOs, the MO is most effective when it overlaps with the start site, which limits the region against which MOs can be designed. This is a particular issue for neural genes, which have many different splice isoforms in the 5′UTR, which govern promoter specificity. Because the Xenopus genome is not yet completely annotated, shifting the target region of the MO around the start site may unwittingly end up targeting alternate transcripts and generate confusing or erroneous results with respect to specificity of knockdown. Similarly, recommendations to validate knockdown by comparing effects of translation-blocking MOs and MOs that interfere with splicing are valid in systems in which genomic information is complete, but cannot be applied where this is not the case. The central nervous system has the highest density of splice variants of an organ. Even as this feature makes studying the molecular genetic control of brain development particularly exciting, it also means that the challenges unique to studying nervous system development and plasticity must be recognized.

When antibodies to the target protein are available, it is quite common to test whether MOs decrease levels of protein expression with immunocytochemistry or Western blots. However, Western blot analysis may not detect protein knockdown in the brain when MOs are administered with electroporation. Unlike MO injection into a fertilized egg or blastomere, where protein knockdown can be widespread, targeted electroporation of the tectum, retina or other areas of the CNS, may result in fewer cells that take up the MO. The sensitivity of the Western blot to reveal knockdown will be hindered by the relatively few MO-containing cells in the tissue that will be homogenized with the surrounding cells that have not taken up the MO. Despite these caveats, this method has been shown to verify MO knockdown [31, 33, 35].

Lastly, alternative methods to evaluate knockdown of genes of interest in the development of the CNS include functional analysis, such as electrophysiological recordings from neurons electroporated with MOs against neurotransmitter receptor subunits [33, 36].

4 Notes

1. Design of experimental MOs. The MO vendor, GeneTools, provides a free design service. Discussion of MO design can be found here [37].
2. MO concentration. The MO vendor, GeneTools, provides detailed methods for storing and reconstituting the MO. Briefly, MOs are stable when reconstituted in pure water and stored at room temperature. Fluorescently tagged MOs should be stored in the dark. As stated on the GeneTools website, the lissamine tag decreases the solubility of the MO and lissamine-tagged MOs in particular should be heated to 65 °C for 10 min and vortexed for 30 min until the MO solution is fully resuspended. MO concentration can be measured using the protocol on the GeneTools website. Considerations for long-term storage and dilution of MOs are also discussed in [27].
3. Control MOs, choices, and their design. Though the knockdown efficacy will be attenuated, MOs are capable of complementing a target sequence that shares 21/25 bases [3, 38]. "Specific" control MOs that are incapable of interacting with the target sequence consist of a 5/25 base mispair and can be designed for each experimental MO.
4. There are two signs to look for as you electroporate as a way to verify that the settings are correct and the equipment is wired correctly. The first is the formation of small electrolysis-induced bubbles (>10 bubbles). These will appear where the electrode makes contact with the moist skin. Large violently rolling bubbles are an indication of a problem with the settings/equipment. The second indicator is a slight twitch of the ocular muscles and movement of the eyes.

5. Positioning tadpoles in Sylgard chamber. For prescreening, this can be done quickly with little regard to the position of the coverslip and tadpole. When z-stacks are to be acquired, take care that no bubbles are transferred with the tadpoles inside of the Sylgard chamber because they might move and shift the tadpole during the image acquisition. Bubbles can usually be brushed out with the fine paintbrush. It is also important that the coverslip gently press on the head of the tadpole. The best image is acquired when there is no space between the tadpole and the coverslip. The coverslip must be secure, as it will have the water droplet for the immersion objective on it. If the coverslip becomes loose during the acquisition, it will shift the tadpole and disrupt the image mid-acquisition.
6. Inspecting tadpoles for damage as a result of the electroporation procedure. We occasionally find tadpoles that do not recover, or do not recover properly, from electroporation. This may result from the tadpole drying out during the procedure, and with experience, this can be minimized. Tadpoles that sustain damage can be spotted because they fail to swim properly, and inspection of the brain may show herniation of cells into the ventricle or signs of bleeding. The evidence of bleeding usually clears up within 48 h, but these are signs that indicate that the electrode pressed too hard on the tadpole, the volume injected into the ventricle was too great, and/or voltage level and number of electroporation pulses should be scaled back.

References

1. Robu ME, Larson JD, Nasevicius A, Beiraghi S, Brenner C et al (2007) p53 activation by knockdown technologies. PLoS Genet 3:e78
2. Hardy S, Legagneux V, Audic Y, Paillard L (2010) Reverse genetics in eukaryotes. Biol Cell 102:561–580
3. Eisen JS, Smith JC (2008) Controlling morpholino experiments: don't stop making antisense. Development 135:1735–1743
4. Summerton JE (2007) Morpholino, siRNA, and S-DNA compared: impact of structure and mechanism of action on off-target effects adn sequence specificity. Curr Top Med Chem 7:651–660
5. Chen CM, Chiu SL, Shen W, Cline HT (2009) Co-expression of Argonaute2 enhances short hairpin RNA-induced RNA interference in Xenopus CNS neurons in vivo. Front Neurosci 3:63
6. Lund E, Sheets MD, Imboden SB, Dahlberg JE (2011) Limiting ago protein restricts RNAi and microRNA biogenesis during early development in Xenopus laevis. Genes Dev 25:1121–1131
7. Draper BW, Morcos PA, Kimmel CB (2001) Inhibition of zebrafish fgf8 pre-mRNA splicing with morpholino oligos: a quantifiable method for gene knockdown. Genesis 30:154–156
8. Morcos PA (2007) Achieving targeted and quantifiable alteration of mRNA splicing with Morpholino oligos. Biochem Biophys Res Commun 358:521–527
9. Kloosterman WP, Lagendijk AK, Ketting RF, Moulton JD, Plasterk RH (2007) Targeted inhibition of miRNA maturation with morpholinos reveals a role for miR-375 in pancreatic islet development. PLoS Biol 5:e203
10. Choi WY, Giraldez AJ, Schier AF (2007) Target protectors reveal dampening and balancing of Nodal agonist and antagonist by miR-430. Science 318:271–274
11. Bruno IG, Jin W, Cote GJ (2004) Correction of aberrant FGFR1 alternative RNA splicing through targeting of intronic regulatory elements. Hum Mol Genet 13:2409–2420
12. Heasman J, Kofron M, Wylie C (2000) Beta-catenin signaling activity dissected in the early

Xenopus embryo: a novel antisense approach. Dev Biol 222:124–134

13. Tandon P, Showell C, Christine K, Conlon FL (2012) Morpholino injection in Xenopus. Methods Mol Biol 843:29–46
14. S-i O, Mann F, Boy S, Perron M, Harris WA (2002) Lipofection strategy for the study of Xenopus retinal development. Methods 28: 411–419
15. Ando H, Okamoto H (2006) Efficient transfection strategy for the spatiotemporal control of gene expression in zebrafish. Mar Biotechnol (NY) 8:295–303
16. Sasagawa S, Takabatake T, Takabatake Y, Muramatsu T, Takeshima K (2002) Improved mRNA electroporation method for Xenopus neurula embryos. Genesis 33:81–85
17. Eide FF, Eisenberg SR, Sanders TA (2000) Electroporation-mediated gene transfer in free-swimming embryonic Xenopus laevis. FEBS Lett 486:29–32
18. Bestman JE, Ewald RC, Chiu SL, Cline HT (2006) In vivo single-cell electroporation for transfer of DNA and macromolecules. Nat Protoc 1:1267–1272
19. Falk J, Drinjakovic J, Leung K, Dwivedy A, Regan A et al (2007) Electroporation of cDNA/Morpholinos to targeted areas of embryonic CNS in Xenopus. BMC Dev Biol 7:107
20. Haas K, Jensen K, Sin WC, Foa L, Cline HT (2002) Targeted electroporation in Xenopus tadpoles in vivo–from single cells to the entire brain. Differentiation 70:148–154
21. Haas K, Sin W-C, Javaherian A, Li Z, Cline HT (2001) Single-cell electroporation for gene transfer in vivo. Neuron 29:583–591
22. Javaherian A, Cline HT (2005) Coordinated motor neuron axon growth and neuromuscular synaptogenesis are promoted by CPG15 in vivo. Neuron 45:505–512
23. Ruthazer ES, Li J, Cline HT (2006) Stabilization of axon branch dynamics by synaptic maturation. J Neurosci 26:3594–3603
24. Osterele A (2011) P-1000 & P-97 pipette cookbook, 2011. In:Rev G (ed) Sutter Instrument Company
25. Koster RW, Fraser SE (2001) Tracing transgene expression in living zebrafish embryos. Dev Biol 233:329–346
26. Nieuwkoop PD, Faber J (1956) Normal table of Xenopus laevis (Daudin); a systemical and chronological survey of the development from fertilized egg till the end of metamorphosis. North-Holland Publishing Co, Amsterdam, p 243
27. Bedell VM, Westcot SE, Ekker SC (2011) Lessons from morpholino-based screening in zebrafish. Brief Funct Genomics 10:181–188
28. Kos R, Tucker RP, Hall R, Duong TD, Erickson CA (2003) Methods for introducing morpholinos into the chicken embryo. Dev Dyn 226:470–477
29. Bestman JE, Lee-Osbourne J, Cline HT (2012) In vivo time-lapse imaging of cell proliferation and differentiation in the optic tectum of Xenopus laevis tadpoles. J Comp Neurol 520:401–433
30. Sauka-Spengler T, Barembaum M (2008) Gain- and loss-of-function approaches in the chick embryo. Methods Cell Biol 87:237–256
31. Bestman JE, Cline HT (2008) The RNA binding protein CPEB regulates dendrite morphogenesis and neuronal circuit assembly in vivo. Proc Natl Acad Sci USA 105:20494–20499
32. Chiu S-L, Chen C-M, Cline HT (2008) Insulin receptor signaling regulates synapse number, dendritic plasticity, and circuit function in vivo. Neuron 58:708–719
33. Shen W, McKeown CR, Demas JA, Cline HT (2011) Inhibition to excitation ratio regulates visual system responses and behavior in vivo. J Neurophysiol 106(5):2285–2302
34. Sharma P, Cline HT (2010) Visual activity regulates neural progenitor cells in developing Xenopus CNS through musashi1. Neuron 68:442–455
35. Schwartz N, Schohl A, Ruthazer ES (2011) Activity-dependent transcription of BDNF enhances visual acuity during development. Neuron 70:455–467
36. Ewald RC, Van Keuren-Jensen KR, Aizenman CD, Cline HT (2008) Roles of NR2A and NR2B in the development of dendritic arbor morphology in vivo. J Neurosci 28:850–861
37. Zhao Y, Ishibashi S, Amaya E (2012) Reverse genetic studies using antisense morpholino oligonucleotides. Methods Mol Biol 917: 143–154
38. Rana AA, Collart C, Gilchrist MJ, Smith JC (2006) Defining synphenotype groups in Xenopus tropicalis by use of antisense morpholino oligonucleotides. PLoS Genet 2: 1751–1772

Part V

Zebrafish Protocols

Chapter 12

Sensitive Whole-Mount Fluorescent In Situ Hybridization in Zebrafish Using Enhanced Tyramide Signal Amplification

Gilbert Lauter, Iris Söll, and Giselbert Hauptmann

Abstract

Whole-mount in situ hybridization is the preferred method for detecting transcript distributions in whole embryos, tissues, and organs. We present here a sensitive fluorescent in situ hybridization method for colocalization analysis of different transcripts in whole embryonic zebrafish brains. The method is based on simultaneous hybridization of differently hapten-labeled RNA probes followed by sequential rounds of horseradish peroxidase (POD)-based transcript detection. Sequential detection involves enhancement of fluorescent signals by tyramide signal amplification (TSA) and effective inactivation of the antibody–POD conjugate prior to the following detection round. We provide a detailed description of embryo preparation, hybridization, antibody detection, POD–TSA reaction, and mounting of embryos for imaging. To achieve high signal intensities, we optimized key steps of the method. This includes improvement of embryo permeability by hydrogen peroxide treatment and efficacy of hybridization and TSA–POD reaction by addition of the viscosity-increasing polymer dextran sulfate. The TSA–POD reaction conditions are further optimized by application of substituted phenol compounds as POD accelerators and use of highly efficient bench-made tyramide substrates. The obtained high signal intensities and cellular resolution of our method allows for co-expression analysis and generation of three-dimensional models. Our protocol is tailored to optimally work in zebrafish embryos, but can surely be modified for application in other species as well.

Key words Fluorescent in situ hybridization, FISH, Tyramide signal amplification, TSA, Peroxidase, Zebrafish

1 Introduction

With the rapid progress of high-throughput sequencing and improved cloning technologies, available sequence data and clone collections of genomic DNA and cDNA have been steadily increasing. The use of these resources allows producing gene expression data to construct genomic atlases of different tissues and organs. Vertebrate forebrain molecular maps of the telencephalon [1, 2], hypothalamus [3, 4], thalamus [3, 5], and pretectum [6, 7] have been generated to potentially reveal novel aspects of the underlying basic brain organization. Such gene expression data are routinely produced by in situ hybridization technologies to allow

Simon G. Sprecher (ed.), *Brain Development: Methods and Protocols*, Methods in Molecular Biology, vol. 1082, DOI 10.1007/978-1-62703-655-9_12, © Springer Science+Business Media, LLC 2014

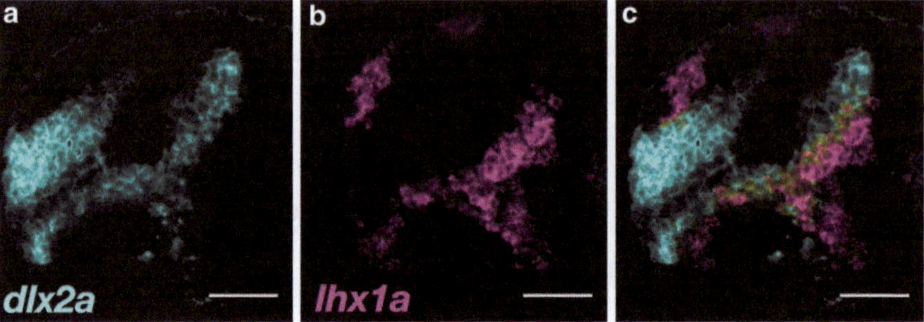

Fig. 1 Two-color whole-mount FISH using enhanced tyramide signal amplification. Lateral views of *dlx2a* and *lhx1a* expression in a 28-hpf zebrafish forebrain with anterior oriented to the left. Transcript signals are visualized in *cyan* and *magenta* as indicated in the panels. Single-channel detection and the overlay of both channels in the same confocal plane are shown. (**a**) Expression of *dlx2a* was visualized by using a DNP-labeled RNA probe together with the DyLight 633-tyramide. The TSA reaction was allowed to proceed for 20 min in the presence of 0.15 mg/ml 4-iodophenol. (**b**) Expression of *lhx1a* was visualized by using a DIG-labeled RNA probe together with the FAM-tyramide. The TSA reaction was allowed to proceed for 30 min in the presence of 0.15 mg/ml 4-iodophenol. (**c**) Overlay of both single channels shows the expression of *dlx2a* and *lhx1a* in the developing forebrain with areas of overlap indicated in *yellow*. Scale bars = 50 μm

global and detailed views on gene expression patterns [8]. Because of the complexity of the brain, it is however often essential to relate a distinct expression domain of a gene of interest to well-known tissue- or cell-type-specific molecular markers. For this purpose, in situ hybridization methods have been developed to detect two or more transcript patterns in different colors simultaneously in the same tissue or embryo [9–12].

In the original whole-mount in situ hybridization protocol, a digoxigenin-labeled nucleic acid probe specific for a unique transcript is detected by anti-digoxigenin antibodies conjugated to alkaline phosphatase. The transcript distribution is visualized by formation of an insoluble cellular purple precipitate during the chromogenic substrate reaction [8]. Protocols for two-color detection add a second nucleic acid probe with another hapten label such as biotin or fluorescein to the hybridization mix and use two different chromogenic substrates to obtain contrasting color precipitates. These protocols were first developed in zebrafish and flies and included alkaline phosphatase, horseradish peroxidase, and ß-galactosidase as reporter enzymes [9, 10, 12–15].

One drawback of chromogenic detection, however, can be compromised cellular resolution especially in case of overlapping expression domains. To overcome this caveat, fluorescent detection methods with increased sensitivity have been developed, so that two or more different transcripts could be detected with high signal strength and signal-to-noise ratio. The problem of weak fluorescence signal was largely solved by implementation of tyramide signal amplification (TSA) [16] into fluorescent in situ hybridization (FISH) protocols [17]. TSA dramatically enhances the FISH signals and makes multiplex mRNA detection feasible in *Drosophila* [18], *Platynereis* [19], zebrafish [20, 21], frog [22], and chicken [23]. One problem with TSA however can be the amplification of background staining in parallel to specific signal enhancement, which may significantly reduce achievable signal-to-noise ratios. Strong autofluorescence of zebrafish and other embryos may additionally lower signal-to-noise ratios to a nonacceptable level for multiplexing. Therefore, further optimization of embryo preparation, hybridization, and detection steps have been necessary to achieve optimal signal strengths and results in zebrafish embryos (Fig. 1) [24].

We describe here an optimized protocol for colocalization analysis of different mRNA transcripts in whole embryonic zebrafish brains. For best results we optimized key steps of the procedure [21, 24]. After fixation, permeability of embryos is improved by treatment with hydrogen peroxide aside standard proteinase K digestion. Hybridization efficacy is enhanced by addition of the viscosity-increasing polymer dextran sulfate. Differently hapten-labeled RNA probes (e.g. dinitrophenol and digoxigenin) are simultaneously hybridized and sequentially detected by their respective anti-hapten antibodies, each coupled to horseradish peroxidase (POD). Inclusion of dextran sulfate in the TSA reaction and addition of substituted phenol compounds as accelerators strongly enhance achievable fluorescent signal intensity. Between detection rounds consisting of antibody–POD conjugate incubation and TSA reaction, the POD activity of the prior round has to be efficiently removed by acidic pH. Finally, we provide a short protocol for synthesis of tyramides, as we find that bench-made substrates are more efficient than commercially available tyramides in producing high signal sensitivity [24]. Our FISH protocol is tailored to work optimally in whole zebrafish embryos and embryonic brains (Fig. 1), but may be easily adapted for application to other model organisms.

2 Materials

2.1 General Buffers

1. Paraformaldehyde (PFA): 4 % (w/v) paraformaldehyde in 1× PBS, pH 7.3. Dissolve 4 g PFA in 100 ml PBS and stir in a fume hood at about 65 °C until everything has dissolved (1–2 h). Let solution cool to room temperature (RT) and adjust to pH 7.3 with NaOH. Store the fixative in 5–10 ml aliquots at −20 °C.
2. Phosphate-buffered saline (1× PBS): 8 % (w/v) NaCl, 0.2 % (w/v) KCl, 16 mM Na_2HPO_4, 4 mM NaH_2PO_4, pH 7.3.
3. Phosphate-buffered saline plus Tween (PBST): 1×PBS, 0.1 % (v/v) Tween-20.
4. Methanol (MeOH).

5. 30 % hydrogen peroxide (H_2O_2), stabilized (Sigma 31642).
6. Rehydration series: 75 %, 50 %, and 25 % (v/v) of methanol in PBST.
7. Proteinase K: 20 mg/ml stock in TE (10 mM Tris–HCl pH 8.0, 1 mM EDTA), store in aliquots at −20 °C. Prepare working solution of 10 μg/ml proteinase K in PBST just prior to use.
8. Glycine: 2 mg/ml glycine in PBST. Store glycine as a 100 mg/ml stock in ddH_2O at −20 °C in aliquots.

2.2 Tyramide Synthesis

1. Succinimidyl esters: 5-(and-6)-carboxyfluorescein succinimidyl ester (FAM-SE; Molecular Probes), DyLight 633 N-hydroxysuccinimide ester (DyLight 633-SE; Pierce). Prepare a 10 mg/ml stock solution of each succinimidyl ester in dimethylformamide (Sigma) just before use.
2. Tyramine hydrochloride: Prepare a 10× stock solution at a concentration of 100 mg/ml in dimethylformamide just before use. Dilute with dimethylformamide to yield the 1× working solution and add 10 μl triethylamine (Sigma) per 1 ml solution.
3. Absolute ethanol (EtOH).

2.3 Hybridization

1. Prehybridization buffer (HB): 50 % (v/v) deionized formamide, 5× SSC, 5 mg/ml torula RNA (Sigma), 50 μg/ml heparin sodium salt, 0.1 % (v/v) Tween-20. Store in aliquots at −20 °C. Store heparin as a 50 mg/ml stock in ddH_2O at −20 °C.
2. Dextran sulfate (Sigma): Prepare a 50 % (w/v) stock solution in water. When used for hybridization, autoclave stock solution for 30 min at 110 °C, otherwise store in aliquots at −20 °C.
3. Hybridization buffer (HBD5): HB including 5 % (w/v) dextran sulfate.
4. Hybridization wash: 50 % (v/v) deionized formamide (AppliChem), 2× SSC, 0.1 % (v/v) Tween-20. Store in aliquots at −20 °C.
5. 20× SSC: 3 M NaCl, 300 mM trisodium citrate, pH 7.0.
6. 2× SSCT: 2× SSC, 0.1 % (v/v) Tween-20.
7. 0.2× SSCT: 0.2× SSC, 0.1 % (v/v) Tween-20.

2.4 Antibody Detection

1. Blocking solution: 8 % (v/v) normal sheep serum in 1× PBST. Heat inactivate sheep serum stock (Sigma) at 56 °C for 30 min and store in aliquots at −20 °C.
2. Sheep-anti-digoxigenin-POD Fab fragments (Roche): Prepare 1:500 working dilution in blocking solution prior to use.

3. Anti-dinitrophenyl-POD (PerkinElmer TSA Plus DNP System): Prepare 1:100 working dilution in blocking solution prior to use.

2.5 Fluorogenic Reaction

1. 4-Iodophenol (Fluka 58020): Prepare a 150 mg/ml stock in DMSO. Store tightly sealed at 4 °C (*see* **Note 1**).
2. Vanillin (Sigma V110-4): Prepare a 150 mg/ml stock in DMSO. Store tightly sealed at 4 °C.
3. TSA reaction buffer: 100 mM borate buffer pH 8.5, 2 % (w/v) dextran sulfate, 0.1 % (v/v) Tween-20, 0.003 % H_2O_2.
4. Borate buffer: 100 mM borate pH 8.5 plus 0.1 % (v/v) Tween-20.
5. POD inactivation solution: 100 mM glycine-HCl pH 2.0 plus 0.1 % (v/v) Tween-20.

2.6 Mounting

1. Glycerol series: 25 %, 50 %, and 75 % (v/v) of glycerol in PBST, 40 mM $NaHCO_3$.
2. Mounting gel: 1 % (w/v) low melting agarose in 75 % (v/v) glycerol in PBST, 40 mM $NaHCO_3$. Keep the solution stirring at 50 °C, as it will turn yellow after repeated heating.
3. Mounting slides: Two stacks of coverslips are glued to a microscope slide with Eukitt (Fluka) leaving a gap, which can be easily bridged by a 24 × 32 mm coverslip. To mount samples of varying thickness, prepare a series of mounting slides using one to several coverslips for each stack.

3 Methods

3.1 Tyramide Synthesis

Despite the advantage of being cost-efficient, bench-made tyramides are also highly concentrated and offer the possibility to specifically match the requirements of the microscope in use.

1. Mix the freshly prepared tyramine working solution and the respective succinimidyl esters at a 1:1.1 equimolar ratio.
2. Allow the reaction to proceed for 2 h in the dark without agitation.
3. The resulting tyramide product is diluted with absolute ethanol to a concentration of 1 mg/ml and stored protected from light at −20 °C (*see* **Note 2**).

3.2 Embryo Pretreatment

1. Transfer dechorionated embryos of the desired developmental stage to a 2 ml microcentrifuge tube. Aspirate supernatant while taking care that embryos remain immersed in liquid. Fixate embryos in 1 ml of 4 % PFA for 24 h at 4 °C (*see* **Note 3**).
2. Rinse embryos four times for 5 min with 1× PBST. Transfer embryos into 100 % MeOH. Exchange 100 % MeOH after

5 min and incubate embryos at −20 °C for at least 30 min. Alternatively, embryos can be kept in 100 % MeOH at −20 °C for long-term storage.

3. Incubate embryos in a 2 % hydrogen peroxide solution in 100 % MeOH for 20 min. Gradually rehydrate embryos by going through a series of 5-min washing steps of 75 %, 50 %, and 25 % MeOH in PBST followed by two PBST washes (*see* **Note 4**).
4. Embryos developed to tail bud stage or further have to be digested with proteinase K in order to increase permeability. The optimal digestion times for different stages have to be determined experimentally. After treatment with proteinase K solution (10 μg/ml in PBST) at RT and under gentle agitation, stop the reaction by rinsing twice with 2 mg/ml glycine in PBST. Postfix embryos for 20 min in 4 % PFA. Afterwards, wash four times for 5 min in PBST. Transfer embryos into 0.5 ml prehybridization buffer (HB) and exchange with 1 ml HB after 5 min. Embryos are now ready to use or can be stored in HB at −20 °C.
5. Transfer up to 25 embryos into 200 μl of HB in a 2 ml microfuge tube with round bottom. Incubate in a water bath at 60 °C for 1 h for prehybridization.

3.3 Hybridization

1. Prepare probe mix by adding dinitrophenol (DNP)- and digoxigenin (DIG)-labeled RNA probes together in 150 μl HBD5 (*see* **Note 5**). Denature the probe mix for 5 min at 80 °C and equilibrate to 60 °C.
2. Carefully aspirate prehybridization solution from embryos and add the pre-warmed probe mix instead. Incubate in a water bath at 60 °C overnight (min. 15 h).
3. For all post-hybridization washes, solutions have to be pre-warmed to 60 °C.
4. Incubate embryos twice in 1 ml SSCT hybridization wash solution for 30 min.
5. Wash once with 1.5 ml 2× SSCT for 15 min.
6. Wash twice with 1.5 ml 0.2× SSCT for 30 min.
7. Add fresh 0.2× SSCT and let cool down to RT.
8. Rinse twice with PBST.

3.4 Antibody Detection

The differently hapten-labeled RNA probes are sequentially detected using peroxidase (POD)-conjugated antibodies directed against the respective hapten. Typically DNP-labeled probes are detected first (for reasons *see* **Note 6**) using an anti-DNP-POD antibody (1:100) followed by detection of DIG-labeled probes with sheep-anti-DIG-POD (1:500) (*see* **Note 7**).

1. Incubate embryos in 100 μl blocking solution for 30 min on a gently rocking table at RT.

2. Carefully remove the blocking solution and add antibody solution. Incubate overnight at 4 °C without agitation.
3. To remove excess antibody, wash six times with PBST for 20 min at RT with gentle agitation.

3.5 Fluorogenic Reaction

The TSA reaction can be significantly enhanced by the use of a POD accelerator. Use 4-iodophenol or vanillin at a concentration of 0.15 mg/ml and 0.45 mg/ml, respectively (*see* **Note 8**).

1. For 1 ml reaction buffer, combine 500 μl borate buffer pH 8.5 (200 mM), 40 μl 50 % dextran sulfate, 10 μl of 10 % Tween-20, 6 μl of 0.5 % H_2O_2, and an appropriate accelerator and adjust the volume with water to 1 ml (*see* **Note 9**).
2. Addition of 4-iodophenol will result in the appearance of a cloudy smear. Mix the reaction buffer well by pipetting up and down until the dextran sulfate has dispersed and the solution appears opaque.
3. Dilute the desired tyramide with the reaction buffer and mix well by pipetting. Use 6 μl of DyLight 633-tyramide or 4 μl of FAM-tyramide stock solution per 1 ml reaction buffer (*see* **Note 10**).
4. Rinse embryos twice with 100 mM borate buffer pH 8.5 containing 0.1 % Tween-20. Remove supernatant borate buffer as closely as possible and add 90 μl of the tyramide solutions to the embryos. Gently mix by pipetting using a cut tip. Incubate for the desired length of time in the dark at RT without agitation (*see* **Note 11**).
5. To stop the TSA reaction, wash for four times with PBST. For each washing step, fill the entire tube with PBST. After inverting the tubes for several times, wait until the embryos are at the bottom and then remove excess buffer carefully. From this step onward, embryos should be protected from light.
6. To inactivate POD activity of the first antibody, incubate samples in 100 mM glycine-HCl pH 2.0 for 10 min followed by four 5-min washes in PBST under agitation (*see* **Note 12**).
7. For the second antibody detection round, repeat the steps described under Subheading 3.4 using sheep-anti-DIG-POD Fab fragments (1:500).
8. Repeat **steps 1–5** for fluorogenic detection (*see* **Note 13**).

3.6 Mounting

1. To avoid shrinkage, gradually transfer the embryos through 5-min washing steps in 25 %, 50 %, and 75 % glycerol in PBST, 40 mM $NaHCO_3$ (*see* **Note 14**).
2. After embyo preparation under a dissecting microscope, immerse the sample in mounting gel solution.

3. Transfer the sample onto a microscope slide between spacers of respective heights and apply coverslip.
4. By gently moving the coverslip, the sample can be rotated into the desired orientation.
5. Put the mounted sample into the fridge until the agarose has solidified.

4 Notes

1. Be *careful* when working with 4-iodophenol, as it is a highly aggressive substance and should *only* be handled under a fume hood wearing appropriate protections at all times (even when highly diluted).
2. Tyramide reagents diluted in absolute ethanol and stored at −20 °C are at least stable for 3 years.
3. The optimal fixing conditions for different stages have to be determined experimentally by adjusting fixation time and temperature. Embryos that are older than 1 day are usually fixed for 24 h at 4 °C.
4. Hydrogen peroxide treatment promotes embryo permeabilization properties resulting in improved signal detection.
5. As a rule of thumb, use comparable concentration of RNA probes as with chromogenic BCIP/NBT detection. Exaggerated probe concentrations will result in a decreased signal-to-noise ratio. In case three different mRNA transcript patterns are to be compared in the same embryo, in addition to DIG- and DNP-, a fluorescein-labeled mRNA probe is included in the probe mix. Labeling probes with biotin is not recommended because the biotin label may produce high background signals in zebrafish embryos [13]. However, biotin-labeled probes work well in *Drosophila* [10]. For detailed descriptions of RNA probe preparation, *see* [11].
6. The anti-DNP-POD antibody shows cross-reactivity with the periderm, which surrounds the zebrafish embryo as a thin cell layer during early stages of development. This usually results in visualization of the outline of the embryo. As the first round of detection usually requires only short staining times resulting in low background, this effect is minimized when detecting the DNP RNA first.
7. In case a fluorescein-labeled probe is included in the hybridization mix, it is detected by rabbit-anti-fluorescein/Oregon Green 488-POD diluted 1:500 in blocking solution prior to use.
8. In general 4-iodophenol is a more potent enhancer than vanillin. Higher concentrations of accelerator maybe used to

further increase signal intensity, although adverse effects on the signal-to-noise ratio should be kept in mind.

9. Always make fresh TSA reaction buffer just before use. Preparing 1 ml reaction buffer in a 2 ml tube helps to minimize eventual spill of aggressive 4-iodophenol-rich solution during mixing.
10. Since acidic POD inactivation greatly diminishes the signal intensity of FAM-tyramide staining [25], DyLight633-tyramide is usually used in the first and FAM-tyramide in the second round of detection. In case of three-color detection, the third transcript is visualized with carboxytetramethylrhodamine (TAMRA) tyramide [24].
11. Since the 50 % dextran sulfate stock solution is very viscose, use a cut 200 μl tip for pipetting. Take special care not to transfer embryos by accident from one tube to another. In order to avoid a decline in the signal-to-noise ratio, the reaction time should not exceed 30 min.
12. The acidic inactivation step is pivotal to avoid that the first detected probe is visible in two fluorescence detection channels. Incubation in hydrogen peroxide solutions bears the risk of partial POD inactivation [24, 26]. This may cause false-positive colocalization signals.
13. In case a third fluorescein-labeled probe is included in the hybridization mix, the second antibody–POD conjugate is also inactivated (repeat **step 6**) and a third antibody detection (repeat the steps described under Subheading 3.4) and fluorogenic reaction (repeat **steps 1–5** of Subheading 3.5) round is added. The fluorescein-labeled probe is detected with rabbit-anti-fluorescein/Oregon Green 488-POD and TAMRA-tyramide is used as third substrate for the fluorogenic reaction. We found that best results in a three-color experiment are obtained when the fluorescein label is detected first with DyLight 633 followed by DNP with TAMRA and DIG with FAM.
14. Using 40 mM $NaHCO_3$ ensures that the pH is above 8 (*see* also **Note 10**). Stained embryos stored in 75 % glycerol in PBST, 40 mM $NaHCO_3$ are stable for several months and the use of special anti-fading agents is usually not necessary.

Acknowledgments

The Stockholm Greater Council and the Karolinska Institutet supported this work. Imaging was in part performed at the Live Cell Imaging Unit, Department of Biosciences and Nutrition, Karolinska Institutet, Huddinge, Sweden, supported by grants from the Knut and Alice Wallenberg Foundation, the Swedish Research Council, and the Centre for Biosciences.

References

1. Abellan A, Medina L (2009) Subdivisions and derivatives of the chicken subpallium based on expression of LIM and other regulatory genes and markers of neuron subpopulations during development. J Comp Neurol 515(4):465–501
2. Puelles L, Kuwana E, Puelles E, Bulfone A, Shimamura K, Keleher J, Smiga S, Rubenstein JL (2000) Pallial and subpallial derivatives in the embryonic chick and mouse telencephalon, traced by the expression of the genes Dlx-2, Emx-1, Nkx-2.1, Pax-6, and Tbr-1. J Comp Neurol 424(3):409–438, doi:10.1002/1096-9861(20000828)424:3<409::AID-CNE3>3.0.CO;2-7 [pii]
3. Diez-Roux G, Banfi S, Sultan M, Geffers L, Anand S, Rozado D, Magen A, Canidio E, Pagani M, Peluso I, Lin-Marq N, Koch M, Bilio M, Cantiello I, Verde R, De Masi C, Bianchi SA, Cicchini J, Perroud E, Mehmeti S, Dagand E, Schrinner S, Nurnberger A, Schmidt K, Metz K, Zwingmann C, Brieske N, Springer C, Hernandez AM, Herzog S, Grabbe F, Sieverding C, Fischer B, Schrader K, Brockmeyer M, Dettmer S, Helbig C, Alunni V, Battaini MA, Mura C, Henrichsen CN, Garcia-Lopez R, Echevarria D, Puelles E, Garcia-Calero E, Kruse S, Uhr M, Kauck C, Feng G, Milyaev N, Ong CK, Kumar L, Lam M, Semple CA, Gyenesei A, Mundlos S, Radelof U, Lehrach H, Sarmientos P, Reymond A, Davidson DR, Dolle P, Antonarakis SE, Yaspo ML, Martinez S, Baldock RA, Eichele G, Ballabio A (2011) A high-resolution anatomical atlas of the transcriptome in the mouse embryo. PLoS Biol 9(1):e1000582
4. Shimogori T, Lee DA, Miranda-Angulo A, Yang Y, Wang H, Jiang L, Yoshida AC, Kataoka A, Mashiko H, Avetisyan M, Qi L, Qian J, Blackshaw S (2010) A genomic atlas of mouse hypothalamic development. Nat Neurosci 13(6):767–775
5. Suzuki-Hirano A, Ogawa M, Kataoka A, Yoshida AC, Itoh D, Ueno M, Blackshaw S, Shimogori T (2011) Dynamic spatiotemporal gene expression in embryonic mouse thalamus. J Comp Neurol 519(3):528–543
6. Ferran JL, Sanchez-Arrones L, Sandoval JE, Puelles L (2007) A model of early molecular regionalization in the chicken embryonic pretectum. J Comp Neurol 505(4):379–403. doi:10.1002/cne.21493
7. Lauter G, Söll I, Hauptmann G (2013) Molecular characterization of prosomeric and intraprosomeric subdivisions of the embryonic zebrafish diencephalon. J Comp Neurol 521(5): 1093–1118
8. Tautz D, Pfeifle C (1989) A non-radioactive in situ hybridization method for the localization of specific RNAs in Drosophila embryos reveals translational control of the segmentation gene hunchback. Chromosoma 98(2):81–85
9. Hauptmann G, Gerster T (1994) Two-color whole-mount in situ hybridization to vertebrate and Drosophila embryos. Trends Genet 10(8):266
10. Hauptmann G, Gerster T (1996) Multicolour whole-mount in situ hybridization to Drosophila embryos. Dev Genes Evol 206(4): 292–295
11. Hauptmann G, Gerster T (2000) Multicolor whole-mount in situ hybridization. Methods Mol Biol 137:139–148, doi:1-59259-066-7-139 [pii] 10.1385/1-59259-066-7:139
12. Jowett T, Lettice L (1994) Whole-mount in situ hybridizations on zebrafish embryos using a mixture of digoxigenin- and fluorescein-labelled probes. Trends Genet 10(3):73–74
13. Hauptmann G (1999) Two-color detection of mRNA transcript localizations in fish and fly embryos using alkaline phosphatase and beta-galactosidase conjugated antibodies. Dev Genes Evol 209(5):317–321
14. Hauptmann G (2001) One-, two-, and three-color whole-mount in situ hybridization to Drosophila embryos. Methods 23(4):359–372, doi:10.1006/meth.2000.1148 S1046-2023(00)91148-4 [pii]
15. O'Neill JW, Bier E (1994) Double-label in situ hybridization using biotin and digoxigenin-tagged RNA probes. Biotechniques 17(5):870, 874, 875
16. Bobrow MN, Harris TD, Shaughnessy KJ, Litt GJ (1989) Catalyzed reporter deposition, a novel method of signal amplification. Application to immunoassays. J Immunol Methods 125(1–2):279–285, doi:0022-1759(89)90104-X [pii]
17. van de Corput MP, Dirks RW, van Gijlswijk RP, van Binnendijk E, Hattinger CM, de Paus RA, Landegent JE, Raap AK (1998) Sensitive mRNA detection by fluorescence in situ hybridization using horseradish peroxidase-labeled oligodeoxynucleotides and tyramide signal amplification. J Histochem Cytochem 46(11):1249–1259
18. Kosman D, Mizutani CM, Lemons D, Cox WG, McGinnis W, Bier E (2004) Multiplex detection of RNA expression in Drosophila embryos. Science 305(5685):846, doi:10.1126/science.1099247 305/5685/846 [pii]

19. Tessmar-Raible K, Steinmetz PR, Snyman H, Hassel M, Arendt D (2005) Fluorescent two-color whole mount in situ hybridization in Platynereis dumerilii (Polychaeta, Annelida), an emerging marine molecular model for evolution and development. Biotechniques 39(4):460, 462, 464. doi:000112023 [pii]
20. Clay H, Ramakrishnan L (2005) Multiplex fluorescent in situ hybridization in zebrafish embryos using tyramide signal amplification. Zebrafish 2(2):105–111. doi:10.1089/zeb.2005.2.105
21. Lauter G, Söll I, Hauptmann G (2011) Two-color fluorescent in situ hybridization in the embryonic zebrafish brain using differential detection systems. BMC Dev Biol 11:43
22. Vize PD, McCoy KE, Zhou X (2009) Multichannel wholemount fluorescent and fluorescent/chromogenic in situ hybridization in Xenopus embryos. Nat Protoc 4(6):975–983, doi:nprot.2009.69 [pii] 10.1038/nprot.2009.69
23. Denkers N, Garcia-Villalba P, Rodesch CK, Nielson KR, Mauch TJ (2004) FISHing for chick genes: triple-label whole-mount fluorescence in situ hybridization detects simultaneous and overlapping gene expression in avian embryos. Dev Dyn 229(3):651–657. doi:10.1002/dvdy.20005
24. Lauter G, Söll I, Hauptmann G (2011) Multicolor fluorescent in situ hybridization to define abutting and overlapping gene expression in the embryonic zebrafish brain. Neural Dev 6(1):10, doi:1749-8104-6-10 [pii]10.1186/1749-8104-6-10
25. Fischer AH, Jacobson KA, Rose J, Zeller R (2008) Media for mounting fixed cells on microscope slides. CSH Protoc 2008:pdb ip52
26. Liu G, Amin S, Okuhama NN, Liao G, Mingle LA (2006) A quantitative evaluation of peroxidase inhibitors for tyramide signal amplification mediated cytochemistry and histochemistry. Histochem Cell Biol 126(2):283–291. doi:10.1007/s00418-006-0161-x

Chapter 13

Dynamic Neuroanatomy at Subcellular Resolution in the Zebrafish

Adèle Faucherre and Hernán López-Schier

Abstract

Genetic means to visualize and manipulate neuronal circuits in the intact animal have revolutionized neurobiology. "Dynamic neuroanatomy" defines a range of approaches aimed at quantifying the architecture or subcellular organization of neurons over time during their development, regeneration, or degeneration. A general feature of these approaches is their reliance on the optical isolation of defined neurons in toto by genetically expressing markers in one or few cells. Here we use the afferent neurons of the lateral line as an example to describe a simple method for the dynamic neuroanatomical study of axon terminals in the zebrafish by laser-scanning confocal microscopy.

Key words Neuroanatomy, Zebrafish, Live imaging, Microscopy, Fluorescent protein

1 Introduction

Neural circuits involved in behaviors evoked by environmental signals convey sensory information from peripheral receptors to the central nervous system and return to the musculature via premotor neural pathways [1]. Because sensory-motor circuits perform their main physiological function only in their natural context, explanted tissues rarely provide an ideal model system for their study [2, 3]. Therefore, enormous effort has been invested in developing techniques for imaging and manipulating neurons over extended periods in the intact animal [4–7]. The zebrafish (*Danio rerio*) has emerged as a powerful model system for the implementation of systematic dynamic neuroanatomical studies [8–10]. It compares favorably with other animal models, offering a number of advantages such as its rapid sexual development and fecundity, which facilitates transgenesis and permits the generation of hundreds of animals for quantitative/statistical analyses. The zebrafish embryo develops rapidly, externally, and is optically transparent, which facilitates the visualization of neurons for detailed cellular, molecular, and physiological analyses [11]. Finally, because all vertebrates share many

Simon G. Sprecher (ed.), *Brain Development: Methods and Protocols*, Methods in Molecular Biology, vol. 1082,
DOI 10.1007/978-1-62703-655-9_13,

of the cellular and physiological mechanisms that underlie neural circuit development and performance, the zebrafish provides a highly relevant model to human neurobiology and neuropathology [12].

Small fluorescent proteins can be expressed in genetically defined neuronal types and also targeted to specific subcellular compartments, serving as excellent markers for the entire neuron and its axonal projections or dendritic trees. However, for densely packed axons or overlapping dendritic trees, expressing a fluorescent protein in a large population of neurons may prevent detailed visualization and quantification [13, 14]. Mosaic gene expression to mark a small fraction of neurons within a large population can overcome this limitation [10, 15, 16]. There are several strategies for the systematic expression of fluorescent markers in single or few genetically defined neurons. Some of these methods are based on the generation of a silent transgene in all the neurons, which can later be expressed in cell clones by recombinase-based excision of the silencing/repressing genetic elements in the transgene [17]. Other approaches are based on site-specific chromosomal recombination, such as the Mosaic Analysis with Double Markers (MADM) or Mosaic Analysis with Repressible Cell Marker (MARCM) [18]. Finally, the Brainbow and derived approaches allow the individualization of every neuron within a complex population by the random expression of different combinations of spectral variant of fluorescent proteins [19, 20]. However, all these methods require the generation and maintenance of stable transgenic lines, with a consequent increment of the time and costs of every study.

Here we detail a simple, fast, and inexpensive strategy for dynamic neuroanatomy in the zebrafish [10]. We use the afferent neuronal pathway of the mechanosensory lateral line as an example of its implementation [10, 15, 21, 22]. This protocol is based on transient expression of engineered transgenes coding for fluorescent proteins in single neurons by DNA injection in the fertilized egg. Our procedure only requires a small zebrafish colony, a basic knowledge of molecular biology, an access to a laser-scanning confocal microscope for image acquisition, and finally, a desktop computer equipped with free or inexpensive commercially available software packages for image analysis and rendering. A single change in the mounting of the samples would allow imaging by Selective Plane Illumination Microscopy (SPIM) [23–25].

2 Materials

2.1 Embryos Obtainment and Injections Components

1. E3 embryo medium stock solution (60×): Dissolve 172 g NaCl, 7.6 g KCl, 29 g $CaCl_2 \cdot 2H_2O$, and 49 g $MgSO_4 \cdot 7H_2O$ in 10 L of distilled H_2O (dH_2O).
2. Necessary equipment for raising fish and collecting eggs.

3. Necessary equipment for fish eggs injections: Glass microcapillary, micropipette puller, microinjector using air pressure, and micropipette holder.
4. Necessary equipment to clone DNA and DNA purification kit.

2.2 Mounting Embryos Components

1. Tricaine (MS222 or MESAB: 3-aminobenzoic acid ethyl ester) stock solution (25×): 0.4 g tricaine, 97.9 mL dH_2O. Adjust the pH to ~7 by adding 9.1 mL of 1 M Tris–HCl (pH 9). Store at 4 °C.
2. 1 % LMP (low-melting-point) agarose in E3 to mount embryos for observation.
3. Cover-glass-bottomed culture dishes.

2.3 Observation and Analysis of Neuron Behaviors

1. Fluorescence stereomicroscope equipped with a filter set for GFP and RFP.
2. Confocal microscope equipped with lasers for excitation at 488 nm and 532 nm.
3. Imaris software (Bitplane).

3 Methods

3.1 Expressing Membrane Red Fluorescent Protein in the Zebrafish Lateral Line Afferent Neurons

1. Prepare DNA for injection (*see* Fig. 1). To achieve mosaic expression in single neurons, clone the cDNA encoding the red fluorescent protein (e.g., mCherry or tdTomato) under the control of a neuronal promoter (e.g., HuC). For a better visualization of neurites, the red fluorescent protein has to be fused to a membrane-targeting sequence (e.g., CAAX domain or first 20 codons of GAP43).

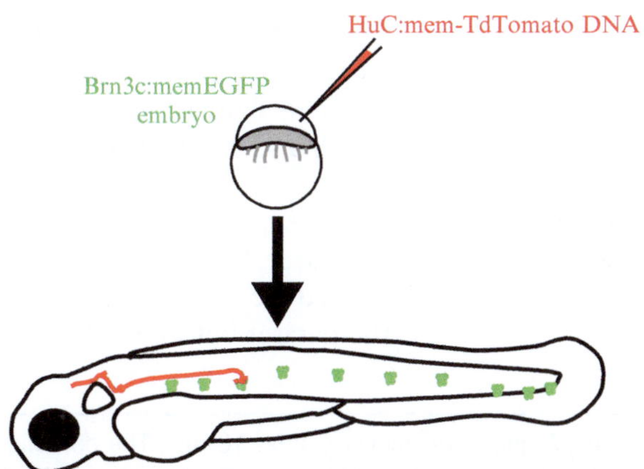

Fig. 1 Scheme of the injection and selection of the samples to image. Overview of the injection strategy to label single lateralis neurons. Injection of cDNA coding with HuC:mem-TdTomato into 1–2-cell Tg[Brn3c:mem-eGFP] host embryos

2. Purify DNA using a plasmid purification kit following the manufacturer's instructions.
3. One day before carrying out injections, set the male and female fish as a pair. Use transgenic fish expression the GFP in the hair cells of the lateral line (e.g., sqET4 or Brn3c:mem-eGFP transgenic lines).
4. For transient expression in single afferent neurons, inject 20 ng of circular DNA (*see* **Note 1**). Dilution of the purified plasmid is carried out in milliQ water.
5. Raise the injected embryos at 28.5 °C in an incubator to obtain a standard developmental speed until the desired stage.
6. Screen and select embryos with a stereomicroscope under ultraviolet light and select those expressing red fluorescence in afferent neurons and GFP in the hair cells for live imaging. The exact time for the start of expression after injection of DNA depends on the promoter/enhancer used in the expression vector.

3.2 Imaging Afferent Arborization

1. Transfer the embryo directly to a cover-glass-bottomed culture dish in a drop of 1 % low-melting-point agarose in E3 medium with tricaine and carefully rotate the embryo with a hair loop so that the desired side of the embryo faces the bottom.
2. Wait for a few minutes. Carefully fill the dish with E3 medium and tricaine.
3. Use excitation light from a mercury lamp passed through the filter set to observe red fluorescence. Select a neuromast innervated by a single red afferent neuron (*see* **Notes 2–4**).
4. Image embryos with a confocal microscope with a 40× oil immersion objective or with a 20× dry objective. Z-stacks are acquired at 0.8 or 0.5 mm intervals, imaging GFP (488 nm excitation, 500–550 nm emission), and TdTomato (532 nm excitation, 570–630 nm emission).
5. Generate three-dimensional images through the entire extent of the neuromast and the neuronal arbor over time for a period of 1 h and every 10 min (6 time points).

3.3 Neuronal-Arbor Tracing (See Fig. 2)

1. Image stacks are rendered three-dimensional using the Imaris software. Under the file menu, chose open and select the desired file. In order to accelerate the analysis process, resample the dataset before loading it (*see* **Note 5**).

Fig. 2 (continued) A and B, respectively, after 10 min. The dendrite C is no more present. (**e**, **f**) Example of tracing of the whole neuronal arbor at T1 (**e**) and T2 (**f**). (**g**) Example of data for four-dimensional quantification. The dendrite length is given by the software in micrometers. To calculate the growth or retraction of a neurite, values obtained at T2 are subtracted from the values obtained at T1 for each neurite. In this case, the dendrite A has extended 4.1 μm whereas the dendrite B has retracted 2.45 μm. In the case of the dendrite C, which was present at T1 but absent at T2, the distance of retraction is calculated from the branching point, here Y

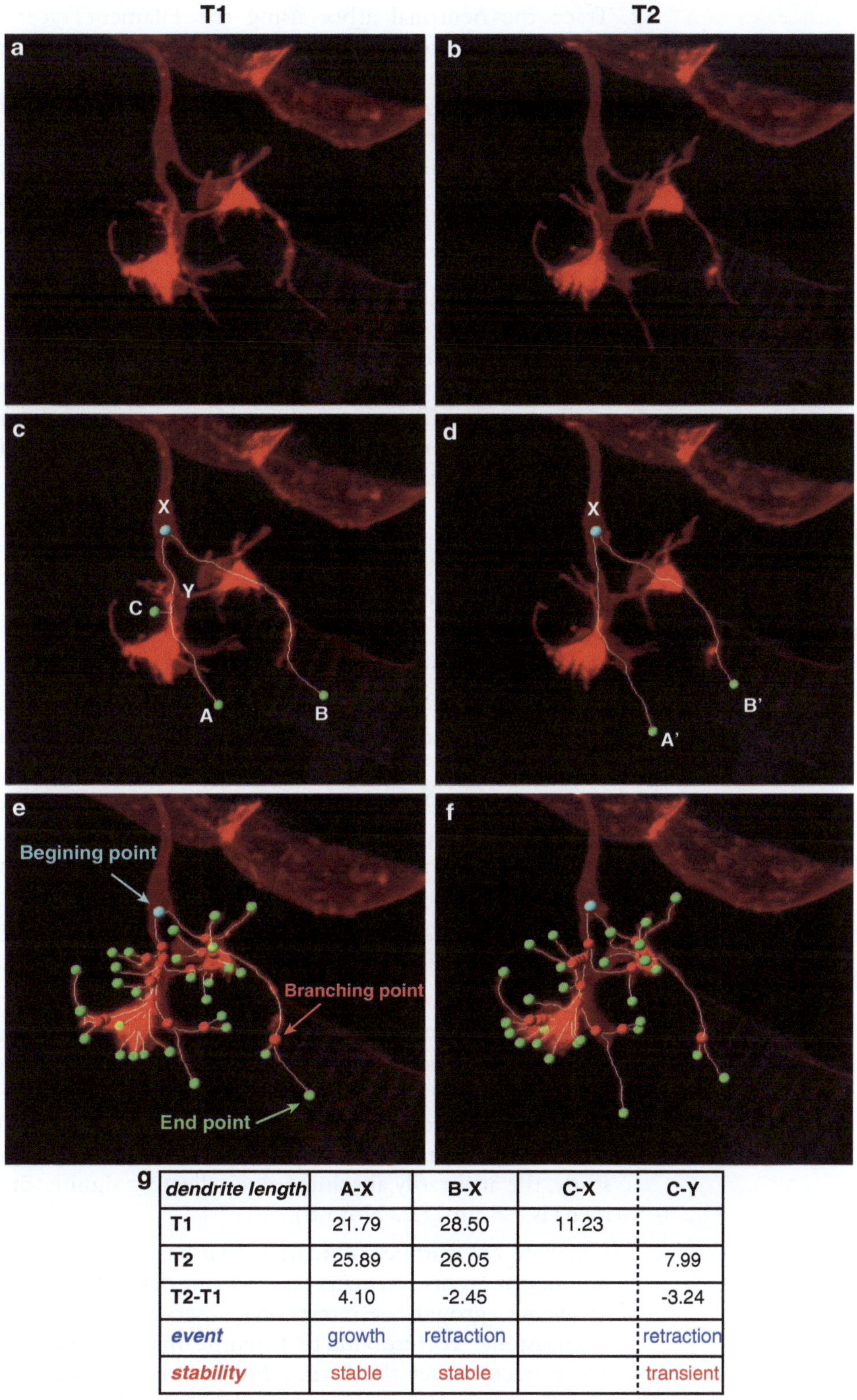

dendrite length	A-X	B-X	C-X	C-Y
T1	21.79	28.50	11.23	
T2	25.89	26.05		7.99
T2-T1	4.10	-2.45		-3.24
event	growth	retraction		retraction
stability	stable	stable		transient

Fig. 2 Neuronal-arbor tracing. (**a**, **b**) Neuronal arbor at time points T1 and T2. (**c**, **d**) Example of dendrite tracing. X represents the beginning point. A, A′; B, B′; and C represent 3 end points and Y represents a branching point. In this case, at T1 (**c**), 3 dendrites have been traced. At T2 (**d**), A′ and B′ represent the position of the dendrites

2. Trace the neuronal arbor using the FilamentTracer Imaris module. Open the *Filament* wizard and select the Autopath (no loops) algorithm. Select the starting point manually at the first bifurcation of the afferent neuron and assign it as *Dendrite Beginning Point* in the *Process Selection* box of the Edit tab. End points are also manually selected and filaments automatically traced. Check the final tracing manually, and if needed, correct the path using the object menu. Repeat the process for every time point (*see* **Note 6**).
3. In order to visualize the different positions of a specific tip over time (*four-dimensional rendering*), chose *Mouse selects point* in the Edit tab and select a terminal point. In the Statistics tab, press *Duplicate Selection to New Filament* and attribute one different color in the Color tab to each terminal point. Repeat for each time point, keeping the same color for the same neurite.

3.4 Four-Dimensional Quantification of Axonal Arbors (See Fig. 3)

1. Select a terminal point. In the Statistics menu, display the *Selection values* and chose *Pt distance* (i.e., dendrite length) (*see* **Note 7**). Copy the value of *Pt distance* and paste it in a new Excel file.
2. For the same neurite, follow the terminal point from one time point to the following one and repeat **step 1**.
3. In the case of transient neurites, select the branching point and repeat **step 15**.
4. To express the *Neurites stability*, calculate the percentage of stable neurites (present throughout all the movie) versus the percentage of transient neurites (neurites visible during less than 6 time points).
5. Calculate the *Neurites persistence* by counting the percentage of neurites present for only one, two, three, four, five, or six time frames.
6. To calculate growth and retraction, subtract the position value of a neurite at a specific time point of the value of the previous time point. Positive values correspond to growth and negative values to retraction. Decide of an arbitrarily absolute value that reflects significant growth or retraction movements (in our study, the arbitrarily absolute value reflecting significant movement is 1.5 μm) (*see* **Note 8**).
7. Calculate the number of neurites that have achieved a defined number of events. *Neurites consistency* is defined as the number of events (growth or retraction) observed for each neurite. "Consistent" corresponds to a neurite that has neither grown nor retracted over the 6 time frames (i.e., no events). "One event" corresponds to a neurite that has only grown or only retracted over the 6 time frames. "Two events" correspond to a

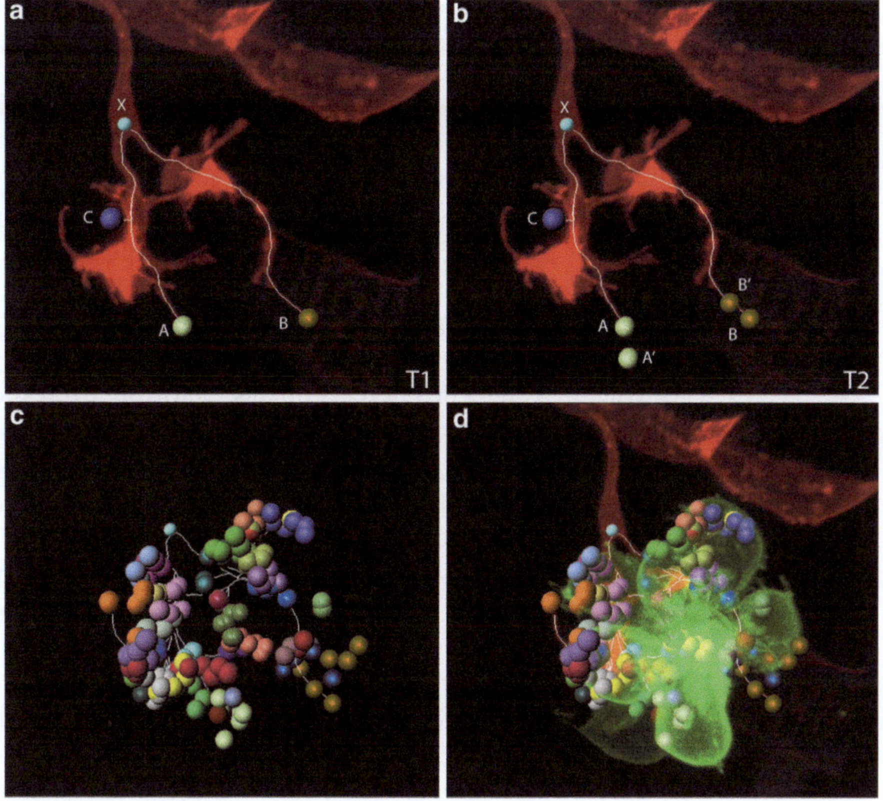

Fig. 3 Four-dimensional representation. (**a**, **b**) One color is attributed to each neurite and end points are duplicated in order to visualize the positions of a specific tip over time. At T1 (**a**), the end points of dendrites A, B, and C have been plotted. At T2 (**b**), A′ and B′ indicate the positions of the tip of the neurites A and B, respectively, after 10 min. (**c**) Four-dimensional representation of the whole arbor after 1 h (six times 10 min). (**d**) Same as in (**c**) with the maximal projection of the mem-tdTomato-expressing neuron and GFP-expressing hair cells

neurite that completed a sequence of two countermovements, namely, growing followed by retraction or the opposite. Similarly, "three events" represent a sequence of three countermovements (e.g., growth followed by retraction followed by growth or vice versa).

8. To calculate the *Neurites presence*, go to the Statistics menu, display the Overall values and select *No. Dendrite Terminal Pts.* Repeat for each time

4 Notes

1. The amount of injected circular DNA may depend on the construct itself. It is recommended to perform a range of injections with different amount of DNA prior to the experiment in order to determine the right amount of DNA for expression of the fluorescent protein in single neurons.

2. Select for analysis only those specimens with single red fluorescent neurons whose soma is localized within the lateralis afferent ganglion, which can be marked using the HGn39D transgenic line.
3. If more than one afferent neuron are expressing the red fluorescent protein, you can try to inject less DNA. Nevertheless, if the neuromast is innervated by only one neuron and if the soma of this neuron can be localized within the lateralis afferent ganglion, the experiment can be carried away.
4. If none of the afferent neurons is expressing the red fluorescent protein, you can try to inject more DNA. Check as well that you are using a promoter/enhancer allowing expression in the afferent neurons.
5. If the sample has moved during the time of the acquisition, align the three-dimensional reconstructions of each time point using the Spot/Correct drift function of Imaris.
6. It is very important to assign the starting point as Dendrite Beginning Point in the Edit menu since every following calculation is based on the position of this point.
7. If, in the statistic menu of Filament tracer, there is no value for the dendrite length, check that the starting point has been assigned as the Dendrite Beginning Point. Check as well that there is no discontinuity in the tracing of the neuronal arbor reconstruction.
8. In the case of transient neurites, the value before appearance or after disappearance is the distance between the beginning point and the branching point.

Acknowledgements

We thank the generosity of C.B. Chien for the Tol2kit and advice and dedicate this article to his memory. We also thank T. Zimmermann and J. Swoger for advice on, respectively, confocal and SPIM microscopy. The original research that encouraged the development of this methodology was supported by a grant from the European Research Council (ERC-2007-StG SENSORINEURAL) and by the Ministerio de Ciencia e Innovación of Spain to H.L. S.A.F. was supported by an Intra-European Marie Curie Fellowship from the European Union.

References

1. Arber S (2012) Motor circuits in action: specification, connectivity, and function. Neuron 74:975–989
2. Seelig JD, Jayaraman V (2011) Studying sensorimotor processing with physiology in behaving *Drosophila*. Int Rev Neurobiol 99:169–189

3. Koch M (1999) The neurobiology of startle. Prog Neurobiol 59:107–128
4. Piggott BJ, Liu J et al (2011) The neural circuits and synaptic mechanisms underlying motor initiation in *C. elegans*. Cell 147:922–933
5. Anikeeva P, Andalman AS et al (2011) Optetrode: a multichannel readout for optogenetic control in freely moving mice. Nat Neurosci 15:163–170
6. Ahrens MB, Li JM et al (2012) Brain-wide neuronal dynamics during motor adaptation in zebrafish. Nature 485:471–477
7. Moser EI, Kropff E, Moser MB (2008) Place cells, grid cells, and the brain's spatial representation system. Annu Rev Neurosci 31:69–89
8. Gahtan E, Baier H (2004) Of lasers, mutants, and see-through brains: functional neuroanatomy in zebrafish. J Neurobiol 59:147–161
9. Friedrich RW, Jacobson GA, Zhu P (2010) Circuit neuroscience in zebrafish. Curr Biol 20:R371–R381
10. Faucherre A, Baudoin JP, Pujol-Martí J et al (2010) Multispectral four-dimensional imaging reveals that evoked activity modulates peripheral arborization and the selection of plane-polarized targets by sensory neurons. Development 137:1635–1643
11. Simmich J, Staykov E, Scott E (2012) Zebrafish as an appealing model for optogenetic studies. Prog Brain Res 196:145–162
12. Santoriello C, Zon LI (2012) Hooked! Modeling human disease in zebrafish. J Clin Invest 122:2337–2343
13. Shu X, Lev-Ram V, Olson ES et al (2011) Spiers Memorial lecture. Breeding and building molecular spies. Faraday Discuss 149:63–77
14. Detrich HW 3rd (2008) Fluorescent proteins in zebrafish cell and developmental biology. Methods Cell Biol 85:219–241
15. Pujol-Martí J, Baudoin JP, Faucherre A et al (2010) Progressive neurogenesis defines lateralis somatotopy. Dev Dyn 239:1919–1930
16. Collins RT, Linker C, Lewis J (2010) MAZe: a tool for mosaic analysis of gene function in zebrafish. Nat Methods 7:219–223
17. Hans S, Kaslin J, Freudenreich D et al (2009) Temporally-controlled site-specific recombination in zebrafish. PLoS One 4:e4640
18. Luo L (2007) Single-neuron labeling using the genetic MARCM method. CSH Protoc 2007: pdb.prot4789
19. Livet J, Weissman TA, Kang H et al (2007) Transgenic strategies for combinatorial expression of fluorescent proteins in the nervous system. Nature 450:56–62
20. Pan YA, Livet J, Sanes JR et al (2011) Multicolor brainbow imaging in zebrafish. CSH Protoc 2011(1):pdb.prot5546
21. Raible DW, Kruse GJ (2000) Organization of the lateral line system in embryonic zebrafish. J Comp Neurol 421:189–198
22. Metcalfe WK, Kimmel CB, Schabtach E (1985) Anatomy of the posterior lateral line system in young larvae of the zebrafish. J Comp Neurol 233:377–389
23. Swoger J, Muzzopappa M, López-Schier H et al (2011) 4D retrospective lineage tracing using SPIM for zebrafish organogenesis studies. J Biophotonics 4:122–134
24. Kaufmann A, Mickoleit M, Weber M et al (2012) Multilayer mounting enables long-term imaging of zebrafish development in a light sheet microscope. Development 139: 3242–3247
25. Huisken J, Stainier DY (2009) Selective plane illumination microscopy techniques in developmental biology. Development 136: 1963–1975

Chapter 14

Anatomical Dissection of Zebrafish Brain Development

Katherine J. Turner, Thomas G. Bracewell, and Thomas A. Hawkins

Abstract

Zebrafishbrain.org is an online neuroanatomical atlas of the embryonic zebrafish. The atlas uses high-resolution confocal images and movies of transgenic lines to describe different brain structures. This chapter covers detail of materials and protocols that we employ to generate data for the atlas.

Key words Zebrafish, Neuroanatomy, Brain atlas, Brain dissection, Labeling

1 Introduction

Increasingly zebrafish are becoming a model system for the study of behavior [1–5]. To truly understand zebrafish behavior, we must first understand the connectivity of the neuronal circuits driving behavior. This requires detailed neuroanatomical characterization of the zebrafish brain. Zebrafish are a relatively new model system; as such, the zebrafish does not benefit from the extensive neuroanatomical descriptions that have been undertaken in other model species [6, 7]. Most neuroanatomical studies in the zebrafish and other teleosts focus on adult neuroanatomy [8, 9]. These studies mainly use serial sections of adult brains which can be challenging to interpret by non-experts and can be difficult to apply to the embryonic brain [10]. Specific description of the neuroanatomy of the embryonic and larval zebrafish brain is not comprehensive. To address this issue, in collaboration with other laboratories, we are in the process of building an online neuroanatomical resource (an atlas) called zebrafishbrain.org.

Zebrafishbrain.org is being built to communicate current knowledge about the neuroanatomy of the developing zebrafish brain, and this is achieved in two ways. The first is to provide a hub for community experts to provide data and write descriptions of neuroanatomical structures, and the second is to provide data and descriptions ourselves. To do this we are mining currently available collections of transgenic zebrafish (and making a few of our own) to generate a

Simon G. Sprecher (ed.), *Brain Development: Methods and Protocols*, Methods in Molecular Biology, vol. 1082, DOI 10.1007/978-1-62703-655-9_14,

collection of transgenic lines with thoroughly described embryonic/larval (and sometimes adult) brain-specific expression patterns. We employ the descriptions of the expression patterns of these lines in the construction of tutorials describing brain structures.

This chapter details the core methods that are currently employed by us when studying transgenic zebrafish embryos and larvae for zebrafishbrain.org. The protocols describe in detail how we undertake brain dissection, immunohistochemistry, and the several available methods we employ to mount embryos for confocal microscopy. We also list software we have found useful for processing the raw data.

The protocols are based upon methods that are widely used in the zebrafish field but with a few adjustments to adapt to characterization of zebrafish developmental neuroanatomy. We also specify some reagents that we have found to be reliable for use in the zebrafish. Many of these protocols have flexible aspects to them, and we have tried to include detail of this flexibility where possible as an indication of how optimization for particular conditions can be undertaken.

2 Materials

2.1 Equipment for Embryonic Culture and Fixation

1. Petri dishes.
2. Watchmakers forceps for dechorionation: Dumont #5 (Fine Science Tools).
3. 7 ml bijou tubes: Appleton Woods.

2.2 Embryonic Culture and Fixation Reagents

1. Fish/embryo water (filter-sterilized aquarium system water) or embryo medium (zebrafish book).
2. 1-Phenyl 2-thiourea (PTU) 3 mg/ml: Keep frozen 40 ml aliquots of 25× stock solution (75 mg/ml). One aliquot makes 1 l of PTU when mixed with fish water. Fresh PTU should be made every 2 days.
3. Sweet fix: 4 % PFA with 4 % sucrose. Make from 20 % PFA stock: 40 ml H_2O + 10 g paraformaldehyde. Heat to 60 °C with stirring. Add 10 drops of 10 M NaOH. Cool and filter through funnel with filter paper, add 10× phosphate-buffered saline (PBS) (pH 7.3), sucrose, and H_2O and adjust pH with HCl to give a final concentration of 4 % paraformaldehyde, 4 % sucrose, and 1× PBS at pH 7.3. Aliquot into 5 ml aliquots and freeze. Defrost fresh fix at room temperature (RT) just prior to fixation. Do not use fix that has been defrosted for longer than 48 h.
4. 2 % TCA in PBS. 10 % TCA (in H_2O) stock can be stored at −20 °C. Defrost and mix with 10× PBS and H_2O to make final concentration of 2 % in 1× PBS.
5. PBS: phosphate-buffered saline pH 7.3.

2.3 Brain Dissection Equipment

1. Glass petri dish.
2. Bunsen burner.
3. Oven (60 °C).
4. Minute pins A1.
5. Superfine Dumont forceps (Fine Science Tools).
6. Needle holders (Fine Science Tools or VWR)
7. Tungsten wire 0.125 mm.
8. 12v DC power supply.
9. Two wires with crocodile clips.
10. Glass jar with lid.
11. Electrode.
12. Microcentrifuge tubes.

2.4 Brain Dissection Reagents and Solutions

1. Sylgard: 1 kit 184 from Dow Corning WPI Sylgard http://www.dowcorning.com/applications/search/products/Details.aspx?prod=01064291&type=&country=GBR.
2. Saturated NaOH.
3. PBS.

2.5 Brain Dissection Sylgard Dishes

1. Sylgard is an elastomer curing agent. The Sylgard kit contains two solutions that set to a hard rubberlike texture when mixed. A layer of Sylgard in a glass petri dish makes a great dissecting dish as you can pin the embryos using insect/minute pins to the Sylgard, immobilizing them while you dissect. You also minimize damage to forceps and dissecting needles.
2. Mix Part A liquid from the Sylgard 184 kit with Part B liquid in a ration of 10:1 in a beaker. You will need 40–50 ml per petri dish.
3. Thoroughly mix the two solutions together using a disposable stirrer such as a tongue depressor. Be careful not to stir too vigorously so as not to create unwanted bubbles in the mixture.
4. After mixing, pour the Sylgard slowly into the glass petri dishes laid on a flat, non-vibrating surface where they can remain undisturbed for several days. Fill each petri dish to 1/3–1/2 full.
5. After the Sylgard has settled for a few minutes, you will see that bubbles have risen to the surface of the liquid. Burst these using a Bunsen burner by passing the flame lightly close to the surface of the Sylgard.
6. Put the petri dish lids on top to prevent dust settling on the plates while they set.
7. Sylgard dishes need at least 24 h to set and improve if left for several days. It is possible to speed up this process by heat curing them at 60 °C for several hours in a mini oven or incubator.

Fig. 1 Sharpening dissection needles using an electric current and saturated sodium hydroxide(NaOH). (**a**) Connect an electrode to the negative terminal of a 12v DC power supply. Connect your dissection needle to the positive terminal. (**b**) Submerge the negative electrode into the saturated NaOH. (**c**) Dip the tip of the tungsten wire into the NaOH and slowly draw back up out of the solution. Repeat this motion several times until you have a very sharp point. When making a new dissecting needle, you will have to repeat the dipping action several times. When you are sharpening a dissecting needle that has already been sharpened before, you only need to dip the needle two or three times

2.6 Brain Dissection Needles

To remove the skin and eyes from zebrafish embryos, we use dissecting needles fashioned from fine tungsten wire. The tungsten wire is clamped into a needle holder and cut using strong scissors so it protrudes from the tip of the holder by about 1 cm. The wire at this point is blunt but can be sharpened to a very fine sharp point using an electric current and saturated sodium hydroxide (Fig. 1).

2.7 Immunohistochemistry Equipment

1. 1.5 ml microcentrifuge tubes.
2. Plastic fine tip Pasteur pipettes or aspirator.
3. 50 ml Falcon tubes.

2.8 Immunohistochemistry Reagents and Solutions

1. PBS pH 7.3 + 0.5–0.8 % Triton-X100 (PBTr).
2. 50 % MeOH in PBS.
3. 100 % MeOH.
4. Proteinase K(PK) 10 mg/ml (this is a 1,000× stock) store in aliquots at −20 °C.
5. Trypsin (0.25 % in PBS). Trypsin stock is 2.5 % (10×).
6. Immunohistochemistry blocking solution (IB): For 1 ml of IB, 100μl normal goat serum (NGS), 10μl of DMSO, 0.89 ml PBTr.
7. Anti-GFP: polyclonal rabbit α-GFP from AMS Biotechnology gives great results. Use 1:1,000.
8. Anti-acetylated tubulin antibody: mouse monoclonal (IgG2b) from Sigma. Use 1:500 [11].
9. Anti-SV2: mouse monoclonal (IgG1) DSHB. Use 1:500 [12, 13].
10. Anti-GFP(Rat): rat monoclonal (IgG2a) from Nacalai Tesque. Use 1:1,000.
11. Chk pAb to GFP: chicken polyclonal from Abcam. Use 1:1,000 [15].
12. Anti-RFP: rabbit polyclonal from MBL. Use 1:2,500 [16].
13. Anti-DsRed (Living Colors): rabbit polyclonal from Clontech. Use 1:300 [14].
14. Anti-Kaede (M-125-3): mouse monoclonal from MBL. Use 1:200 (does not distinguish between red and green forms of protein).
15. Anti-Kaede (PM 102): rabbit polyclonal from MBL. Use 1:1,000 (does not distinguish between red and green forms of protein).
16. Secondary antibody depends upon primary antibody and detection method: For fluorescence, Molecular Probes (www.probes.com) Alexa Fluor 488 goat α-rabbit (highly cross-adsorbed) IgG. Use at 1:200. To detect anti-acetylated tubulin and anti-SV2 in the same sample, use isotype-specific secondaries: Alexa Fluor 568(IgG2b) and Alexa Fluor 633(IgG1). Use at 1:200. Molecular Probes also make fluorescent secondaries against rat and chicken primary antibodies conjugated with various fluorescent dyes.

2.9 Confocal Mounting and Imaging Equipment

1. Silicone grease: Fisher Scientific.
2. Glass rings for mounting.
3. Disposable Pasteur pipettes, glass, short tip.
4. Araldite Rapid: Fisher Scientific.
5. Rubber bulbs for pipettes 1 ml (VWR).

6. Small coverslips 22 × 22 (VWR).
7. Microscope slides (VWR).
8. Heat block set to 40 °C.
9. Microwave.
10. Beaker with microcentrifuge tube plastic stand to melt agarose.
11. Dissecting microscope.
12. Confocal microscope with appropriate laser lines.
13. Long working distance objective lens. Highest possible numerical aperture (NA). Approximately 500 μm (±250) working distance (WD) is required. For example, Leica HCX IRAPO L 25×/0.95 W is a water immersion lens with working distance of 2.4–2.5 mm, and coverglass and non-coverglass corrected versions are available.

2.10 Confocal Mounting and Imaging Reagents and Solutions

1. Fish water without methylene blue.
2. PBS.
3. Low gelling temperature agarose (Sigma): 1 ml aliquots of 1 % agarose in filter-sterilized fish water (with no methylene blue) or embryo medium (E3) for live time-lapse imaging. For fixed embryos use 1 % agarose in PBS or 1 % regular (not low melt) agarose in PBS/80 % glycerol. The latter solution can be challenging to make. Use a water bath and stirrer to make a 5 % agarose solution, and then add 4× volume of glycerol.
4. Tricaine:(3-amino benzoic acid ethyl ester) (Sigma) 400 mg tricaine powder, 97.9 ml DD H_2O, ~2.1 ml 1 M Tris (pH 9). Adjust pH to ~7. Freeze this solution into aliquots and use 4.2 ml per 100 ml of fish water or E3.
5. CyGEL Sustain (500 μl) (BioStatus Limited).
6. 60× E3 embryo medium: NaCl 3 M, KCl 0.1 M, CaCl 0.2 M, $MgSO_4$ 0.2 M dissolved in deionized water.
7. Ice pack or ice bucket.

2.11 Image Processing Software

1. Volocity (Perkin Elmer).
2. XuvTools.
3. ImageJ/Fiji.

3 Methods

3.1 Embryonic Culture and Fixation

1. Embryos should be collected and raised at 28.5 °C in fish water or embryo medium without methylene blue. This medium minimises the auto-fluorescence of the skin caused by

methylene blue that is not desirable with fluorescent imaging methods.

2. At 24 hpf transfer embryos to fish water containing PTU again without methylene blue. PTU inhibits pigmentation of the embryos.
3. Defrost sweet fix just prior to fixation at room temperature. Embryos should be dechorionated using forceps prior to fixation. Fix embryos in large volumes of sweet fix. Transfer embryos to 7 ml bijou tube; remove as much fish water as possible before adding sweet fix. Up to 200 embryos can be fixed per 5 ml of sweet fix (*see* **Note 1**).
4. Fix embryos at 25 °C (employ an incubator if necessary). 1 h of fixation per day of development, e.g., for 24 hpf embryos fix for 1 h at 25 °C.
5. After fixation, remove most but not all sweet fix and top up tube with PBS (do not use any detergent (e.g., triton) at this stage).
6. Leave embryos overnight at 4 °C before starting dissection. Embryos can be stored for several days at 4 °C before dissection but will deteriorate in that time so prompt dissection is advised. If the embryos are not going to be dissected prior to performing the immunohistochemistry, then dehydrate the embryos using these steps.
7. Transfer to 1.5 ml tubes, rinse 2×, and wash 3×10 min in PBTr (phosphate-buffered saline + 0.5–0.8 % Triton-X100) on side on shaker.
8. Transfer to MeOH—1×5 min wash in 50 % MeOH/50 % PBTr and 3×5 min in 100 % MeOH. NB Mixing PBTr and MeOH is exothermic—this solution must be made up at least 15 min before use.
9. Store at −20 °C for at least 6 h (or up to 6 months+) to delipidate.
10. For most antibodies formaldehyde fixation with sweet fix gives the best results; however, some antibodies work better with TCA fixation. This is often true of older embryos (5dpf+). For TCA fixation, simply replace 2 % TCA for sweet fix in the protocol above. TCA-fixed embryos should be permeabilized using trypsin (see below); TCA-fixed embryos are also more opaque and will benefit from clearing in glycerol at the end of the protocol (see below).

3.2 Brain Dissection

Dissecting the skin and eyes off of zebrafish embryos prior to antibody staining vastly improves the penetration of antibodies and reduces auto-fluorescence of the skin common to many transgenic lines. Although this technique requires some dexterity, with practice it becomes routine and the results justify the effort (Fig. 2).

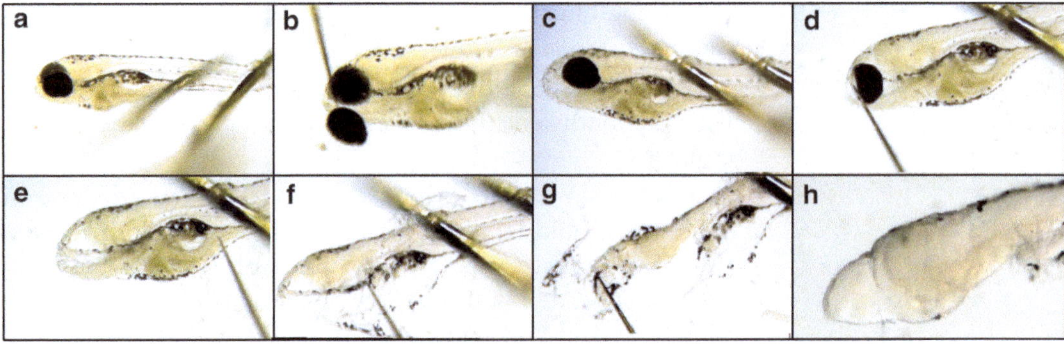

Fig. 2 Dissection. (**a**) Pin embryo laterally to a Sylgard dish. (**b**) Remove eye using a sharpened tungsten needle. (**c**) Remove pins and flip the embryo so the other side now faces upwards. (**d**) Remove second eye using sharpened tungsten needle. (**e**) Cut away the yolk and jaw from the ventral surface of the embryo. (**f**) At the level of the otic vesicle using either a sharpened tungsten needle or fine forceps, grip a piece of skin and tease away from the brain. (**g**) Pull this flap of skin rostrally, removing all of the skin from the surface of the brain. (**h**) Close up showing completely dissected brain ready to be transferred into methanol for immunohistochemistry or in situ hybridization

1. Under a dissecting microscope, transfer embryo to Sylgard dish.
2. Using forceps (a low-grade type), pin the embryo with minute pins. Embryo should be pinned on its side (laterally). Pin the embryo twice through the tail at the level of the notochord. Push down on the pins so they stick into the Sylgard.
3. Using a dissecting needle, gently cut the skin surrounding the eye and tease the eye off the head. The eye should pop off easily at stages post-48 hpf but is trickier to remove at earlier stages.
4. Unpin embryo, taking care not to break the tail. Turn the embryo to lie on the opposite side and re-pin to the Sylgard.
5. Remove second eye using the same procedure as above.
6. Cut skin between anterior yolk and heart/head using dissecting needle. Then cut away the skin attaching the yolk sack to the body. The yolk should come off as an intact structure.
7. Remove the jaw and other connective tissue from the ventral surface of the brain.
8. Make a shallow cut at the level of the otic vesicle. Using fine forceps tease a flap of skin off and pull this flap rostrally removing all of the skin from the surface of the brain.
9. Using fine forceps (*see* **Note 2**) or a dissecting needle, pull off any remaining skin, leaving just the brain attached to the tail.
10. Unpin the embryo and transfer to a 1.5 ml tube filled with PBTr using normal forceps.

11. After all embryos are dissected and transferred to 1.5 ml tube. Wash several times with PBTr, then transfer to MeOH (*see* Subheading 3.1), and store at −20 °C until you want to start the immunohistochemistry (*see* **Note 3**).

3.3 Fluorescent Immunohistochemistry Protocol

1. Rehydrate embryos: Wash embryos 5 min in 50 % MeOH/50 % PBTr, and wash 3 × 5 min in PBTr.
2. Permeabilize embryos (PFA-fixed embryos): Proteinase K (pK) digestion times vary with embryo age, and also pK batches vary. The times below are a guide, and tests should be carried out with the particular batch employed (store pK at −20 °C at 10 mg/ml—this is 1,000× stock). Dilute pK stock 1/1,000 (1×) in PBTr.
   - Up to tailbud, no PK
   - 2–10ss, in and out of 1× PK
   - 10–15ss, 1 min 1× PK
   - 16–26ss, 2 min 1× PK
   - 24 h, 10–15 min 1× PK
   - 30 h, 20 min 1× PK
   - 36–48 h, 30–40 min 1× PK
   - 2.5d, 30–40 min 1.5× PK
   - 3d, 30–40 min 2× PK
   - 4d, 30–40 min 3× PK
   - 5d, 30–40 min 4× PK.

   Digestions are at room temperature (18–22 °C) with tube lying on its side. Dissected embryos should be treated as somite-stage embryos (minimal permeabilization). Rinse 3× in PBTr. Postfix in 4 % PFA for 20 min at room temperature (also denatures pK). Wash 3 × 5min in PBTr.
3. For TCA-fixed embryos: Rinse embryos 3 × 5 min in PBS. Prechill trypsin solution (0.25 % in PBS) and 5 ml per sample of PBTr on ice until cold. Incubate embryos in trypsin, on ice for 5–10 min (according to age 36 h to 5 days), longer for older embryos, depending on trypsin batch; titrate upon first use. Rinse 2× in cold PBTr then 3 × 10 min in cold PBTr, bring to RT.
4. Block endogenous binding sites: Incubate in IB for at least 1 h at room temperature on shaker.
5. Primary antibody incubation: Incubate in IB + primary antibody overnight at 4 °C on shaker. Some antibodies work better with longer incubations or room temperature incubations, and for room temperature incubation, consider adding 2 mM sodium azide to the IB to inhibit mold growth. For longer incubations increase the number and length of post-incubation washes.

6. Postprimary incubation washes: Remove primary antibody (can be kept at 4 °C for reuse within a week). Rinse 3× in PBTr. Wash at least 4 × 30 min in PBTr on shaker, can be longer/more washes.
7. Secondary antibody incubation: Incubate in IB + secondary antibody overnight on shaker at 4 °C or room temperature.
8. Postsecondary incubation washes: Rinse 3× in PBTr. Wash at least 4 × 30 min on shaker. Fluorescent-stained embryos are now ready to be imaged. Transfer either to 80 % glycerol (through 25 % and 50 % glycerol/PBS solutions) and mount. Alternatively, keep in PBTr and mount in agarose for imaging. Keep at 4 °C in the dark and image as soon as possible.

3.4 Antibodies for Neuroanatomy

Zebrafishbrain.org focuses mainly on the characterization of transgenic zebrafish lines. Many hundreds of transgenic and enhancer trap lines have been created by the zebrafish community and can be generated easily using protocols provided elsewhere. To generate data for zebrafishbrain.org, high-resolution confocal

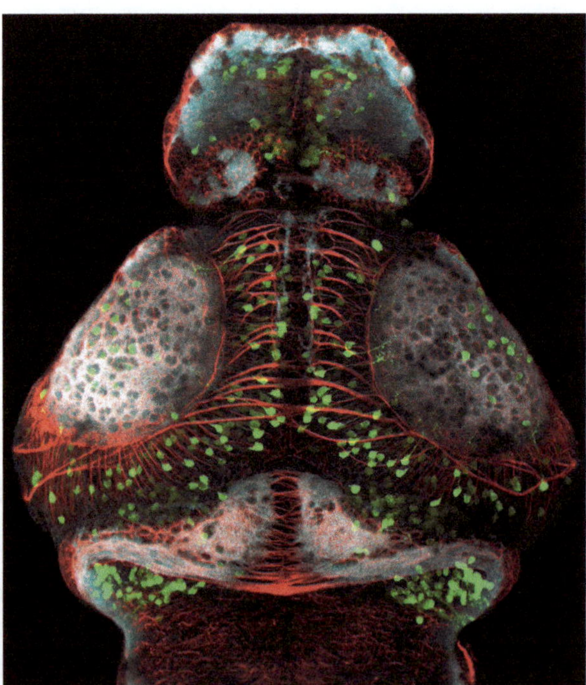

Fig. 3 Transgenic zebrafish embryo labelled with anti-GFP (*green*), anti-SV2 (*blue*), and anti-acetylated tubulin (*red*). This is a dorsal view of a 4dpf Tg(dlx 4/6:GFP) embryo. In this image we can see GFP-positive cells in the telencephalon, optic tectum, and the cerebellum. The SV2 staining labels the synaptic neuropil and axonal tracts are visualized by labeling with anti-acetylated tubulin antibody. This is an example of the type of images we use to populate the zebrafishbrain.org database (Image generated by Monica Folgueira)

imaging is performed on suitable transgenics generated in-house and externally that have fluorescent expression in specific brain regions or nuclei (Fig. 3). With the majority of transgenic lines expressing GFP, the primary antibody used most frequently is a polyclonal rabbit anti-GFP from AMS Biotechnology (TP401). This antibody has excellent penetration and works equally well in whole-mount embryos at all stages and on sections.

To aid anatomical orientation transgenic specimens can also be labelled with anti-acetylated tubulin (IgG2b, Sigma) and/or anti-SV2 (IgG1, DSHB). These antibodies label beautifully the axonal connections and neuropil, respectively. In addition to these antibodies being informative from a neuroanatomical perspective, they are also invaluable as a tool for anatomical localization. These antibodies can be used as a framework to easily compare the expression patterns of different transgenic lines and locate GFP-positive structure in the context of the brain. Both antibodies are mouse monoclonals; fortunately, they are different subtypes and can be detected in the same specimen using subtype-specific secondary antibodies.

We have trialed many neurotransmitter antibodies in whole-mount zebrafish embryonic/larval preps with little success. Colocalization to check which neurotransmitters a particular cell is expressing in a transgenic normally needs to be done using immunohistochemistry on cryosections. There is a protocol for this on zebrafishbrain.org. Some exceptions of antibodies that have worked well in whole-mount can also be seen there. Different fixation methods can improve the efficacy of some of these antibodies for immunohistochemistry. Many antibodies that do not work after PFA fixation will work better after fixation with TCA or other fixatives such as glutaraldehyde.

3.5 Cell Dyes for Neuroanatomy

Using a nuclear label in conjunction with a fluorescent immunostaining can be very useful to delineate brain nuclei, neuropil, and ventricles through the tissue (Fig. 4). Nuclear staining has also been employed by the Driever lab to produce a 3D reference brain for their ViBE-Z software. They have also used acetylated tubulin immunohistochemistry with their reference brain [17, 18].

For nuclear staining, use SYTOX Orange or Green at a concentration of 1:10,000 or Topro3 at a concentration of 1:5,000 depending upon the wavelengths required. These dyes can be added with the secondary antibody incubation. Staining with nuclear dyes works best following room temperature incubation in the dye so it is advisable to add 2 mM azide with the secondary antibody/nuclear dye IB mixture to inhibit mold formation. The dyes bleach very easily so minimize exposure to light and keep the laser intensity on the confocal as low as possible when imaging, and also, timely imaging following the staining process produces the best results.

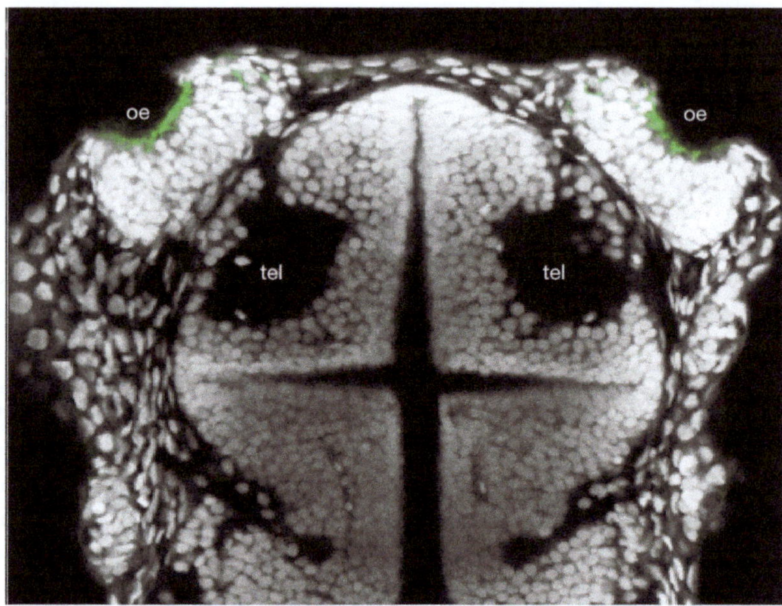

Fig. 4 Nuclear dyes. Nuclear dyes show brain nuclei and also expose morphological boundaries between brain areas. Neuropil is not marked and is revealed as dark spaces. In this confocal slice through the forebrain of this enhancer trap line, we can see expression of GFP in the olfactory epithelia (Image generated by Tim Geach and TAH)

3.6 Confocal Mounting with Glass Ring

1. Mounting media: The choice of mounting medium depends upon the experimental procedure that is being undertaken. For live imaging (e.g., timelapse), embryos can be mounted in agarose (made with fish water) or CyGEL Sustain (*see* Subheading 3.9 below). For imaging of fixed preparations, embryos can be mounted in agarose (made with PBS) or glycerol agarose. Agarose in simple PBS leaves embryos slightly opaque; depth penetration can be improved if the embryo is mounted in glycerol agarose (as described in the protocol associated with the recent Driever lab paper: [17, 18]). The methods of ring mounting described below can be achieved using either aqueous or glycerol agarose. Alternatively, embryos can be mounted between stacked coverslips. Aqueous agarose is slightly preferable as a mountant when using water immersion lenses, and glycerol immersion lenses should be employed where available when mounting in glycerol agarose. Water immersion lenses can produce acceptable results with glycerol agarose as a mountant despite being an optically suboptimal system.
2. Mounting in glass rings (Fig. 5): The advantage of this preparation is that the embryos are securely mounted in a large volume of agarose that permits the orientation of the embryos as is required for imaging. It can be achieved using either aque-

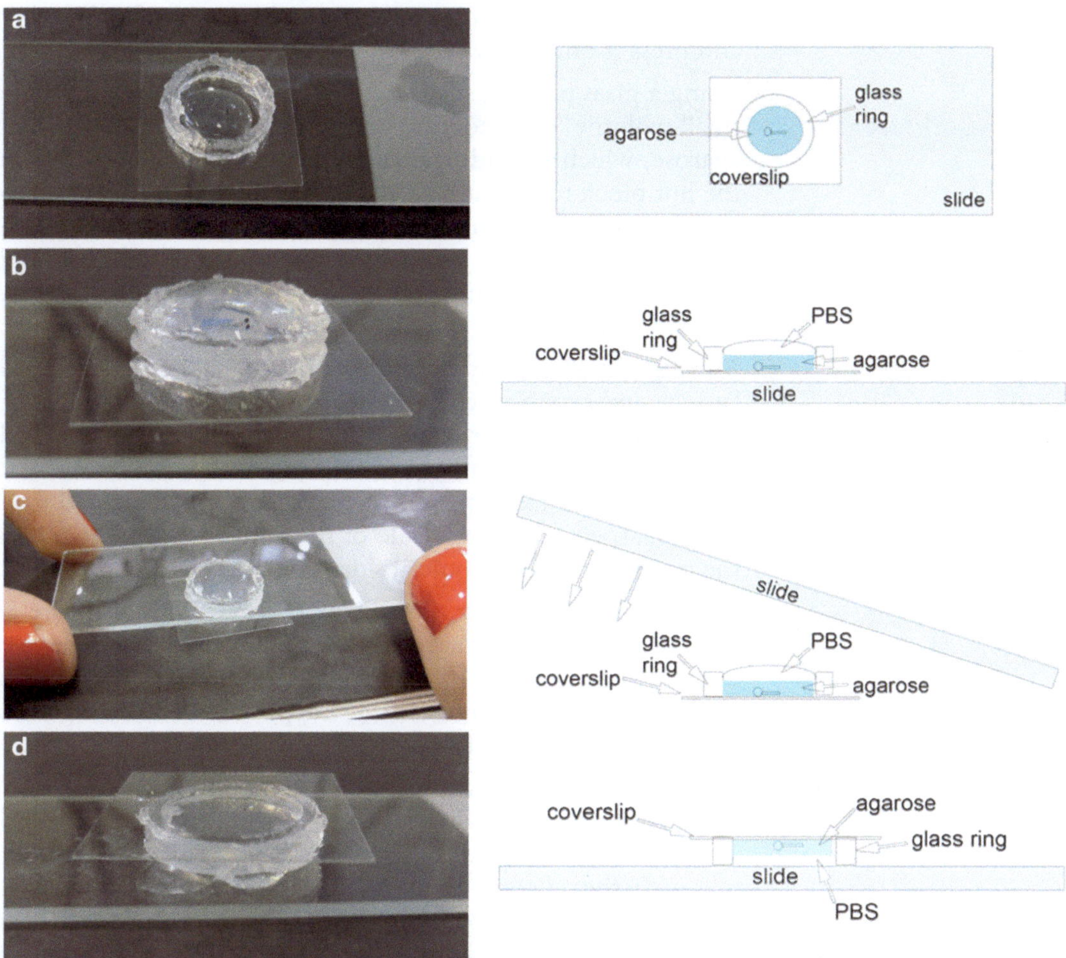

Fig. 5 Mounting in glass rings. (**a**) On a dissecting microscope, place a 22 × 22 coverslip onto a slide. Smear silicone grease on both sides of a glass ring, and press it onto the coverslip forming a watertight seal. Pipette the embryo into the agarose and then onto the coverslip in the center of the glass ring. Orientate the embryo so the surface of the embryo you wish to image is against the coverslip. (**b**) Fill the ring 2/3 full with agarose and let it set. Once the agarose has set firm, pipette enough PBS onto the surface of the agarose to form a convex meniscus over the top of the ring. Overfilling the ring with PBS means that no bubbles will be trapped when you place the slide on top. (**c**) Take another slide and press down onto the top of the glass ring expelling the excess PBS. Make sure the slide and the glass ring have formed a watertight seal with the silicone grease. (**d**) The prep can now be inverted with the coverslip on top and slide on the bottom. Your embryo will now be at the top of the prep just under the coverslip ready for imaging

ous or glycerol agarose as the mounting medium although glycerol agarose should only be used for fixed preparations. Before beginning this protocol, several tubes (as many as required) of agarose should be melted in the microwave and transferred to a heat block set at 45 °C.

3. Place a slide on the dissecting microscope.
4. Place a 22 × 22 coverslip onto the slide.

5. Smear silicone grease on to the top and bottom of a glass ring, and press it onto the coverslip to form a watertight seal.
6. Using a glass pipette, suck up the embryo to be mounted in a small volume of medium, and pipette it into the molten agarose, which should be around 40 °C (slightly cooled from the hot block temperature). Refill the pipette with 0.5–1 ml of agarose with the embryo. Expel the pipette contents (agarose and embryo) onto the coverslip in the center of the glass ring. There should be enough agarose to fill the glass ring around two-thirds full.
7. Moving quickly, using forceps or other suitable tool, manipulate the embryo to place in the correct orientation for imaging. Bear in mind that after **step 7,** the preparation will be inverted; thus the side of the embryo to be imaged should be closest to the coverslip; for example, if the embryo is to be imaged from a dorsal aspect, the dorsal side of the embryo will be touching the coverslip with the ventral side facing upwards at this point.
8. Once the agarose has set firm, pipette some PBS (*see* **Note 5**) onto the surface of the agarose to form a convex meniscus over the top of the ring (it is important to slightly overfill the ring as this prevents bubbles getting trapped inside the ring once the top slide is secured).
9. Take another slide and press down onto the top of the glass ring starting from a slight angle, expelling the excess PBS. Make sure the slide and the glass ring have formed a watertight seal with the silicone grease. The prep can now be inverted with the coverslip on top and slide on the bottom. Use a tissue to remove the excess liquid. Your embryo will now be at the top of the prep just under the coverslip, ready for imaging.

3.7 Confocal Mounting Without Glass Ring

If glass rings are not available, a similar mounting method can be used where a well of silicone grease is constructed directly onto a slide (Fig. 6). A syringe filled with silicone grease should be used to squeeze the grease onto the slide. The embryo is mounted on a coverslip as above in a large drop of agarose (the diameter and depth of this drop should not exceed the width and depth of the silicone grease well). The silicone grease well is filled with PBS to form a convex meniscus. The coverslip with the embryo attached is inverted and pressed gently down into the well expelling the excess PBS. Make sure the coverslip forms a watertight seal with the top of the silicone grease well. The embryo is now ready for imaging. This method also has the advantage of allowing for a small adjustment in the orientation of the sample.

3.8 Mounting of Live Embryos for Confocal Imaging

For imaging of live embryos (particularly for timelapse), embryos can be mounted in a large epoxy resin chamber or well. This allows for constant gaseous exchange and thus improves the health of the embryo. These chambers are easily made using Araldite or similar

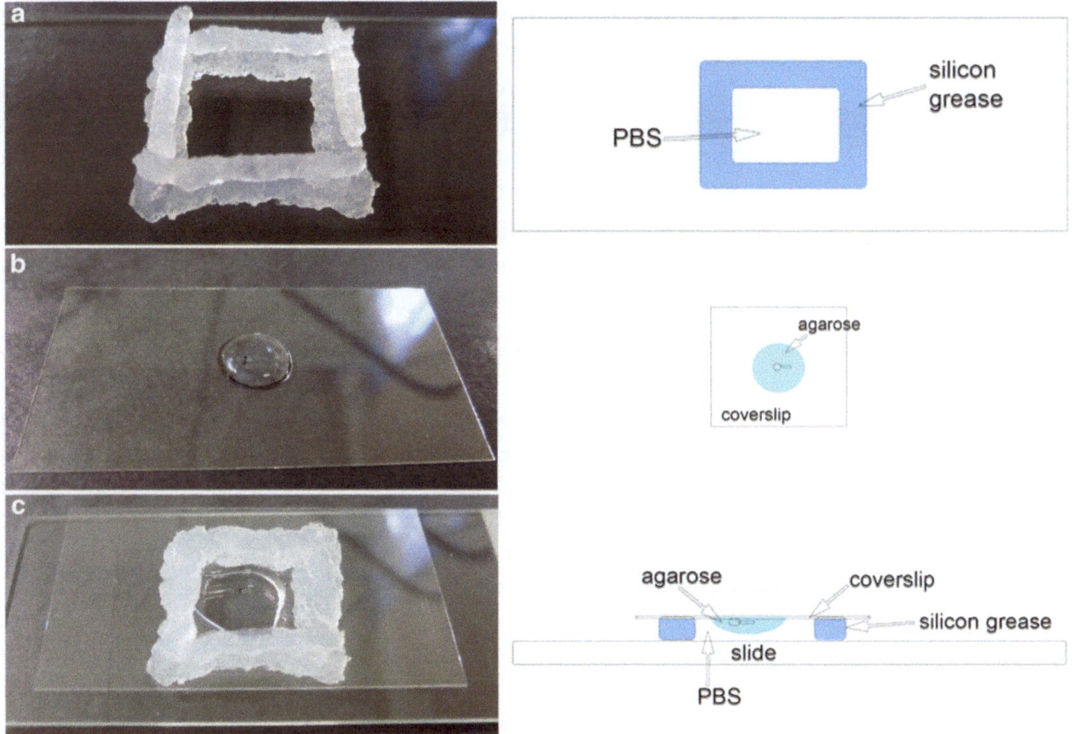

Fig. 6 Mounting fixed embryos in silicone grease wells. (**a**) Using a syringe filled with silicone grease, pipe a well onto a microscope slide. (**b**) On a coverslip orientate your embryo in a drop of 1 % agarose. The surface of the embryo you wish to image should be touching the coverslip. Half fill the silicone grease well with PBS. (**c**) Once the agarose is set, invert your coverslip so that the embryo is submerged in PBS inside the silicone grease well. Lightly push down on the coverslip to create a sealed chamber

epoxy compounds. Mix the two components and make a rectangular well, a few millimeters deep on a clean slide. Make sure the well is continuous so the liquid won't leak out. Allow the resin to polymerize until hard before using. These large wells mean that it is possible to mount several embryos on the same slide. This is particularly useful for multiple timelapses on a confocal with a motorized X/Y stage. Once made, these slides can also be reused many times. Mounting protocol:

1. Anesthetize embryo(s) in a petri dish by adding tricaine (final concentration 1.6 mg/l) to the fish water. Wait until the embryo has completely stopped twitching before trying to mount. Check that heart beat of the embryo is still strong under the dissecting scope.
2. Place resin chamber slide onto the dissecting scope.
3. Remove tube of agarose from the heat block and allow to cool down to below 37 °C to avoid heat-shocking the embryo.
4. Pipette the anesthetized embryo into the warm molten agarose then refill the pipette with agarose and the embryo. Pipette

embryo and agarose onto the slide with enough agarose to form a small bubble around the embryo.

5. Orientate the embryo very gently using forceps or other appropriate implement. This prep will not be inverted so the side of the embryo to be imaged should face upmost; it should also not be too deep in the agarose.
6. Repeat this procedure for two or three other embryos per slide. Do this as quickly as possible so the embryos mounted earlier do not dry out.
7. When the agarose is set, flood the chamber with fish water containing tricaine.
8. The embryos are now ready for imaging. Embryos mounted in this way should be imaged using a non-coverslip corrected water immersion lens. These lenses can be dipped straight into the fish water as long as it does not contain methylene blue. For the modifications required to adapt this method to live embryos (*see* **Note 4**).

3.9 Mounting of Live Embryos Using CyGEL Sustain

An alternative to agarose for mounting live embryos is a compound called CyGEL Sustain. CyGEL is liquid at low temperatures and changes from a sol to a gel at 23–24 °C. The sol–gel conversion can be reversed, by simply placing the sample on ice for a few seconds. This is particularly useful for short procedures that require a fast and easy way of recovering the embryo undamaged after manipulation, such as electroporation or Kaede photoconversion. CyGEL Sustain is a compound commercialized by BioStatus Limited (www.biostatus.com). They provide protocols on their web page, but these are optimized for the growth of cells. Here we provide a protocol adapted for zebrafish mounting:

1. Place vial of CyGEL Sustain on ice and make sure it is a sol.
2. Add E3 so that the final concentration will be 1× (8.4 μl of 60XE3 for 500 μl of CyGEL). Keep it on ice.
3. Prepare a chamber on a slide with silicone grease, or use a glass ring as described in Subheading 3.6.
4. Warm the vial slightly using hands so that it is not too cold (but not too much because it gels).
5. Pipette the anesthetized embryo into the CyGEL then refill the pipette with the embryo and enough CyGEL to fill the silicone chamber or glass ring.
6. Orientate embryo as required the consistency of CyGEL quickly becomes dense, becoming a tight gel within minutes. At that point, the embryo is ready for microscopy.

No additional liquid should be placed on the gel as it immediately becomes a sol again. If the embryo needs to be imaged using a water immersion lens for a long period, it should be mounted upside down in a glass ring on a coverslip filled completely with CyGEL and tightly sealed with a slide. This ensures that the preparation stays moist for the length of the procedure and that no extra liquid disrupts the gel. Tricaine should be added to the E3 medium (4.2 ml tricaine per 100 ml E3) used to prepare the CyGEL, to ensure that the embryo remains anesthetized for the length of the procedure.

3.10 Confocal Microscopy Setup

Setup of the imaging system for optimal imaging depends very much on the individual sample employed. High-NA, long working distance objectives (water and/or glycerol immersion) are available from most microscope manufacturers. These lenses are ideally suited to imaging the zebrafish brain as they permit imaging through the whole depth of the brain at high resolution. There are many tradeoffs in the setup of the imaging parameters. For high-quality three-dimensional reconstructions when imaging fixed, antibody-stained preparations, it is best to aim for approximately isometric voxels at resolutions approaching the limit of diffraction; however, this should be traded off against the time taken and the file size. For live imaging of fluorophores, it is best to minimize bleaching of the (usually dim) fluorescent proteins. Thus minimizing the dwell time at any particular voxel is advisable.

3.11 Image Processing

To process/visualize images, there are a number of free and commercial packages available. To produce zebrafishbrain.org, we mainly use Volocity (PerkinElmer), but other software or a combination of packages can be used for processing/visualization. Examples of other packages are Imaris (Bitplane), Amira, FluoRender, OsiriX, Drishti, and ImageJ/Fiji. There are also some web-based processing servers that are available to do various tasks (e.g. XuvTools and ViBE-Z).

4 Notes

1. Depending on the antigen certain antibodies may work better after fixation with TCA: 2 % TCA in PBS for exactly 3 h at RT in 5 ml bijous. Transfer to 1.5 ml tubes, rinse 2×, and wash 3×5 min PBS (on side). Store at 4 °C for a week, but add 20 mM azide if storing for longer (to prevent mold). In azide PBS embryos should keep for a month.
2. Take great care of superfine grade forceps. Do not use them to touch anything except the embryo, and try not the stick them into the Sylgard as even this can bend and blunt them. To pin and manipulate the embryo, use normal dechorionation-grade forceps. Superfine forceps are expensive and not strictly neces-

sary for dissecting. A combination of dissecting needle and normal forceps can be sufficient. Forceps can be rehoned using fine pliers and a whetstone.

3. It is possible to dissect an embryo without turning it halfway through. See http://zebrafishbrain.org/movies/brain_dissection.
4. To mount live embryos in glass rings for timelapse. Follow the procedure described in Subheading 3.4.2 with the following changes. Embryos must first be anesthetized by adding tricaine to the fish water prior to mounting. Use 0.5–08 % low melt agarose in filtered fish water or E3. Use fish water with tricaine instead of PBS to flood the ring.
5. Glycerol agarose sets very slowly. When using glycerol agarose, replace the PBS used to top up the ring with 80 % glycerol/PBS.

References

1. Norton W, Bally-Cuif L (2010) Adult zebrafish as a model organism for behavioural genetics. BMC Neurosci 11:90
2. Ahrens MB, Li JM, Orger MB et al (2012) Brain-wide neuronal dynamics during motor adaptation in zebrafish. Nature 485:471–477
3. Stewart A, Gaikwad S, Kyzar E et al (2012) Modeling anxiety using adult zebrafish: a conceptual review. Neuropharmacology 62:135–143
4. Mathur P, Guo S (2010) Use of zebrafish as a model to understand mechanisms of addiction and complex neurobehavioral phenotypes. Neurobiol Dis 40:66–72
5. Prober DA, Rihel J, Onah AA et al (2006) Hypocretin/orexin overexpression induces an insomnia-like phenotype in zebrafish. J Neurosci 26:13400–13410
6. Lein ES, Hawrylycz MJ, Ao N et al (2007) Genome-wide atlas of gene expression in the adult mouse brain. Nature 445:168–176
7. Hawrylycz MJ, Lein ES, Guillozet-Bongaarts AL et al (2012) An anatomically comprehensive atlas of the adult human brain transcriptome. Nature 489:391–399
8. Rink E, Wullimann MF (2002) Connections of the ventral telencephalon and tyrosine hydroxylase distribution in the zebrafish brain (Danio rerio) lead to identification of an ascending dopaminergic system in a teleost. Brain Res Bull 57:385–387
9. Mueller T, Dong Z, Berberoglu MA et al (2011) The dorsal pallium in zebrafish, Danio rerio (Cyprinidae, Teleostei). Brain Res 1381:95–105
10. Wullimann MF, Rupp B, Reichert H (eds) (1996) Neuroanatomy of the zebrafish brain: a topological atlas. Basel, Switzerland: Birkhäuser Verlag, p. 160
11. Wilson SW, Ross LS, Parrett T et al (1990) The development of a simple scaffold of axon tracts in the brain of the embryonic zebrafish, Brachydanio rerio. Development 108(1):121–145
12. Buckley K, Kelly RB (1985) Identification of a transmembrane glycoprotein specific for secretory vesicles of neural and endocrine cells. J Cell Biol 100:1284–1294
13. Hendricks M, Jesuthasan S (2007) Asymmetric innervation of the habenula in zebrafish. J Comp Neurol 502:611–619
14. Miyasaka N, Morimoto K, Tsubokawa T et al (2009) From the olfactory bulb to higher brain centers: genetic visualization of secondary olfactory pathways in zebrafish. J Neurosci 29:4756–4767
15. Cerveny KL, Cavodeassi F, Turner KJ et al (2010) The zebrafish flotte lotte mutant reveals that the local retinal environment promotes the differentiation of proliferating precursors emerging from their stem cell niche. Development 137:2107–2115
16. Collins RT, Linker C, Lewis J (2010) MAZe: a tool for mosaic analysis of gene function in zebrafish. Nat Methods 7:219–223
17. Ronneberger O, Liu K, Rath M et al (2012) ViBE-Z: a framework for 3D virtual colocalization analysis in zebrafish larval brains. Nat Methods 9:735–742
18. Meta R, Roland N, Alida F, Olaf R et al (2012) Generation of high quality multi-view confocal 3D datasets of zebrafish larval brains suitable for analysis using Virtual Brain Explorer (ViBE-Z) software. Nat Protoc Exchange. doi:10.1038/protex.2012.031

Part VI

Chicken Protocols

Chapter 15

Immunohistochemistry and In Situ Hybridization in the Developing Chicken Brain

Richard P. Tucker and Qizhi Gong

Abstract

One of the first steps in studies of gene function is the spatiotemporal analysis of patterns of gene expression. Indirect immunohistochemistry is a method that allows the detection of a protein of interest by incubating a histological section with an antibody or antiserum raised against the protein and then localizing this primary antibody with a tagged secondary antibody. To determine the cellular source of a protein of interest, or if a specific antibody is not available, specific transcripts can be localized using in situ hybridization. A histological section is incubated with a labeled RNA probe that is complementary to the target transcript; after hybridization with the target transcript, the labeled RNA probe can be identified with an antibody. Here we describe materials and methods used to perform basic indirect immunohistochemistry and in situ hybridization on frozen sections through the developing chicken brain, emphasizing controls and potential problems that may be encountered.

Key words Immunohistochemistry, In situ hybridization, Riboprobe, Antibody, Fluorescence, Protocol, Technique, Cryosection

1 Introduction

Immunohistochemistry was first used in the 1940s to identify bacterial antigens in mouse tissues by applying fluorescein-labeled antibodies to frozen histological sections [1]. The technique was modified in the following decades to amplify the signal by "indirectly" identifying the so-called primary antibodies with labeled secondary antibodies [2]. The basic method of indirect immunohistochemistry is unchanged to this day: a histological section is incubated with a primary antibody (e.g., a mouse monoclonal antibody specific to an antigen of interest), then the section is incubated with a tagged secondary antibody (e.g., fluorescently tagged rabbit polyclonal antibodies against mouse antibodies), and the protein of interest is then observed using a microscope fitted with special illumination and optics. The technique is widely used in the

Simon G. Sprecher (ed.), *Brain Development: Methods and Protocols*, Methods in Molecular Biology, vol. 1082,
DOI 10.1007/978-1-62703-655-9_15,

neurosciences for determining patterns of expression of novel gene products as well as for identifying specific types of neurons and supporting cells in brain sections.

In situ hybridization is used to detect transcripts in their native tissue environment. The technique was first used in the early 1980s to study gene expression in *Drosophila* embryos [3, 4], but it was quickly adapted to studies of vertebrate brain [5]. In short, a tissue section is treated to make it amenable to hybridization, and then it is incubated with a labeled nucleic acid probe with sequences complementary to the target transcript. Excess probe is removed and the label is detected. There are several established methods for tissue in situ hybridization [6]. One of the major differences between the published methods is the choice of hybridization probes; in general, RNA probes provide more sensitive detection and cleaner background [7].

Here we present detailed protocols for fixing and cryosectioning embryonic chicken brains as well as methods for performing indirect immunohistochemistry and in situ hybridization with RNA probes on these frozen sections. The methods are easily adapted to other tissues and to tissues from other species. For illustrative purposes we have localized the extracellular matrix glycoprotein tenascin-C, which plays critical roles in brain development [8], and its transcript.

2 Materials

The required materials are listed below in Subheadings 2.1, 2.2, and 2.3.

2.1 Cryosectioning Components

1. Phosphate-buffered saline, pH 7.4 (PBS): 10 g NaCl, 2 g KCl, 11.5 g $Na_2HPO_4 \cdot 7H_2O$, and 2 g KH_2PO_4 in 800 mL of ddH_2O. Bring volume to 1 L and adjust pH.
2. Fixative: 4 % paraformaldehyde in PBS, pH 7.4, made fresh (same day). Add 2 g of reagent grade paraformaldehyde powder to 40 mL of PBS in a 50 mL conical tube. Add two NaOH pellets, cap, and allow the pellets to dissolve with gentle shaking. When the paraformaldehyde is in solution, adjust the pH (using pH-sensitive paper) with 2 N HCl, and then top off with PBS to a final volume of 50 mL. Store on ice (*see* **Note 1**).
3. Cryoprotection: sucrose.
4. Embedding medium: Tissue-Tek O.C.T. Compound (Sakura Finetek).
5. Embedding molds: disposable base molds (various sizes, e.g., 15 × 15 × 5 mm).

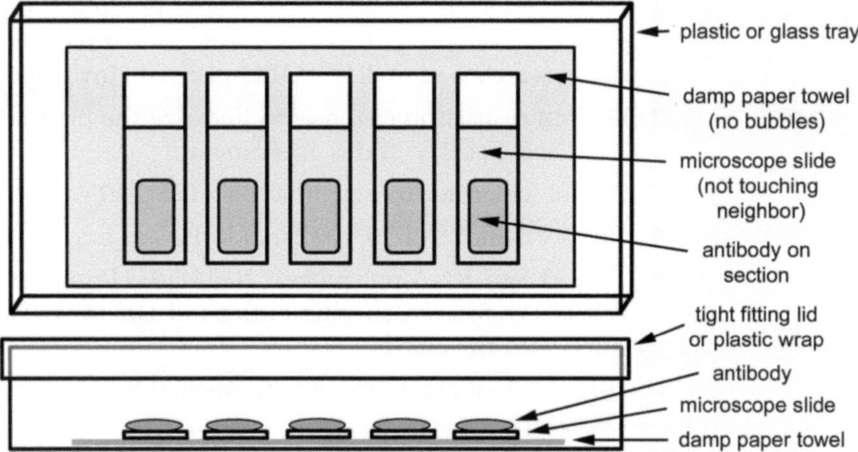

Fig. 1 Schematic illustration of an immunohistochemistry incubation chamber, viewed from above (*top*) and the side (*bottom*). The chamber should have a flat bottom and either a securely fitting lid or a lid fashioned from plastic wrap. As the slides rest directly on the moist paper towel, care should be taken not to let the slides touch each other, and the "puddle" of antibody should not extend to the very edge of the slide

6. Freezing solution: 2-methylbutane. Just before use, break up a 10 cm × 10 cm × 3 cm block of dry ice into powder and chunks approximately 1 cm^3. Put the dry ice into a glass dish, and add the 2-methylbutane to make a slurry 1–2 cm deep.
7. Slides: Superfrost Plus precleaned microscope slides (25 × 75 × 1 mm).
8. Cryostat microtome: for example, Leica CM3050.

2.2 Immunohistochemistry Components

1. Buffer: PBS (see above).
2. Blocking agent: 0.5 % bovine serum albumin (BSA) in PBS.
3. Coplin jars: polypropylene Coplin staining jar.
4. Staining tray: any flat-bottomed glass or plastic tray, approximately 5 cm deep. Place a layer of paper towels on the bottom and dampen with water, pushing out excess water and air bubbles to keep the surface flat. Cover with sticky plastic wrap (Fig. 1).
5. Antibodies: The primary antibody used as an example in this protocol is M1-B4 (mouse anti-chicken tenascin-C [9]; Developmental Studies Hybridoma Bank). Secondary antibodies should be against the animal source of the primary antibody. The secondary antibody used here is Alexa 594 rabbit anti-mouse IgG (Invitrogen).
6. Coverslips: various sizes (e.g., 22 × 30 mm).
7. Nuclear stain: Hoechst (Sigma-Aldrich). Make a 2 mg/mL 100× stock and store in a foil-wrapped tube at 4 °C.
8. Wet coverslip mounting medium: 5 mL PBS/5 mL glycerol.

2.3 In Situ Hybridization Riboprobes

1. cDNA clones in an RNA polymerase binding site containing plasmid: Example given in this protocol is a tenascin-C cDNA clone, which is in pCRII-TOPO plasmid (Invitrogen).
2. Desired restriction enzymes to linearize the plasmid: We used Spe I and EcoR V in this experiment.
3. Phenol/chloroform/isoamyl alcohol (25:24:1) and chloroform.
4. 100 % ethanol and 70 % ethanol in RNase-free H_2O.
5. T7 and Sp6 RNA polymerase 20 U/μL (Promega, Madison, WI, USA). 5× transcription buffer and 100 mM DTT are provided with the enzyme.
6. DIG RNA labeling mix (Roche Applied Science, Indianapolis, IN, USA).
7. RNase-free DNase 10 U/μL (Promega).
8. RNase inhibitor 20 U/μL (New England Biolabs, Ipswich, MA, USA).
9. 4 M LiCl: Dissolve 17 g of LiCl in 100 mL of ddH_2O. Add 100 μL of diethyl pyrocarbonate (DEPC) and mix well. Let the solution stand overnight and then autoclave for at least 15 min.

2.4 In Situ Hybridization Components

1. Hair dryer with cool air setting.
2. Slides.
3. Plastic storage box with a tight-fitting lid.
4. Microscope cover glass (e.g., 40×22 or 50×22 mm; Fisher Scientific or Corning).
5. DEPC-treated H_2O: 1 mL of DEPC, 1 L of ddH_2O. Mix well. Let stand at room temperature overnight or 37 °C for 2 h. Autoclave at least 15 min to inactivate the DEPC.
6. DEPC-treated PBS: 100 mL 10× PBS (see above), 900 mL of ddH_2O, 1 mL of DEPC. Mix well by shaking, and let stand in room temperature overnight. Autoclave to inactivate the DEPC.
7. 1 M Tris–Cl, pH 8.0: 60.5 g Tris base in 400 mL DEPC-treated H_2O. Adjust pH to 8.0 with concentrated HCl, and bring volume up to 500 mL with DEPC-treated H_2O. Filter sterilize (*see* **Note 2**).
8. 1 M Tris–Cl, pH 7.5 and pH 9.5: Same as above; adjust pH to 7.5 and 9.5, respectively.
9. 0.5 M ethylenediaminetetraacetic acid (EDTA), pH 8.0: 19 g of EDTA, 80 mL DEPC-treated H_2O, 10 N NaOH (about 4 mL). Mix to dissolve, adjust pH to 8.0 with NaOH, and bring volume to 100 mL with DEPC-treated H_2O. Filter sterilize.

10. TE buffer, pH 8.0: 5 mL of 1 M Tris–Cl, pH 8.0, and 1 mL of 0.5 M EDTA. Bring to 500 mL with DEPC-treated H_2O.
11. Proteinase K: 12.8 μL of 15.6 mg/mL proteinase K, 20 mL of TE buffer, pH 8.0.
12. 2 N HCl.
13. 1 M triethanolamine–HCl stock: 66.25 mL triethanolamine, 11.25 mL of HCl, 500 mL of DEPC H_2O. Dilute with DEPC-treated H_2O to make 0.1 M triethanolamine–HCl.
14. Acetic anhydride.
15. Ethanol series: Dilute ethanol with DEPC-treated H_2O to make 50 %, 75 %, and 90 % ethanol.
16. Hybridization solution:

Stock	For 10 mL solution	Final concentration
Formamide	5 mL	50 %
Yeast tRNA, 10 mg/mL	200 μL	0.2 mg/mL
Dextran sulfate, 50 %	2 mL	10 %
1 M Tris–Cl, pH 8.0	100 μL	10 mM
0.5 M EDTA, pH 8.0	20 μL	1 mM
Denhardt's, 50×	100 μL	0.5×
5 M NaCl	1.2 mL	600 mM
10 % SDS	250 μL	0.25 %
DEPC-treated H_2O	To 10 mL	

17. 20× SSC: 175.3 g NaCl, 88.2 g sodium citrate ($Na_3C_6H_5O_7{\cdot}2H_2O$), 800 mL DEPC-treated H_2O. Adjust pH to 7.0 with 1 M HCl, and add DEPC-treated H_2O to 1 L.
18. RNase A stock solution 10 mg/mL.
19. 5 M NaCl: 146.1 g NaCl, 450 ddH_2O. Mix by stirring. Add ddH_2O to 500 mL.
20. TNE buffer: 5 mL 1 M Tris–Cl, pH 7.5, 50 mL 5 M NaCl, 1 mL 0.5 M EDTA pH 8.0, and 444 mL ddH_2O.
21. Buffer 1: 100 mM Tris–Cl pH7.5, 150 mM NaCl
22. Blocking reagent (Roche Applied Science).
23. Alkaline phosphatase (AP)-conjugated anti-DIG antibody (Roche Applied Science).
24. Buffer 2: 4 mL 1 M Tris–Cl, pH 9.5, 2 mL 1 M $MgCl_2$, 800 μL 5 M NaCl, ddH_2O to make 40 mL. Make fresh for each use.
25. NBT/BCIP stock solution (Roche Applied Science).
26. Flouromount-G (SouthernBiotech).

3 Methods

If studying a range of developmental stages, it can be convenient to put the fertilized chicken eggs into the incubator at different intervals and then collect and process all of the embryos at once. Be sure to follow appropriate institutional and regional ethical guidelines for the treatment of research animals. The fixation and sectioning protocol described below is used to make frozen sections suitable for both immunohistochemistry and in situ hybridization.

3.1 Cryosectioning

1. Remove the embryo from the egg by cracking the egg into a small bowl containing a small volume of PBS (room temperature). After trimming put the whole embryo (embryonic day [E]4–E7) or the head (E8–E16) into a 50 mL conical centrifuge tube with ice-cold 4 % paraformaldehyde in PBS (pH 7.6). Cap the tube and place it on ice on a rotary shaker for 1–6 h, and then store the tube at 4 °C overnight (*see* **Note 3**).
2. Pour off and dispose of the fixative appropriately and add ice-cold PBS to the tube. Cap the tube and place on a rotary shaker in an ice bath for 30 min. Pour off the PBS and repeat twice with fresh PBS.
3. Add sucrose to the third PBS rinse to a final concentration of 20–25 %. Cap the tube and place on rotating shaker until the sucrose has gone into solution. The tissues should "float" on the sucrose. Store the tube with floating tissues at 4 °C overnight (*see* **Note 4**).
4. Put the sucrose infiltrated (cryoprotected) tissues into a petri dish or onto a piece of Parafilm©, and trim them with a fresh razor blade such that the trimmed surface will match the orientation of the sections to be cut in the cryostat. Transfer the trimmed tissues to a "puddle" of embedding compound to minimize the presence of any sucrose solution, and then transfer the tissue to an embedding mold partially filled with embedding medium. With forceps orient the tissue so that the cut surface is resting along the bottom of the mold (*see* **Note 5**). Float the mold in a shallow slurry of 2-methylbutane and crushed dry ice, taking care not to let the 2-methylbutane come into direct contact with the embedding medium. When the embedding medium is solid white, transfer the mold with the frozen tissue to dry ice (*see* **Note 6**).
5. Cut sections with a cryostat using appropriate protocols, and collect the sections on pre-subbed slides (*see* **Note 7**). Allow the sections to air-dry for 4–6 h, and then either proceed with the protocol or store the slides at −20 °C to −70 °C (*see* **Note 8**).

3.2 Immunohistochemistry

All of the immunohistochemistry procedures described below are carried out at room temperature.

1. Allow slides to come to room temperature (if stored in the freezer), and then place them into a Coplin jar or slide staining rack (depending on the number of slides being processed) with PBS for 10–30 min. Pour off the PBS and add PBS with BSA blocking solution (*see* **Note 9**). Incubate the slides in blocking solution for 15–30 min (longer blocking can reduce background, but typically is not necessary).
2. Dilute the primary antibody in blocking solution in a 1.5 mL snap cap tube. It may be necessary to use a range of dilutions (e.g., 1:10, 1:100, and 1:1,000) with antibodies that have not been characterized before. Spin the diluted antibodies in a tabletop microfuge for 5 min. Reserve a few mL of blocking solution for the secondary antibody (store at 4 °C until used).
3. As the primary antibody is spinning, remove the slides from the blocking solution. Wipe away excess solution using a Kimwipe© taking care not to disturb the section and taking care not to let the section dry out (*see* **Note 10**). Label the slide and place it on the moist paper towel in the incubation chamber; add an appropriate volume of diluted and centrifuged primary antibody until the section is completely covered (about 200 μL). The antibody should stay on the region that is wet and not spread onto the parts of the slide that were dried with the Kimwipe©.
4. Cover the chamber (e.g., with a clingy plastic wrap) and leave undisturbed on the bench top overnight (Fig. 1).
5. Gently tap the primary antibody from the slides onto a Kimwipe©, and rinse the slides three times for 10 min in PBS.
6. Dilute the secondary antibody (following the manufacturer's instructions) in the reserve blocking solution in a 1.5 mL snap cap tube, and centrifuge in a tabletop microfuge for 5 min.
7. As the secondary antibody is spinning, remove the slides from the PBS rinse and carefully dry them (as above), leaving the area where the section is found wet. Place the slides back into the incubation chamber, and add the secondary antibody until it covers the section. Take the same precautions noted above for adding the primary antibody.
8. Cover the chamber and incubate the slides with the secondary antibody for 1–3 h (longer incubations are possible, but usually not necessary).
9. Gently tap the secondary antibody off each slide onto a Kimwipe©, and rinse the slides three times for 10 min in PBS. Optional: Sections can be counterstained with a fluorescent nuclear dye at this stage. Between the second and third PBS

rinse, put the slides into a Coplin jar with nuclear stain (Hoechst) diluted to the 1× working concentration in PBS for 1 min.

10. Coverslip carefully, without introducing air bubbles, using either the PBS/glycerol wet mount or a permanent mounting medium suited for fluorescence microscopy (e.g., Fluoromount-G) and appropriately sized glass coverslips (*see* **Note 11**).
11. One or more control incubations should also be run. If the antibody is well characterized, it may be sufficient to incubate a section in the blocking solution overnight and then treat it with the secondary antibody the following day. Such "secondary antibody only" controls will reveal background fluorescence or nonspecific binding of the secondary antibody. When illustrating this type of control (Fig. 2), be sure to use the same camera settings that were used when imaging the experimental sections. If a polyclonal antiserum is used, one should run a control that includes similarly diluted pre-immune serum.

3.3 In Situ Hybridization

In situ hybridization is a multistep procedure that can take 3–4 days. For better planning, make all solutions ahead of time. Frozen sections can be prepared days to weeks ahead and stored at −20 °C. DIG-labeled RNA probes need to be made and the quality of the probe evaluated before starting in situ hybridization.

3.4 Preparation of DIG-Labeled Riboprobes

For this protocol we used riboprobes generated against tenascin-C to match the immunohistochemistry described above.

1. cDNAs of tenascin-C (GENE ID: 396440 TNC, 2,761–3,179 bp) were obtained by RT-PCR using primers, 5′- acaggactaccattgacctctctg -3′ and 5′- atcagtgccagcattaatggtagc -3′, and cloned into pCRII-TOPO. Orientation and the sequence were confirmed by sequencing (*see* **Note 12**).
2. 10 μg of cDNA-containing plasmids were linearized either at the 5′ end for an antisense probe or at the 3′ end of the cDNA for a control sense probe, with appropriate restriction endonucleases (Fig. 3). The restriction digest should be done overnight to ensure completeness (*see* **Notes 13** and **14**).
3. Take out 1/20 of the reaction to run on an agarose gel to confirm that the restriction digest is complete.
4. Purify the restriction digest reactions with phenol extraction: Bring the volume to 100 μL with ddH_2O in a 1.5 mL tube. Add 100 μL of phenol/chloroform/isoamyl alcohol (25:24:1); vortex for 1 min and centrifuge at 16,000×*g* for 5 min; carefully remove the top aqueous layer and add it to a new tube; add an equal volume of chloroform to the aqueous layer and vortex for 1 min and centrifuge at 16,000×*g* for 5 min. Again, collect the top aqueous layer and transfer to a new tube.

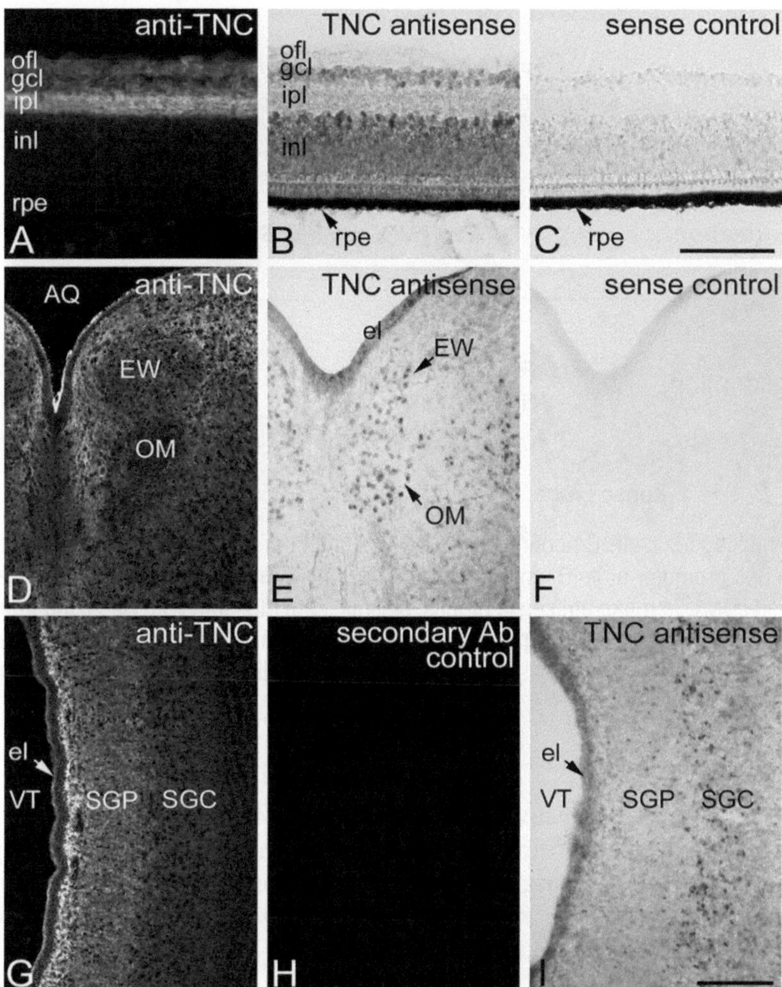

Fig. 2 Examples of indirect immunohistochemistry and in situ hybridization on adjacent sections through the E12 chicken brain. The primary antibody recognizes tenascin-C, and the antisense riboprobe recognizes tenascin-C transcripts. (**a**) Anti-tenascin-C labels the optic fiber layer (ofl) and inner plexiform layer (ipl) of the E12 retina. gcl, ganglion cell layer; inl, inner nuclear layer; rpe, retina pigment epithelium. (**b**) An adjacent section through the E12 retina subjected to in situ hybridization with a tenascin-C antisense probe. The dark reaction product is found in the gcl and inl. (**c**). An adjacent section incubated with a control sense probe, illustrating the specificity of the hybridization signal in **b**. The rpe is dark due to the presence of melanosomes and not the target transcript. Scale bar: 100 μm (**a–c**). (**d**) Anti-tenascin-C immunoreactivity is found around the nucleus of Edinger-Westphal (EW) and oculomotor nucleus (ON). AQ, cerebral aqueduct. (**e**) The tenascin-C antisense probe hybridizes in large cells within the EW and ON as well as in the ependymal layer (el). (**f**). The control section incubated with a sense probe shows that the reaction products seen in E are specific. (**g**). The tenascin-C antibody labels the region underlying the ependymal layer (el) of the optic tectum as well as the stratum griseum periventriculare (SGP). *VT* ventricle of the optic tectum, *SGC* stratum griseum centrale. (**h**) A secondary antibody only control shows the absence of background fluorescence in the same region in an adjacent section. (**i**) An adjacent section incubated with a tenascin-C antisense probe. The reaction product is found in the el and SGC. Scale bar: 500 μm (**d–i**)

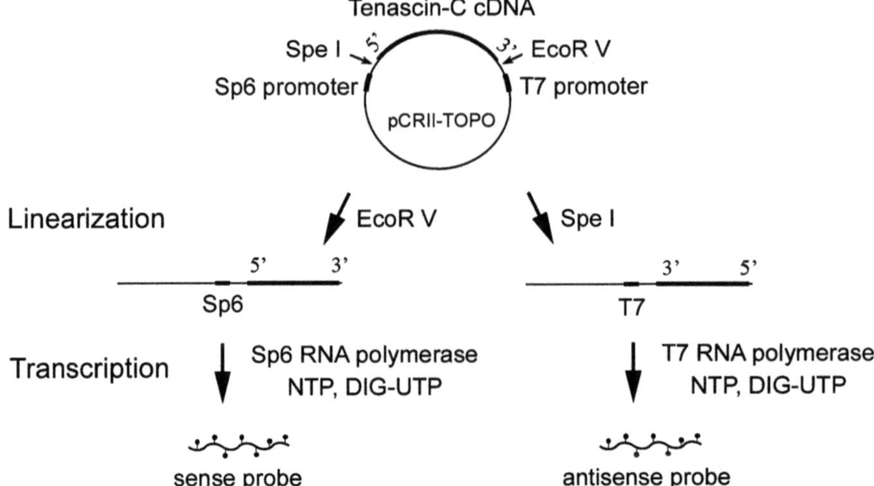

Fig. 3 Steps for making tenascin-C riboprobes. A cDNA fragment of tenascin-C is cloned into the pCRII-TOPO plasmid with an Sp6 promoter at its 5′ end. Restriction enzyme Spe I and EcoRV sites are located at different ends of the cloned cDNA. To make antisense probes, plasmids are digested with Spe I and transcribed with T7 RNA polymerase. To make control sense probes, plasmids are digested with EcoR V and transcribed with Sp6 RNA polymerase. DIG-labeled UTPs are used to incorporate DIG (indicated by *black dots*) into the riboprobes

5. Precipitate the linearized plasmids: Add 1/10 volume of 3 M sodium acetate and 3 volumes of ethanol; mix by briefly vortexing and leave the reaction at −20 °C for 2 h or −80 °C for 1 h; centrifuge at 16,000 × *g* in a refrigerated centrifuge for 20 min.
6. Take out the liquid and wash the pellet with 1 mL of 70 % ethanol (made with DEPC-treated H_2O); spin again in the refrigerated centrifuge for 10 min, empty the tube, and let the pellet dry at room temperature for 10 min.
7. Dissolve the pellet with 20 μL of TE, pH 8.0.
8. To make DIG-labeled RNA probes, assemble the reaction as follows:

Template DNA	1 μg
5× transcription buffer	4 μL
DIG RNA labeling mix	2 μL
0.1 M DTT	1 μL
RNA polymerase	2 μL
RNase inhibitor	0.5 μL
RNase-free water to	20 μL

9. Incubate the reaction at 37 °C for 2 h.
10. Add the following and incubate at 37 °C for 15 min:

RNase inhibitor	1 μL
DNase	1 μL

11. Precipitate RNA probes: Add 1/10 volume of 4 M LiCl and 2.5 volumes of ethanol; incubate at –20 °C for 2 h or –80 °C for 1 h.
12. Pellet RNA probes for 20 min at high speed in a 4 °C centrifuge. Wash with 70 % ethanol (RNase-free) and air dry. Resuspend in 20 μL RNase-free water (*see* **Note 15**).

3.5 In Situ Hybridization: Day 1. Pre-hybridization and Hybridization: RNase-Free

1. Dry sections with cold air using a hair dryer for up to 1 min (*see* **Note 16**).
2. All pre-hybridization and subsequent washing steps are done in a vertical slide mailer that holds five slides. Complete immersion of tissue sections requires 15 mL of solution for each mailer.
3. Place slides into a slide mailer and postfix with 4 % paraformaldehyde in PBS for 15 min.
4. Rinse in DEPC-treated PBS three times, 1 min each.
5. Digest sections with proteinase K at room temperature for 6–12 min (*see* **Note 17**).
6. Inactivate the proteinase K by incubating the sections in 4 % paraformaldehyde in PBS for 10 min at room temperature.
7. Rinse in DEPC-treated PBS three times, 1 min each.
8. Treat the sections with 0.2 N HCl for 10 min at room temperature (*see* **Note 18**).
9. Rinse in DEPC-treated PBS three times, 1 min each.
10. Measure 15 mL of 0.1 M triethanolamine–HCl into a 15 mL tube. Add 37.5 μL of acetic anhydride drop by drop. Mix by inverting and pour into the mailer immediately.
11. Transfer slides into the mailer and incubate sections for 10 min at room temperature.
12. Wash in DEPC-treated PBS three times, 1 min each.
13. Dehydrate sections in increasing concentrations of ethanol (50 %, 75 %, 95 %, 100 %, 100 %) for 2 min each. Sections are subsequently air-dried for 2 min.
14. Add 400 μL of hybridization solution to two 1.5 mL tubes. Heat the hybridization solution at 85 °C for 10 min.

15. Add DIG-labeled sense and antisense probes (~ 1 μg/mL) to the preheated hybridization solution, and incubate at 85 °C for 3 min (*see* **Note 19**).
16. Add 400 μL of probe solution onto each slide. Cover the slide with a glass coverslip. Be sure there are no bubbles and that the hybridization solution covers the entire coverslip area.
17. Place the slides into a sealed, humid chamber. We line the bottom of the plastic box with a piece of paper towel soaked with 50 % formamide in 5× SSC. Instead of placing the slides directly on the paper towel, place a 1.5 mL tube rack upside down to build a platform for the slides. Cover the box with a tight lid.
18. Hybridize at 60 °C overnight (*see* **Notes 20** and **21**).

3.6 In Situ Hybridization: Day 2. Post-hybridization Washes and Immunological Detection: Non-RNase-Free

1. Carefully remove the coverslips by dipping the slides in and out of 2× SSC (*see* **Note 22**).
2. Incubate in 2× SSC and 50 % formamide at 60 °C for 30 min.
3. Transfer sections to TNE buffer at 37 °C for 10 min.
4. Replace the TNE buffer with TNE plus RNaseA (20 μg/mL) and incubate at 37 °C for 30 min.
5. Wash with TNE for 10 min at 37 °C.
6. Further wash the sections in 2× SSC at 60 °C for 30 min.
7. Proceed with high-stringency washes in 0.2× SSC twice for 30 min at 60 °C (*see* **Note 23**).
8. Let the sections stay cool at room temperature for 5 min.
9. Wash with Buffer 1 for 5 min at room temperature.
10. Incubate with 1 % blocking reagent for 1 h at room temperature.
11. Briefly rinse the sections with Buffer 1 plus 0.1 % Tween 20 (B1-T). Put 300 μL of alkaline phosphatase-conjugated anti-DIG antibody (1:1,000 with B1-T) onto each slide. Cover the sections with coverslips and place the slides into a humidified chamber (e.g., a flat-bottomed box with a moist paper towel as in Fig. 1, but with the sections coverslipped to reduce the volume of antibody solution).
12. Incubate overnight at 4 °C.

3.7 In Situ Hybridization: Day 3. Signal Detection: Non-RNase-Free

1. Remove the coverslips by dipping the slides in B1-T. Wash in B1-T three times 10 min each at room temperature.
2. Precondition with Buffer 2 (B2) for 5 min.
3. Replace with 20 mL of B2 with 400 μL of NBT/BCIP. The NBT/BCIP solution is light sensitive. Keep the reaction in the dark.

4. The color reaction may take 30 min to 2 h or longer. Check the staining at 30 min intervals. If the reaction is weak at the end of 2 h, leave the staining overnight in a dark place at room temperature (*see* **Note 24**). The reaction product is blue-purple in color and located in the cytoplasm.
5. Stop reaction by washing with TE, pH 7.5 for 10 min.
6. Rinse quickly with ddH_2O and mount with Fluoromount-G (*see* **Note 25**).
7. Determine appropriate conditions for imaging the sections hybridized with antisense, and then photograph the control sense slides with the same camera settings (Fig. 2).

4 Notes

1. Paraformaldehyde causes skin, eye, and respiratory tract irritation and is a suspect carcinogen. All handling of paraformaldehyde should be done in the fume hood. Read and follow precautions provided by the manufacturer.
2. DEPC inactivates RNase at 0.1 % concentration and is therefore used to treat solutions used for in situ hybridization. It is important to remember that DEPC should not be used to treat Tris buffer. When making RNase-free Tris buffer, dissolve Tris base into DEPC-treated water and filter sterilize.
3. It is important to use freshly prepared formaldehyde as fixative made more than a day or two before use can increase background fluorescence. Do not over fix. Tissues left in fixative for more than 24 h can become so cross-linked that in situ hybridization will not work, and over fixation can also adversely affect epitopes. Some antibodies will not work on formaldehyde-fixed tissues. It may be necessary to experiment with different fixatives, like methanol-, ethanol-, or acetone-based fixatives.
4. The sucrose will prevent tissue disruption from the expansion of water in tissues as it freezes. Tissues that have taken up sufficient sucrose to withstand freezing will sink into the solution over time. Smaller tissues will sink sooner and can be processed the same day. Larger tissues will take more time, but most will sink after 24 h. Air bubbles trapped in the tissues may prevent them from sinking even when they have taken up the sucrose, so avoid introducing air bubbles into the specimens.
5. It is important to orient the material properly before it is frozen. Chose traditional orientations that correspond to orientations found in atlases to help interpret your results.
6. Tissues frozen in embedding medium can typically be stored in ziplock storage bags at –20 °C for several weeks or at –70 °C

for several months. Tissues can usually be stored for longer periods if the sections will be used for immunohistochemistry than if they are to be used for in situ hybridization.

7. Wear gloves when handling slides that will be used for in situ hybridization. It is easier to cut and collect thicker sections, but try to collect sections in the 12–16 μm range. The ideal temperature for cutting sections through chicken brain is usually colder than the ideal temperature for cutting murine tissues; precise block, knife, and chamber tissues will vary from cryostat to cryostat and require practice before working with precious specimens.
8. Store the slides in a clean box and wrap the edges of the box with Parafilm©, and place it in a ziplock bag before refreezing. It is always best to work with fresh slides, but not always practical. Slides kept in the freezer for several years are often suitable for immunohistochemistry, but slides used for in situ hybridization should be used within a few weeks.
9. The BSA-based blocking solution is easy to make and use and works well for most antibodies, but with some antibodies it may be necessary to use a fetal calf serum or condensed milk-based blocking solution.
10. The moisture left behind after drying around the section will help keep the primary antibody from spreading away from the section during the antibody incubation step. There is an increased chance of auto-fluorescence if the section dries out during this step. To avoid this, work with small batches of slides.
11. The wet mounting medium (PBS/glycerol) is easy to use, but care must be taken not to use too much or the medium may spread onto the microscope stage and/or objectives. Dab at the edge of the coverslip after mounting to remove excess medium, and let the coverslipped slides air-dry for several hours before use. Another advantage of this medium is that it is easy to "float off" the coverslip after photography, allowing the section to be counterstained or stained with another primary and secondary antibody (double label immunohistochemistry).
12. The most important factor for probe selection is its sequence specificity to the target gene. When choosing cDNA fragments as template, perform a blast search to find out whether or not they have homology to other transcripts. The optimal probe size is 100–1000 bp. If the probe is too long, it can inhibit tissue penetration.
13. In pCRII-TOPO, T7 and Sp6 polymerase binding sites are located on different sides of the cloned cDNA. In this experiment, sequencing indicated that the tenascin-C cDNA clone was oriented so that the T7 RNA polymerase transcribes the

antisense riboprobe and the Sp6 transcribes the sense (control) probe. When selecting restriction enzymes, it is preferable to use enzymes that result in either a 5′ overhang or a blunt end DNA template. We used Spe I to linearize the template for the antisense probe and EcoR V for the sense probe. Selected restriction enzymes will cut on either ends of the cDNA, respectively, to allow the RNA polymerase to transcribe through the cDNA, but relatively little of the plasmid sequences.

14. When performing in situ hybridization, it is important to incorporate controls. The sense probe often serves as a negative control for nonspecific hybridization signals. Under high-stringency condition, the sense probe should give no in situ signal while the antisense probe hopefully hybridizes specifically to its target. Others suggest to evaluate tissue transcript quality by using labeled poly T probes [10].

15. The yield of the labeled RNA probe is usually 10–15 μg. To evaluate the quantity of the probe, measure the concentration using a nanodrop or conventional spectrophotometer. To determine the quality and the size of the RNA probe, run 0.5–1 μg of the probe on a 2 % agarose formaldehyde gel. To assemble a 2 % agarose formaldehyde gel, combine 2 g of agarose and 73.3 mL of DEPC-treated ddH_2O, and heat to boil; in the fume hood add 16.7 mL formalin (37 % formaldehyde) and 10 mL of 10× MOPS (200 mM MOPS, pH 7.0, 20 mM sodium acetate, 10 mM EDTA, pH 8.0); mix and pour into the tray to set. The probes are diluted with 1× MOPS, 20 % formaldehyde, and 50 % formamide; heat denature at 70 °C for 10 min before loading onto the gel. The gel is run in 1× MOPS.

16. RNases are long lived and difficult to inactivate. Small amount of contamination can destroy the transcripts in the tissue section, so it is important to prevent RNase contamination. On day 1 of in situ hybridization, all solutions should be RNase-free. Use DEPC-treated water to make all solutions. Wear gloves during day 1. It is important to make sure that the slide mailers for pre-hybridization steps are brand new and free of RNase.

17. Proteinase K treatment is used to allow better access of the probe to its target transcripts in tissue sections. Excessive digestion will result in damage to tissue morphology. The concentration, duration, and the temperature of the treatment should be determined empirically. Proteinase K is commonly used at 10 μg/mL. Dependent upon the tissue type, digestion can be done between 25 and 37 °C for 5–15 min. A good starting point is 10 min at room temperature. When treating embryonic

tissue, it is suggested to start with lower concentration of proteinase K (2 μg/mL) or a shorter digestion time.

18. The function of HCl is not entirely known. It is believed that HCl extracts protein and hydrolyzes the target sequence. The treatment may allow better permeability and also appears to reduce background.

19. Dependent upon the abundance of the target transcript, the optimal concentration of the labeled probe should be determined empirically. A good starting range is 0.5–1 μg/mL.

20. The theoretical melting temperature (T_m), which is the temperature at which 50 % of the probe is dissociated from the target, is determined by a number of factors including monovalent cation concentration, the presence of formamide, probe length, and GC content. T_m can be calculated as follows: $T_m = 79.8 + 18.8 \times \log\ [\text{Na}] + 0.58 \times \text{GC}\ \% + 11.8 \times [\text{GC}\ \%]^2 - 0.35 \times \%\text{formamide} - 820/\text{length}$ [11]. Hybridization is often carried out at 20 °C below the T_m.

21. The T_m on tissue (tissue T_m) is different from the theoretical T_m. While theoretical T_m is determined by DNA or RNA behavior in solution, tissue T_m is determined by chromogenic detection of a specific hybridization signal. In general, tissue T_m is around 10° higher than the theoretical T_m. It is important to keep this mind when designing hybridization and washing conditions [12–15].

22. During post-hybridization washes and all subsequent steps, it is not necessary to maintain an RNase-free environment. In general, RNases cleave single-stranded but not double-stranded RNAs. After hybridization, probes are hybridized with their transcript targets and are resistant to degradation by RNase contamination.

23. The stringency of the washes needs to be determined empirically for each probe. Washing stringency is determined by both the salt concentration and the washing temperature. For RNA probes larger than 300 bp, washing temperatures are generally between 55 and 65 °C, and the salt concentrations vary between 1× SSC and 0.1× SSC. As discussed in (*see* **Note 21**), tissue T_m is generally 10–15° higher than theoretical Tm. In this experiment, washing with 0.2× SSC at 60 °C yields the most specific in situ signal for TNC.

24. NBT/BCIP are AP substrate-chromogens. The reaction results in a blue-purple precipitate where hybridization has occurred. If the tissue has endogenous AP activity, it will yield a false positive signal. Adding 0.1 mM levamisole to the NBT/BCIP solution will help inhibit endogenous AP activity.

25. Mounting should be done with a quick wash with water and using an aqueous mount. Do not dehydrate with xylene as the treatment will result in the formation of crystals.

Acknowledgments

We would like to thank Huaiyang Chen for her help in the design of the riboprobe and the cloning of the cDNA, and Chenghao Qian for his technical help with the in situ hybridization.

References

1. Coons AH, Creech HJJ, Jones RN, Berliner E (1942) The demonstration of pneumococcal antigen in tissues by the use of fluorescent antibody. J Immunol 45:159–170
2. Beutner EH, Witebsky E (1963) Studies on organ specificity. Xv. Immunohistologic evaluation of reactions produced by thyroid autoantibodies. J Immunol 91:204–209
3. Akam ME (1983) The location of Ultrabithorax transcripts in Drosophila tissue sections. EMBO J 2:2075–2084
4. Hafen E, Levine M, Garber RL, Gehring WJ (1983) An improved in situ hybridization method for the detection of cellular RNAs in Drosophila tissue sections and its application for localizing transcripts of the homeotic Antennapedia gene complex. EMBO J 2:617–623
5. Garner CC, Tucker RP, Matus A (1988) Selective localization of messenger RNA for cytoskeletal protein MAP2 in dendrites. Nature 336:674–677. doi:10.1038/336674a0
6. Wilkinson DG (1998) In situ hybridization: a practical approach, 1999/01/30th edn. Oxford University Press, Oxford, p 224
7. Zeller R, Rogers M, Haramis AG, Carrasceo AS (2001) In situ hybridization to cellular RNA. Current protocols in molecular biology/edited by Frederick M. Ausubel ... [et al.] Chapter 14, Unit 14 13, doi:10.1002/0471142727.mb1403s55
8. Chiquet-Ehrismann R, Tucker RP (2011) Tenascins and the importance of adhesion modulation. Cold Spring Harb Perspect Biol 3, doi:10.1101/cshperspect.a004960
9. Chiquet M, Fambrough DM (1984) Chick myotendinous antigen. I. A monoclonal antibody as a marker for tendon and muscle morphogenesis. J Cell Biol 98:1926–1936
10. Iezzoni JC, Kang JH, Bucana CD, Reed JA, Brigati DJ (1993) Rapid colorimetric detection of epidermal growth factor receptor mRNA by in situ hybridization. J Clin Lab Anal 7:247–251
11. Bodkin DK, Knudson DL (1985) Assessment of sequence relatedness of double-stranded RNA genes by RNA-RNA blot hybridization. J Virol Methods 10:45–52
12. Herrington CS, McGee JO (1994) Discrimination of closely homologous human genomic and viral sequences in cells and tissues: further characterization of Tmt. Histochem J 26:545–552
13. Herrington CS, Anderson SM, Graham AK, McGee JO (1993) The discrimination of high-risk HPV types by in situ hybridization and the polymerase chain reaction. Histochem J 25:191–198
14. Herrington CS, Graham AK, Flannery DM, Burns J, McGee JO (1990) Discrimination of closely homologous HPV types by nonisotopic in situ hybridization: definition and derivation of tissue melting temperatures. Histochem J 22:545–554
15. Evans MF, Aliesky HA, Cooper K (2003) Optimization of biotinyl-tyramide-based in situ hybridization for sensitive background-free applications on formalin-fixed, paraffin-embedded tissue specimens. BMC Clin Pathol 3:2. doi:10.1186/1472-6890-3-2

Chapter 16

Transplantation of Neural Tissue: Quail-Chick Chimeras

Andrea Streit and Claudio D. Stern

Abstract

Tissue transplantation is an important approach in developmental neurobiology to determine cell fate, to uncover inductive interactions required for tissue specification and patterning as well as to establish tissue competence and commitment. Avian species are among the favorite model systems for these approaches because of their accessibility and relatively large size. Here we describe two culture techniques used to generate quail-chick chimeras at different embryonic stages and methods to distinguish graft and donor tissue.

Key words Chick, Brain, Neural plate, Neural tube, Quail, Transplantation

1 Introduction

During development the entire central nervous system arises from the neural plate, which is induced in the ectoderm by signals from the organizer [1–3]. Shortly thereafter, precursors for the forebrain, midbrain, hindbrain, and spinal cord occupy different, albeit overlapping, territories [4–7]. As the neural plate folds to form the neural tube, anterior–posterior and dorsoventral patterning is established through signals from surrounding tissues but also through the action of local organizers like the midbrain–hindbrain boundary, the floor plate, and roof plate [8]. Many of the molecular mechanisms that control these processes have now been identified, but originally these paradigms were established by transplantation experiments using various ways to distinguish host and donor tissue. In particular, grafting experiments established fate maps of the neural plate and brain [4, 5, 9–12], the location and action of organizers [8, 13], and also the time of competence during which tissue can respond to organizer signals and the time when cells become committed to a particular fate [14–16].

Avian model systems have been very popular for such studies, because the development of their nervous system parallels that of mammalian embryos in many aspects and has been described in great detail. Unlike mammals, however, avian embryos are easily

Simon G. Sprecher (ed.), *Brain Development: Methods and Protocols*, Methods in Molecular Biology, vol. 1082, DOI 10.1007/978-1-62703-655-9_16, © Springer Science+Business Media, LLC 2014

accessible and relatively cheap to obtain, and little specialized equipment is needed for operations and for growing the embryos. Most, if not all, transplantation experiments require a reliable system to distinguish host and donor tissue to locate the graft, to follow its progeny or examine cell behavior like axonal projections, but also to establish whether, e.g., changes in gene expression or neuronal morphology occur cell autonomously (e.g., within the graft) or are induced in surrounding cells (e.g., in neighboring tissues). Many studies have used transplantation of tissues labelled with fluorescent dyes (e.g., [17], infected with retroviral vectors (transplanted into resistant hosts) [18] and more recently tissues from GFP transgenic chickens [19, 20]). However, one of the most extensively used techniques is cross-species transplantations generating chick-quail chimeras to provide permanent cell tracing [9–12]. Quail and chick are closely related species; their early development is fairly similar, but they differ very slightly in timing. The chimeras generated by transplantation of neural tissue and neural crest cells develop normally and are even able to hatch. Early experiments to distinguish quail and chick tissue made use of the fact that quail nucleoli are associated with a fair amount of heterochromatin, which is absent in most other species including the chick. Therefore, histological staining for DNA can differentiate quail and chick tissue [21]. More recently, however, quail-specific antibodies have become available, which recognize either all quail cells or quail neurites [22, 23]. These are now frequently used and their detection can be combined with other techniques like in situ hybridization [24].

This chapter focuses on quail-chick transplantation of neural tissue at early neural plate and at later neural tube stages. After embryonic days 3–4, the brain becomes less accessible due to the formation of blood vessels and extraembryonic membranes, while the spinal cord remains accessible. The procedures described are used to replace neural tissue from chick with the identical tissue from quail (orthotopic) at the same stage of development (isochronic). However, similar strategies can be used for heterotopic or heterochronic grafts as well as to any other tissue.

2 Materials

All procedures described below require two pairs of watchmakers' forceps (number 5); one pair of coarse forceps, about 15 cm long; one pair of small, fine scissors, with straight blades about 2 cm long; Pasteur pipettes (short form), end lightly flamed to remove sharp edges and rubber teats; container for egg waste; and small beakers (50–100 ml). Instruments should be cleaned with lightly soapy water, rinsed in distilled water, and washed in 70 % ethanol before drying on a tissue. You need a good stereomicroscope with transmitted light base and for in ovo work a cold light source (fiber optics) for illumination from the top. Fertile hens' or quails' eggs are incubated in a humidified incubator at 38 °C until they have reached the stage desired; staging of host and donor embryos is performed according to Hamburger and Hamilton [25]. All solutions are diluted from autoclaved stock solutions in distilled water immediately before use; beakers for salines are autoclaved before use.

2.1 Preparing Chick Hosts for New Culture

Operations on primitive streak to early somite stage ($HH3^{+}$-8) chick host embryos are performed in modified New culture [26, 27]; at this stage embryos are fragile and difficult to manipulate in ovo, and survival rate in ovo is poor. On the other hand, in New culture embryos can only be grown for 24–36 h even in an expert's hands. In addition to the above materials, this method requires:

1. Pyrex baking dish about 5 cm deep with 2 l capacity.
2. Watch glasses about 5–7 cm diameter.
3. Rings cut from glass tubing (approx. 27 mm outer diameter, 24 mm inner diameter and 3–4 mm deep; obtained from a local glass blower).
4. 35 mm plastic dishes with lids (bacteriological grade).
5. Plastic box with lid for incubating culture dishes.
6. Pannett-Compton saline is prepared from two stock solutions, which can be kept at 4 °C if autoclaved. Solution A: 121 g NaCl, 15.5 g KCl, 10.42 g $CaCl_2 \cdot 2H_2O$, 12.7 g $MgCl_2 \cdot 6H_2O$, H_2O to 1 l. Solution B: 2.365 g $Na_2HPO_4 \cdot 2H_2O$, 0.188 g $NaH_2PO_4 \cdot 2H_2O$, H_2O to 1 l. To prepare working solution just before use, mix (in order) 120 ml A, 2,700 ml H_2O, and 180 ml B. Do not mix concentrated stocks of A and B.

2.2 Preparing Chick Hosts for In Ovo Operation

After HH8-9, operations are performed in ovo; under perfect conditions embryos can be grown for a long time, even until hatching. Collect the following materials in addition to those listed at the beginning of the methods section.

1. Scalpel with No. 3 handle and No. 11 blades.
2. Plasticine or foam (from packaging) to make a ring for resting eggs on their side.
3. PVC tape to seal the eggs.
4. 5 ml syringe with 21G needle (for removing albumen).
5. 1 ml syringe with 27G (or finer) needle (for ink injection).
6. 1 ml syringe with 21G needle (for antibiotics).
7. Paper tissues.
8. Indian ink (Pelikan Fount India or Winsor and Newton; diluted 1:10 in saline).

9. Silicone grease in a 10 ml syringe (no needle).
10. 10× stock Tyrode's saline: 80 g NaCl, 2 g KCl, 0.5 g $NaH_2PO_4{\cdot}2H_2O$, and 10 g glucose in 1 l H_2O. Prepared in advance, autoclaved, and kept at 4 °C after opening. Dilute this to 1× working solution with autoclaved water (about 100 ml are needed).
11. Antibiotics: 100× penicillin/streptomycin solution (Sigma A9909).
12. 70 % ethanol is also required.

2.3 Preparing Quail Donors

The following materials are required to harvest quail embryos:

1. Glass Petri dish (10 or 15 cm diameter, depending on the number of embryos to be collected).
2. Spoon spatula for collecting embryos.
3. Glass Pasteur pipette cut at the shoulder and fire polished (for embryo transfer).
4. Rubber teats.
5. 500 ml Tyrode's saline.
6. Dissecting microscope with transmitted light base.

2.4 Grafting

To dissect quail and chick tissue for grafting, the following materials are required:

1. 35 mm Sylgard-coated dish for dissecting (this should never come into contact with fixative).
2. Entomological insect pins (A1; steel) for pinning out embryos on the Sylgard dish.
3. Insect pins mounted on Pasteur pipettes or tungsten needles (Goodfellow; 100 μm diameter, mounted on aluminum holders or glass rods using sealing wax; sharpen by repeated exposure of the tip to a very hot Bunsen flame).
4. 30G needles mounted on 1 ml syringes.
5. P20 Gilson pipette and yellow tips.

In addition, for in ovo transplantation in older embryos, the following materials are needed:

6. Micro-knife (e.g., micro-feather microsurgery knives for eye surgery 15E blade angle).
7. Aspirator tube (Sigma A5177).
8. 50 μl borosilicate glass capillaries (for trypsin injection) pulled to fine injection needles using an electrode puller, tips broken off (puller settings need to be determined; needles should be fine enough to avoid fluid uptake by capillary forces but large enough to deliver small amounts of trypsin by air pressure).

9. 50 ml 0.12 % trypsin (Difco) in Tyrode's saline.
10. 5 % serum (any species) in Tyrode's saline for stopping the trypsin.

2.5 Fixing Embryos and Analyzing Results

Embryos are fixed several hours or days after transplantation; they can be analyzed by in situ hybridization or immunostaining to label specific tissues or cell types, followed by labelling with quail-specific antibodies to detect the graft by in situ hybridization using chick- and quail-specific probes or by histological sectioning followed by Feulgen and Rossenbeck staining to reveal nucleoli [21]. Depending on the analysis, different fixatives are used. Fixing requires:

1. Petri dish for collecting embryos (for in ovo).
2. 3.5 or 10 mm Sylgard dish for pinning out embryos.
3. Insect pins (see above).
4. 7 ml glass vials and phosphate-buffered saline (PBS).
5. For whole-mount in situ hybridization, embryos are fixed in 4 % PFA, 1 mM EGTA in PBS, and 0.1 % Tween-20. Embryos are stored in methanol.
6. For Feulgen staining and in situ hybridization on sections [28], embryos are fixed in Zenker's fixative: 50 g $HgCl_2$, 25 g $K_2Cr_2O_7$, 10 g of $Na_2SO_4 \times 10\ H_2O$ in 1 l distilled water; before use, add 5 ml glacial acetic acid to 100 ml of the solution: or fixed in Carnoy's fixative: 50 % ethanol, 11.1 % formaldehyde, 10 % glacial acetic acid [24, 28].

The most important point for further analysis is the detection of quail tissue. While traditionally Feulgen staining has been used to reveal the difference between chick and quail nucleoli [22, 29, 30], now quail-specific antibodies are the favorite method. The monoclonal mouse antibody QCPN (Developmental Studies Hybridoma Bank) labels all quail cells, while QN is specific for neurites [23]. We generally perform QCPN staining after whole-mount in situ hybridization; this procedure requires:

1. Glass vials.
2. Pasteur pipettes.
3. Rubber teats.
4. Rocking platform.
5. PBS, blocking buffer: 1 % goat serum, 0.5 % Triton X-100 in PBS.
6. Anti mouse IgG-HRP coupled (Jackson).
7. 100 mM Tris–HCl pH 7.4.
8. 50 mg/ml 3,3′-diaminobenzidine (DAB) in 100 mM Tris–HCl pH 7.4.
9. 0.3 % H_2O_2 in 100 mM Tris–HCl pH 7.4.

It is also possible to perform in situ hybridization with a probe directed against sequences that differ between chick and quail (most likely the 3′ UTR of a specific gene of interest) to distinguish transcripts produced by cells of the graft and donor [31].

3 Methods

3.1 Preparing Quail Donor Embryos for Grafting

To dissect the donor tissue, quail embryos are first removed from the egg and cleaned using the following steps.

1. Remove quail eggs from incubator. Using fine scissors, gently tap the shell near the blunt end of the egg to penetrate the shell. Use the tip of the scissors to cut off a small cap of shell; avoid damaging the yolk.
2. Pour thin egg white into waste; use the scissors to help and cut through the rather thick albumen if required (*see* **Note 1**).
3. Once most albumen is removed, turn the yolk by stroking it very gently with the sides of the scissors to make the embryo become visible on top of the yolk.
4. Using the scissors make four cuts into the vitelline membrane around the embryo; make sure that all the cuts meet (*see* **Note 2**).
5. Using the spoon, pick up the square of embryo and membrane including a little yolk; try to collect as little yolk as possible.
6. Transfer the embryo including yolk and membrane into the large Petri dish containing Tyrode's saline under a dissecting microscope by sliding it carefully off the spoon. Using fine forceps, turn the square of yolk/membrane/embryo so you can see the embryo.
7. Once all donor embryos have been collected in the Petri dish, separate the embryos from adhering yolk. Work at low magnification; use two pairs of forceps to pick up a corner of the vitelline membrane with one and slowly but steadily fold it back, steadying the yolk with the other. Make sure that the membrane and embryo remain submerged in saline. The embryo will remain attached to the membrane. If not, peel off the membrane completely, and then use the forceps gently to remove the embryo from the underlying yolk.
8. Using the wide-mouth Pasteur pipette, pick up the embryo, with (better) or without membrane, and transfer to a 10 cm dish with clean saline. To clean the embryo use a fire-polished Pasteur pipette and gently blow saline over it; this will remove yolk particles. The embryos are now ready for dissection and grafting and can be kept for 1–2 h before proceeding further.

3.2 Preparing Chick Hosts for New Culture

At primitive streak and neural plate stages, operations in avian embryos are most easily performed in New culture. The method described below is based on New's original technique [27] modified by Stern and Ireland [26]. This modified culture method uses rings cut from glass tubes, instead of rings bent from glass rods and 35 mm plastic dishes instead of glass watch glasses resting inside a large glass Petri dish. The rings cut from tubing generate a slightly rough surface that grips the vitelline membrane and therefore allows easy transfer of the culture into the plastic dish. To set up the cultures, proceed as follows (Fig. 1):

1. Remove eggs from the incubator.
2. Fill the large Pyrex dish with about 1.5 l of Pannett-Compton saline; the volume should be large enough that eggs yolks are submerged completely.
3. To open an egg, tap its blunt egg with the coarse forceps and carefully remove pieces of the shell. Discard the thick albumen into the waste bucket, assisted with the coarse forceps. Collect the thin albumen in a small beaker (*see* **Note 3**).
4. Carefully tip the yolk into the Pyrex dish containing saline, taking care not to damage the membrane on the edges of the shell. Carefully turn the yolk with the side of the coarse forceps so the embryo is facing upwards. Now place a watch glass and a glass ring into the dish.
5. Cut the vitelline membrane enveloping the yolk just below the equator using small scissors; you can use one pair of forceps to push the yolk around gently while continuing to cut all the way around its circumference (*see* **Note 4**).
6. Using both pairs of fine forceps, peel the vitelline membrane with the embryo attached slowly but very steadily off the yolk. Use one pair of forceps to pull the edge of the membrane slightly upwards (about 25–30° angle from the yolk surface) and the other to hold the yolk down. Do not pull tangentially along the yolk: this may detach the embryo from the membrane. Do not stop during this process. The embryo should come off with the membrane (*see* **Note 5**).
7. Turn the vitelline membrane with the inner face containing the embryo pointing upwards and slide it, preserving its orientation, onto the watch glass. Place the ring over it so that the membrane protrudes around the ring and the embryo sits in its center. Remove the watch glass, ring, and embryo from the dish; tilt the assembly gently to pour off some saline while steadying the ring with one finger (*see* **Note 6**).
8. Dry the bottom of the watch glass on some tissue. Using fine forceps, carefully wrap the lose edges of the vitelline membrane

Fig. 1 Setting up New culture. (**a**) Set up. (**b**) Instruments. (**c**) Opening eggs. (**d**) Removing albumen. (**e**) Yolks in Pyrex dish. (**f**) Cutting the vitelline membrane at the equator. (**g**) Removing membrane and embryo from the yolk. (**h**) Membrane with embryo facing upwards on watch glass. (**i**) Placing glass ring on the membrane. (**j**) Culture assembled on ring and watch glass. (**k**) Removing assembly form Pyrex dish. (**l**) Cleaning the culture. (**m**) Primitive streak stage embryo on watch glass. (**n**) Embryo after removal of a piece of neural plate. (**o**) Setting up the culture in a Petri dish with albumen. (**p**) Finished New culture

over the edge of the ring, all the way around its circumference. Pull the membrane slightly so its bottom is smooth and free from wrinkles, but be careful not to pull so tight that it breaks (*see* **Note 7**).

9. Using the fire-polished Pasteur pipette, rinse the outside of the ring to remove yolk particles. If there is a lot of egg albumen remaining under the membrane, lift the ring gently and use the

Pasteur pipette to remove it. Clean the yolk over and around the embryo using clean saline; be careful not to dislodge the embryo form the membrane. Damaged embryos do not grow well or normal. If there is a lot of vitelline membrane inside the ring, trim off the excess with the fine scissors while lifting the edges with fine forceps. At this stage embryos are ready for transplantation and can be kept on the bench for some time; make sure they remain well submerged under saline and there is sufficient saline on the watch glass. If keeping them for a few hours, place them on a wet tissue and cover with a large glass or plastic plate/dish.

10. To finalize the cultures after transplantations, work under the microscope; carefully remove any remaining saline, both inside and outside the ring. Drying helps the graft and host tissue to heal faster. During culture the embryo and the inside of the ring must remain dry.
11. Pour some thin albumen (about 2–3 mm thick layer) on the bottom of a 35 mm Petri dish. Using fine forceps, slide the ring with vitelline membrane off the watch glass, and transfer it to the dish; lower it onto the albumen making sure that no air is trapped underneath. Press the ring lightly onto the bottom of the dish using two forceps to allow it to adhere.
12. Remove the excess albumen if its level comes close to the edge of the ring. At this point, remove any remaining liquid from inside the ring using a fire-polished Pasteur pipette. The vitelline membrane should be slightly dome shaped; this will help to drain off any fluid that accumulates during culture (*see* **Note 8**).
13. To seal the Petri dish, wet the inside of the lid with a thin film of albumen all around the edge, discard the excess, and place onto the bottom part. Press lid down slightly to seal (*see* **Note 9**).
14. Place the dish in a plastic box containing a piece of wet tissue, seal the box, and place it into an incubator at 38 °C.

3.3 Preparing Chick Hosts for In Ovo Culture

Later-stage embryos are generally grown in ovo, which allows embryos to grow for long periods, even until hatching. Eggs must be incubated lying on their side, so the yolk turns with the embryo facing upwards to make it accessible for manipulation. To prepare hosts for in ovo operations, use the following procedure (Fig. 2):

1. Remove the eggs from the incubator, place one egg onto the egg rest, and clean with 70 % ethanol; be careful not to rotate the egg.
2. Hold the 5 ml syringe with 21G needle nearly vertical, and insert the needle into the blunt end of egg until you feel the shell at the bottom. Remove about 1 ml egg albumen and discard. This lowers the embryo away from the top of the shell.

Fig. 2 In ovo culture. (**a**) Eggs are incubated on their side and placed on egg rest. *Circle* labels the blunt end of the egg. (**b**) Blunt end is used to remove albumen with 5 ml syringe. (**c**) Inserting syringe. (**d**) Scoring the window. (**e**) Windowed egg. (**f**) Windowed egg with Vaseline border surrounding the window. (**g**) Embryo before ink injection. (**h**) Ink injection using a 1 ml syringe. (**i**) Ink injection: Embryo is clearly visible. (**j**) Embryo after ink injection. (**k**) Removing albumen to lower the embryo. (**l**) Eggs sealed with tape

3. Using the scalpel score a 1 cm × 1 cm square on the top of the shell and lift it up using the blade or a pair of forceps.
4. Moisten the white membrane under the shell with a little Tyrode's saline and remove it with fine forceps, to the edge of the window. Be careful to avoid damage to the embryo underneath.
5. Add saline to the egg so that the embryo floats up to the level of the window.
6. Take up diluted ink into a 1 ml syringe with a 27G needle; make sure there are no air bubbles in the syringe. Insert the needle under the vitelline membrane almost parallel to the yolk surface; choose a position away from the embryo proper, and point the needle towards and underneath the embryo. Inject about 50–100 μl ink; the amount should be kept as little as possible. Avoid moving the needle after initial penetration; otherwise the hole will become too big and yolk and/or ink

will leak out. The embryo should now be clearly visible on a dark background (*see* **Note 10**).

7. Line the shell window with a shallow edge of Vaseline by ejecting it from the syringe. This will allow you to cover the embryo with a drop of saline during the operation for moisture and good optics. Fill the chamber with saline until there is a good dome of fluid. Adjust the fiber optic light so the light shines tangentially onto the embryo, which is now ready for manipulation (*see* **Note 11**).
8. Once the embryo has received the graft and it is in the correct position, remove the saline very carefully from above the embryo using a Pasteur pipette. Watch under low magnification to ensure the graft does not move. If required, reposition it using a pin mounted on a Pasteur pipette.
9. Carefully insert the 5 ml syringe with 21G needle into the original hole in the blunt end of the shell. Carefully remove about 3 ml egg albumen to lower the embryo to its original position.
10. Add 1–2 drops (50–100 μl) of antibiotic solution (*see* **Note 12**).
11. Use a tissue with 70 % ethanol to wipe the Vaseline off the shell; dry the shell thoroughly with another tissue.
12. Cut a piece of PVC tape about 3–4 cm long, stretch it slightly, and allow it to relax again. To seal the egg, place the tape over the window and carefully smoothen any wrinkles without putting too much pressure on the shell. Make sure the edges of the tape are firmly attached to the shell; if not they will roll up and expose the embryo.
13. Incubate the egg in a well-humidified incubator at 38 °C; next day you can turn the egg window side down which helps to keep the embryo moist and improves their development. Incubate for the desired period; generally the 2–4 day survival rate should be 80–90 % (*see* **Note 13**).

3.4 Grafting Procedure: Quail Neural Plate into Chick Hosts in New Culture

The procedure below describes orthotopic, isochronic neural plate grafts from quail donors into chick hosts at HH3$^+$/4. The same method can be applied for heterotopic and heterochronic grafts or transplantation of other tissues.

1. To prepare the host follow the procedure described under Subheading 3.1 until **step 9**.
2. Replace one of the eyepieces of the microscope with an eyepiece containing a graticule; a protractor, to measure angles, is particularly useful.
3. Place a host embryo from Subheading 3.1, **step 9** (kept on a watch glass) under the microscope and center the graticule on the node.

4. Define the area to be replaced using the graticule coordinates. New culture embryos face ventral side up; to reach the ectoderm lower layers need to be removed. Fold back the endoderm and mesoderm above the area to be replaced by quail tissue using 30G needles on a 1 ml syringe serving as a holder. Use the sharp side of the needle gently to score both layers on three sides (e.g. anterior, posterior and lateral leaving them attached medially); then using the back of the needle carefully peel away the endoderm and mesoderm overlying the area to be grafted.
5. Use tungsten needles, mounted insect pins or 30G needles to cut out the region of the neural plate to be replaced by quail tissue (*see* **Note 14**; Fig. 1n). Set aside the host embryo and turn to the quail donor.
6. Pin out a quail embryo of the same stage, ventral side up, on a Sylgard dish containing Tyrode's saline; use insect pins through the extraembryonic region to stretch the embryo slightly.
7. Locate the area to be grafted using the graticule, remove the endoderm and mesoderm as described in **step 4** and excise the underlying neural plate as described in **step 5**.
8. Working under low magnification, use a Gilson pipette to pick up the graft in 3–5 μl saline.
9. Move the host embryo back on the stage and, working under low magnification, place the graft close to the target site.
10. Use mounted insect pins to move the graft into the hole cut previously. It is crucial to maintain the apical-basal orientation of the graft; after excision the ectoderm generally contracts basally and the tissue curves slightly.
11. Carefully remove all the liquid outside the ring and most of the fluid inside the ring.
12. Flip back the mesoderm and endoderm to secure the graft in its position and carefully remove all remaining liquid inside the ring. Excess liquid around the grafted area should be removed using pulled capillaries on aspirator tubes (*see* **Note 15**).
13. Now finish setting up the cultures by following **steps 10–13** in Subheading 3.2. Make sure that the dome of albumen is rather flat; a high dome causes too much tension and the grafts do not integrate properly. Leave embryos on the bench for 30–60 min to let the grafts heal and then proceed to **step 14** in Subheading 3.2.

3.5 Grafting Procedures: Quail Neuroectoderm into Chick Host In Ovo

This section describes orthotopic and isochronic neural tube grafts in embryos older than HH9; as with transplantations in younger embryos described above, the same techniques can be used for heterotopic and heterochronic experiments. First, prepare the donor embryo, and then turn to the host.

1. To prepare the quail donor, pin out the embryo dorsal side up on a Sylgard dish in Tyrode's saline and place under a dissecting microscope with transmitted light base.
2. Using tungsten needles or a micro-knife, make a longitudinal incision into the ectoderm dorsal to the neural tube on both of its sides.
3. Replace the Tyrode's with trypsin solution; working at high magnification peel the ectoderm away from the neural tube using the back of a 30G needle. In the same way, gently scrape any loosely attached cells (neural crest depending on stage) off the neural tube and free it from the adjacent tissues (*see* **Note 16**).
4. Progressively separate the neural tube from the underlying notochord using a micro-knife to push it from side to side, allowing the Trypsin to penetrate; when completely detached, cut the neural tube transversely at its anterior and posterior ends to free it.
5. Remove the excised neural tube using a Gilson pipette set to 3–5 μl and place into a 35 mm Petri dish containing saline with 5 % serum (*see* **Note 17**). Graft can be kept on ice until use.
6. Prepare the chick host by following **steps 1–7** in Subheading 3.2. Use a mounted insect pin make a small hole into the vitelline membrane just over the area to be operated. The hole should be as small as possible.
7. Replace the drop of Tyrode's saline with trypsin solution and follow **steps 2–4** above to excise the same section of the neural tube as in the quail donor.
8. Remove the excised neural tube using a Gilson pipette set to 3–5 μl and replace the trypsin solution with fresh Tyrode's saline twice.
9. Pick up the graft using a Gilson pipette; rinse in Tyrode's saline without serum before transferring it to the host.
10. Using a Gilson pipette and working under low magnification, transfer the graft into the saline bubble over the host embryo.
11. Use a mounted insect pin or 30G needle to place the graft into the hole made by removal of the host neural tube. Preserve anterior-posterior and dorsoventral orientation (*see* **Note 18**).
12. Once the transplant is in position, carefully remove the saline using a Pasteur pipette while observing under low magnification. If needed, reposition the graft using a mounted pin.
13. Finish the egg by following **steps 8–13** in Subheading 3.2.

3.6 Detecting Quail Tissue

As outlined above, grafted embryos can be analyzed in various ways depending on the question; these include whole-mount or section in situ hybridization and tissue- or cell-specific immunohistochemistry.

All of these techniques can be combined with the antibody staining using the quail-specific antibody QCPN. We generally perform whole-mount in situ hybridization for embryos up to embryonic day 3 and section in situ for older embryos followed by QCPN staining. The protocol below describes the whole-mount procedure; for other applications, *see* [19, 20, 28]. For embryos incubated to HH13 or older, it is a good idea to treat them with 6 % H_2O_2 in PBS, 0.1 % Tween after fixation or rehydration after storing in methanol (*see* [24]); this reduces background for the in situ hybridization and immunostaining signal.

1. After developing the in situ hybridization color reaction, wash and fix embryos as normal. Remove fixative by washing in PBS three times for 10–30 min depending on the age of the embryo.
2. Block embryos for 1–3 h in blocking buffer at room temperature on a rocking platform.
3. Replace blocking solution with QCPN antibody solution (dilute antibody in blocking buffer; determine concentration for each batch of antibody) and incubate at 4 °C on a rocking platform for two nights.
4. Remove antibody and wash embryos in PBS for five to seven times 1 h each; for older embryos leave the final wash overnight at 4 °C.
5. Incubate embryos in secondary antibody (generally 1:1,000 in blocking buffer, but may need titration) over one or two nights at 4 °C.
6. Wash as in **step 4**.
7. Wash twice for 15 min in 100 mM Tris–HCl pH 7.4 (*see* **Note 19**); in the second wash measure the volume (generally 1 ml is sufficient).
8. Add DAB from the stock to vial with embryos to a final concentration of 0.5 mg/ml; incubate for 10–15 min rocking in the dark (*see* **Note 20**).
9. Add the appropriate amount of H_2O_2 from the 0.3 % stock to the embryos to make a final concentration of 0.003 %.
10. Incubate in the dark until brown color develops, normally within 5–10 min; check occasionally using illumination from the top on a white background.
11. Stop reaction by rinsing several times in distilled H_2O to remove residual substrate and postfix embryos in 4 % formaldehyde.
12. Embryos can now be cleared, photographed, and embedded for paraffin or vibratome sectioning as required.

4 Notes

1. Remove as much albumen as possible; the yolks move less in the next step.
2. Make sure scissors are cleaned after each egg; crusts of egg yolk make the vitelline membrane stick to the scissors and the embryos tend to sink into the yolk. Do not hesitate when making the cuts; work rapidly so the embryos do not move.
3. Try to remove as much albumen as possible; albumen adhering to the vitelline membrane makes the following steps more difficult.
4. Make sure to cut at or slightly below the equator; otherwise it will be difficult to fit the membrane around the ring in the next step.
5. Occasionally embryos remain attached to the yolk particularly at early primitive streak stages. Keep the membrane because it can be used for other embryos, in case you accidentally punctured a membrane. The embryo can also be retrieved from the yolk but requires thorough cleaning: use a pair of fine forceps, close them, and gently push the edges of the extraembryonic region away from the yolk. Work all the way around the edge of the embryo. Transfer the embryo using a wide-mouthed pipette into a dish with fresh saline, ventral side up gently blow the attached yolk plug off the embryo using a Pasteur pipette. The embryo can now be returned to the membrane; make sure to keep it orientation: embryos do not grow with the ventral side on the membrane.
6. Make sure you do not turn the membrane inside out; embryos do not grow on the outer surface.
7. Be careful not to make any holes into the membrane; this will allow albumen to accumulate inside the ring and prevent the embryo from growing.
8. Be careful not to use too much albumen for grafted embryos; this will increase the tension and prevent healing.
9. Sealing is important to prevent condensation on the lid during incubation.
10. Too much or too high concentration of ink is toxic for the embryos. Recently Pelikan Indian ink has become difficult to obtain; if you are using other brands they need to be tested for toxicity.
11. A dome of liquid considerably improves the optics and also prevents the embryo from drying. The latter is critical as drying out reduces the survival rate.

12. Antibiotics are generally only required for long culture periods. Normally embryos survive well for 2–3 days without this.
13. Low survival rates can be due to a number of factors. The most common problems are dehydration, damage to critical blood vessels, and infection. Working in a drop of saline helps to alleviate dehydration. To avoid infection ensure you use clean solutions; instruments can be cleaned periodically while working in distilled H_2O and ethanol (make sure it evaporates before using to operate), while tungsten needles are flamed periodically to keep them sharp.
14. Be careful not to make any holes in the vitelline membrane.
15. Drying the area surrounding the graft greatly improves healing; be careful not to suck up the graft into the capillary.
16. Be patient; the trypsin works almost by itself, and it generally sufficient to push the adjacent tissues away using your instruments. Avoid going too deep and cut a hole into the endoderm; this will make ink and yolk leak out.
17. This inactivates trypsin; this is important because the tissue should be exposed to proteolytic enzymes for as short as possible to avoid disintegration.
18. If you encounter difficulties to preserve orientation, mark one end of the neural tube with a small crystal of carmine powder.
19. Make sure the pH is properly adjusted to 7.4; the reaction is pH sensitive.
20. DAB is carcinogenic; make sure to wear appropriate protective clothing (lab coat, gloves) and consult the local health and safety regulations for inactivation and disposal.

Acknowledgment

This work is supported by the BBSRC, NIH, and ERC. We thank Anneliese Norris for assistance with New culture photography.

References

1. Stern CD (2005) Neural induction: old problem, new findings, yet more questions. Development 132:2007–2021
2. Waddington CH (1934) Experiments on embryonic induction. Part I: the competence of the extra-embryonic ectoderm. Part II: experiments on coagulated organisers in the chick. Part III: A note on inductions by chick primitive streak transplanted to the rabbit embryo. J Exp Biol 11:211–227
3. Spemann H, Mangold H (1924) Über Induktion von Embryonalanlagen durch Implantation artfremder Organisatoren. Wilhelm Roux' Arch Entwicklungsmech Org 100:599–638
4. Cobos I, Shimamura K, Rubenstein JL, Martinez S, Puelles L (2001) Fate map of the avian anterior forebrain at the four-somite stage, based on the analysis of quail-chick chimeras. Dev Biol 239:46–67

5. Fernandez-Garre P, Rodriguez-Gallardo L, Gallego-Diaz V, Alvarez IS, Puelles L (2002) Fate map of the chicken neural plate at stage 4. Development 129:2807–2822
6. Garcia-Martinez V, Alvarez IS, Schoenwolf GC (1993) Locations of the ectodermal and non-ectodermal subdivisions of the epiblast at stages 3 and 4 of avian gastrulation and neurulation. J Exp Zool 267:431–446
7. Eagleson GW, Harris WA (1990) Mapping of the presumptive brain regions in the neural plate of Xenopus laevis. J Neurobiol 21: 427–440
8. Vieira C, Pombero A, Garcia-Lopez R, Gimeno L, Echevarria D, Martinez S (2010) Molecular mechanisms controlling brain development: an overview of neuroepithelial secondary organizers. Int J Dev Biol 54:7–20
9. Couly G, Le Douarin NM (1988) The fate map of the cephalic neural primordium at the presomitic to the 3-somite stage in the avian embryo. Development 103:101–113
10. Couly G, Le Douarin NM (1990) Head morphogenesis in embryonic avian chimeras: evidence for a segmental pattern in the ectoderm corresponding to the neuromeres. Development 108:543–558
11. Couly GF, Le Douarin NM (1985) Mapping of the early neural primordium in quail-chick chimeras. I. Developmental relationships between placodes, facial ectoderm, and prosencephalon. Dev Biol 110:422–439
12. Couly GF, Le Douarin NM (1987) Mapping of the early neural primordium in quail-chick chimeras. II. The prosencephalic neural plate and neural folds: implications for the genesis of cephalic human congenital abnormalities. Dev Biol 120:198–214
13. Martinez S, Wassef M, Alvarado-Mallart RM (1991) Induction of a mesencephalic phenotype in the 2-day-old chick prosencephalon is preceded by the early expression of the homeobox gene en. Neuron 6:971–981
14. Baker CV, Stark MR, Marcelle C, Bronner-Fraser M (1999) Competence, specification and induction of Pax-3 in the trigeminal placode. Development 126:147–156
15. Bhattacharyya S, Bronner-Fraser M (2008) Competence, specification and commitment to an olfactory placode fate. Development 135:4165–4177
16. Groves AK, Bronner-Fraser M (2000) Competence, specification and commitment in otic placode induction. Development 127: 3489–3499
17. Guthrie S, Prince V, Lumsden A (1993) Selective dispersal of avian rhombomere cells in orthotopic and heterotopic grafts. Development 118:527–538
18. Fekete DM, Cepko CL (1993) Retroviral infection coupled with tissue transplantation limits gene transfer in the chicken embryo. Proc Natl Acad Sci USA 90:2350–2354
19. Barraud P, Seferiadis AA, Tyson LD, Zwart MF, Szabo-Rogers HL, Ruhrberg C, Liu KJ, Baker CV (2010) Neural crest origin of olfactory ensheathing glia. Proc Natl Acad Sci USA 107:21040–21045
20. Sabado V, Barraud P, Baker CV, Streit A (2012) Specification of GnRH-1 neurons by antagonistic FGF and retinoic acid signaling. Dev Biol 362:254–262
21. Feulgen R, Rossenbeck H (1924) Mikroskopisch-chemischer Nachweis einer Nukleinsaeure vom Typus der Thymonukleinsaeure und die daruf beruhende elecktive Faerbung von Zellkernen in miroskopischen Praeparaten. Hoppe-Seyler's Z Physiol Chem 135:203–248
22. Teillet MA, Ziller C, Le Douarin NM (2008) Quail-chick chimeras. Methods Mol Biol 461:337–350
23. Tanaka H (1990) Selective motoneuron outgrowth from the cord in the avian embryo. Neurosci Res Suppl 13:S147–S151
24. Streit A, Stern CD (2001) Combined whole-mount in situ hybridization and immunohistochemistry in avian embryos. Methods 23:339–344
25. Hamburger V, Hamilton HL (1951) A series of normal stages in the development of the chick embryo. J Morph 88:49–92
26. Stern CD, Ireland GW (1981) An integrated experimental study of endoderm formation in avian embryos. Anat Embryol (Berl) 163: 245–263
27. New DAT (1955) A new technique for the cultivation of the chick embryo in vitro. J Embryol Exp Morph 3:326–331
28. Lassiter RN, Dude CM, Reynolds SB, Winters NI, Baker CV, Stark MR (2007) Canonical Wnt signaling is required for ophthalmic trigeminal placode cell fate determination and maintenance. Dev Biol 308:392–406
29. Storey KG, Selleck MA, Stern CD (1995) Neural induction and regionalisation by different subpopulations of cells in Hensen's node. Development 121:417–428
30. Le Douarin N, Dieterlen-Lievre F, Creuzet S, Teillet MA (2008) Quail-chick transplantations. Methods Cell Biol 87:19–58
31. Izpisua-Belmonte JC, De Robertis EM, Storey KG, Stern CD (1993) The homeobox gene goosecoid and the origin of organizer cells in the early chick blastoderm. Cell 74:645–659

Chapter 17

RNAi-Based Gene Silencing in Chicken Brain Development

Irwin Andermatt and Esther T. Stoeckli

Abstract

The mouse is the most commonly used vertebrate model for the analysis of gene function because of the well-established genetic tools that are available for loss-of-function studies. However, studies of gene function during development can be problematic in mammals. Many genes are active during different stages of development. Absence of gene function during early development may cause embryonic lethality and thus prevent analysis of later stages of development. To avoid these problems, precise temporal control of gene silencing is required.

In contrast to mammals, oviparous animals are accessible for experimental manipulations during embryonic development. The combination of accessibility and RNAi-based gene silencing makes the chicken embryo a powerful model for developmental studies. Depending on the time window during which gene silencing is attempted, chicken embryos can be used for RNAi in ovo or cultured in a domed dish for easier access during ex ovo RNAi. Both techniques allow for precise temporal control of gene silencing during embryonic development.

Key words Neural development, Chicken embryo, Artificial miRNA, RNA interference, Electroporation, Gene silencing

1 Introduction

The chicken embryo has been used to study developmental processes for a long time. The advantage of the chicken as a model is its easy accessibility for experimental manipulation during embryonic development. Compared to the mouse, however, the chicken cannot be manipulated with the same powerful genetic tools. Although discovered a couple of years ago, chicken embryonic stem cells cannot be maintained and manipulated in vitro. Furthermore, size of the adult animals and long generation time would also be unfavorable features of the chicken with respect to genetic approaches.

However, the development of RNAi-based loss-of-function approaches has overcome these problems and made the chicken embryo a great model organism for developmental studies [1–4].

Simon G. Sprecher (ed.), *Brain Development: Methods and Protocols*, Methods in Molecular Biology, vol. 1082,
DOI 10.1007/978-1-62703-655-9_17,

In fact, the precise temporal control and the cell-type specificity of gene silencing provide unique opportunities for the analysis of gene function during development.

Since many genes involved in neural development are also involved in the development of other tissues, such as the heart, classical genetic knockout strategies can lead to early lethality. One approach to overcome this restriction in the mouse model is a conditional, tissue-specific knockout of a gene of interest provided that suitable Cre lines are available to excise the floxed target gene. Conversely, in chicken, organ- or even cell-type specificity is easy to achieve and can be combined with temporal control of gene silencing.

We have developed approaches for temporally and spatially controlled gene silencing based on RNAi in chicken embryos [4–6]. Temporal control can be achieved with both in ovo and ex ovo RNAi by using long dsRNA or plasmid-based miRNA/shRNA for gene silencing [1, 4]. The latter, miRNA-based RNAi comes with the additional advantage of cell-type specificity of gene silencing. Both dsRNA- and miRNA-based RNAi can be used to silence several genes at the same time.

For gene analysis during early stages, when embryos can easily be accessed through a window in the eggshell, in ovo RNAi is the method of choice, as it has higher survival rates compared to ex ovo RNAi. To analyze brain function during late developmental stages, embryos can be cultured ex ovo (shell-less) in a dish which enables direct access to the desired brain areas for injection and electroporation [3, 7].

Timing is important! Keep in mind that RNAi does not remove the preexisting protein. Therefore, electroporation has to be done before the protein of interest has accumulated. It is thus necessary to carefully analyze the temporal expression of the target gene. We routinely use in situ hybridization to analyze gene expression during embryonic development [8]. Timing of injection and electroporation of dsRNA or miRNA is determined by the temporal expression of the target gene, not by the time point of analysis or the developmental milestones of the part of the nervous system that is analyzed. In general, injection and electroporation are easier at younger stages. However, cell proliferation will dilute the active RNA-induced silencing complex (RISC) loaded with the specific siRNA produced from the injected dsRNA or the miRNA. Thus, gene silencing will be very effective over long periods of time in neurons but will be less efficient in cells that keep proliferating after electroporation.

Once manual skills for handling embryos and efficient injection and electroporation have been acquired, RNAi will provide results on gene function very rapidly. For a first approach, it is often easiest to just use long dsRNA prepared by in vitro transcription from cDNA fragments. For thorough functional analysis of a target gene, miRNA-based plasmids can be used in a cell-type-specific manner. Since these miRNAs are coupled to a fluorescent protein, efficiency of electroporation is directly visualized [4, 9]. These vectors can be used in combinatorial knockdowns with different tissue-specific promoters or in combination with misexpression constructs resistant to the miRNA to perform rescue experiments.

An excellent example demonstrating the power of precise temporal control of gene silencing is provided by our discovery of a role of the morphogen Shh in axon guidance [10–12]. Morphogens are required for cell differentiation and patterning of the nervous system during early stages of development. Thus, precise temporal control of gene silencing was key to our finding of a direct and an indirect effect of Shh on post-crossing commissural axon guidance [10–12]. Loss of Shh function during the morphogenesis phase of neural development would have prevented these findings as cell types in the neural tube would not have been induced properly.

Here, we describe how to culture chicken embryos in ovo or ex ovo for subsequent electroporation of miRNA constructs (described in [4]) or long dsRNA [1, 3] to analyze neural development. In ovo RNAi is well-suited for manipulation of embryos at young stages and limited to about the fourth day of embryonic development [10, 13]. After these stages the brain of the embryo is no longer easily accessible. For manipulations at older stages, chicken embryos can be transferred from the egg to a plastic dish enabling injection and electroporation of diverse brain regions at late stages [3, 7].

To study neural crest derivatives, injection and electroporation have to be carried out in the first 2.5 days of development. Injections into the eye are easier in the first 3–4 days, as the poor elasticity of the sclera makes good injections without leakiness more difficult at later stages. The cerebellum, for instance, starts to emerge very late, only after about 1 week of embryonic development. At this stage, the head needs to be fixed for injection in a more upright position, and the blood vessels at the back of the head can be used as landmarks to guide the injection needle [3].

We include two protocols in this chapter, one for injections and electroporation of the neural tube within the first 3–3.5 days of development and the second one for electroporation of the cerebellum at late stages (HH34–36; Hamburger and Hamilton stages 34–36 [14]).

2 Materials

2.1 Windowing the Eggs

1. Fertilized eggs from a local hatchery.
2. Incubator set at 38.5 °C and 45 % humidity. We use two incubators, one to incubate eggs before they are windowed and a second one for incubation during the experiment (e.g., Heraeus/Kendro Model B12, Kendro Laboratory Products, Germany, or Juppiter 576 Setter + Hatcher; FIEM, Italy).
3. Facial tissues.
4. 70 % ethanol.
5. Paraffin wax (Paraplast tissue embedding medium).
6. Heating plate set at 80 °C to melt paraffin.
7. Paint brush.
8. Scalpel.
9. Scotch tape or coverslips (24 × 24 mm).
10. Soldering iron (when eggs are closed with coverslips).
11. Sterile syringe with 18G needle.
12. Fine scissors.

2.2 Ex Ovo Culture

1. Fertilized eggs from a local hatchery.
2. Domed dish with a diameter of 80 mm and a depth of 40 mm (these dishes are produced for the food industry from oriented polystyrene (OPS; Bellaplast, Altstaetten, Switzerland)).
3. Lid for domed dish (we use the lid of a 10 cm Ø petri dish).
4. Incubator set at 38.5 °C and 45 % humidity. We use two incubators, one to incubate eggs and a second one for incubation of embryos in the dishes (e.g., Heraeus/Kendro Model B12, Kendro Laboratory Products, Germany, for dishes; Juppiter 576 Setter + Hatcher; FIEM, Italy, for eggs).
5. Facial tissues.
6. 70 % ethanol.

2.3 In Ovo and Ex Ovo Electroporation

1. Borosilicate glass capillaries (outer Ø/inner Ø: 1.2 mm/0.68 mm; World Precision Instruments; 1B120F-4).
2. Glass needle puller (Narishige, Japan; PC-10).
3. Square wave electroporator (BTX ECM 830).
4. Spring scissors (Fine Science Tools; 15003-08).
5. Dumont #5 forceps (Fine Science Tools; 11252-20).
6. *For* in ovo *electroporation*: Platinum electrodes (4 mm length, 4 mm distance between cathode and anode).

7. *For* ex ovo *electroporation*: Platelet electrode of 7 mm diameter (Tweezertrodes Model #520, BTX Instrument Division, Harvard Apparatus, Holliston, MA, USA, also see http://www.btxonline.com/tweezertrodes/).
8. Spritz bottle filled with ddH_2O.
9. Polyethylene tubing (Ø 1.24 mm).
10. 0.2-μm filter (Sarstedt, Switzerland).
11. Trypan blue solution, 0.4 % (Invitrogen).
12. 20× phosphate-buffered saline (PBS).
13. Sterile PBS.
14. Spatula bent to a hook (ex ovo electroporation).

3 Methods

3.1 Windowing Eggs

To obtain access to the developing embryo, a window is cut into the eggshell (Fig. 1). Independent of the time of electroporation, eggs are windowed at the second or third day of incubation (E2 and E3, respectively; *see* **Note 1**).

1. Incubate eggs at 38.5 °C and 45 % humidity (*see* **Notes 2–4**).
2. Let embryos develop until they have reached the desired stage for the experimental manipulation. Staging of the embryo is done according to Hamburger and Hamilton [14]; *see* **Note 5**.
3. About 20 min before windowing, place the egg on its side to allow the embryo to reposition on top of the egg yolk.
4. Wipe the eggshell using facial tissue and 70 % ethanol to avoid contamination (*see* **Note 6**).
5. Place a strip of Scotch tape to cover the area of the planned window. This will prevent pieces of eggshell from falling onto the embryo when cutting the window.
6. Use a scalpel to drill a hole into the corner of the intended window and into the blunt end of the egg (*see* Fig. 1 and **Note 7**).
7. Remove ~3 ml of albumen by pushing the needle of the syringe at an angle greater than 45° into the hole at the blunt end of the egg. This will avoid damage to the egg yolk (*see* **Note 8**).
8. Use a paint brush to seal the hole at the blunt end of the egg with melted paraffin.
9. Cut a window into the eggshell. To avoid damaging the embryo, carefully hold scissors horizontally.
10. Seal the window with a coverslip and paraffin. Use a brush to apply melted paraffin to the edges of the window and carefully press a coverslip onto the hot paraffin (alternatively, use Scotch

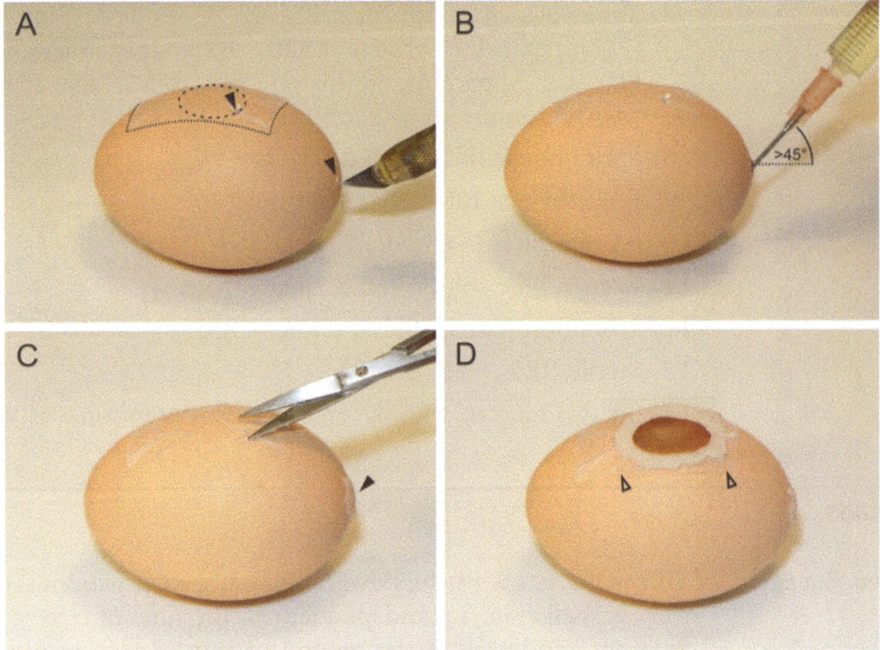

Fig. 1 Egg windowing. (**a**) After the egg is cleaned with 70 % ethanol, a strip of sticky tape is put on the top of the egg, and two holes are drilled into the edge of the presumptive window and the blunt end of the egg (*arrowheads*). (**b**) The needle is inserted at an angle greater than 45° in order not to damage the egg yolk. Removal of about 3 ml of albumen will detach the embryo and the blood vessels from the eggshell. (**c**) The hole at the blunt end is sealed (*arrowhead*), and a window is cut into the shell starting at the previously drilled hole. (**d**) With a brush melted paraffin is applied to the edges of the window and immediately covered with a coverslip (*open arrows*)

tape to seal the window; *see* **Note 9**). As paraffin cools down quickly, carefully pressing a soldering iron on the coverslip will remelt the paraffin below the glass and lead to proper sealing.

11. Put the windowed egg in the incubator at 38.5 °C and 45 % humidity until further use (*see* **Notes 3** and **4**).

3.2 In Ovo Electroporation

1. Clean the working space with 70 % ethanol and autoclave your working tools (*see* **Note 6**).
2. Prepare capillaries to make injection needles using the glass needle puller (*see* **Note 10**).
3. Remove the coverslip or the tape covering the window. To remove the coverslip briefly press the hot soldering iron onto the glass.
4. To get direct access to the embryo, carefully remove the extra-embryonic membranes with forceps and spring scissors (Fig. 2).
5. Break off the tip of the previously pulled needles to obtain a diameter of 5–7 μm. Plug the needle into the polyethylene tubing and fill the tip of the needle with your injection mix (*see* **Note 10**).

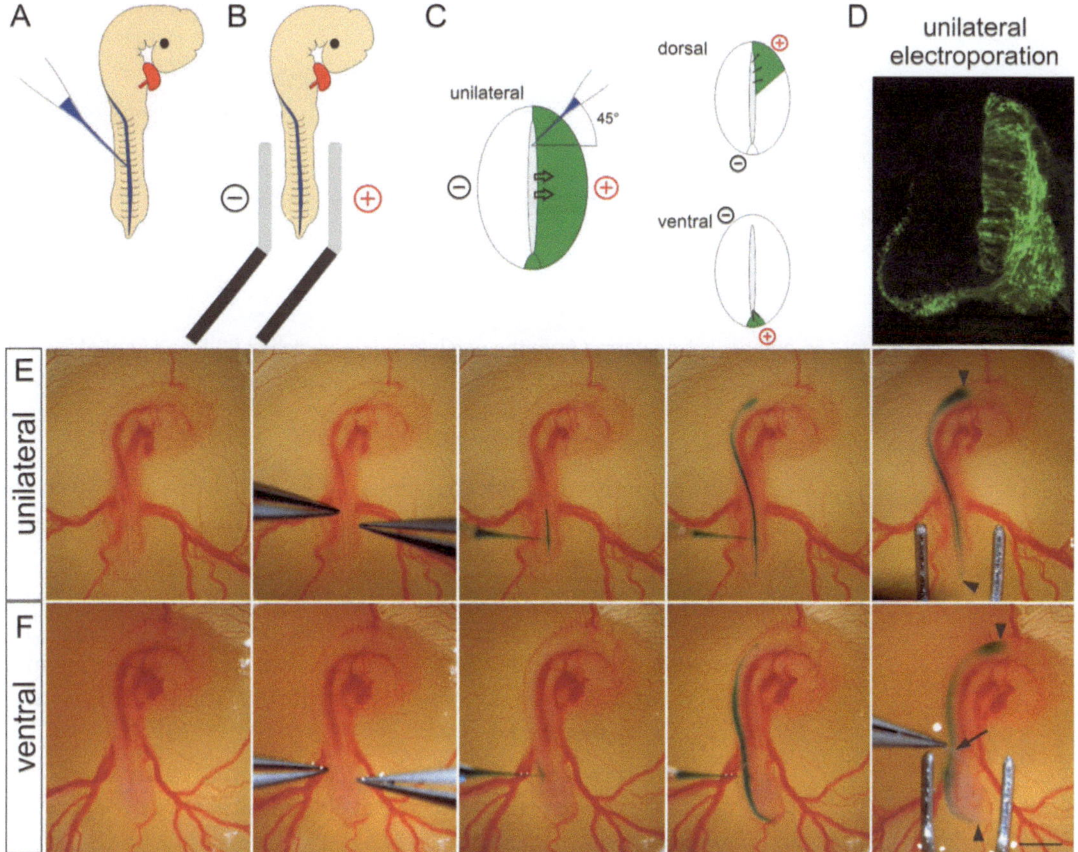

Fig. 2 (**a**, **b**) Schematic drawing of in ovo electroporation of the caudal neural tube. (**a**) Injection of dsRNA or miRNA into the central canal of the spinal cord. (**b**) Electroporation with anode and cathode positioned parallel to the body axis. (**c**) Depending on the positioning of the electrodes, electroporation can be spatially controlled. (**d**) Successful unilateral transfection of EGFP in the right hemisegment is visualized on a cryosection of a HH25 embryo. (**e**) Unilateral in ovo injection and electroporation of an E3 chicken embryo. Using forceps, the extra-embryonic membranes are removed to obtain access to the spinal cord. The injection mix is injected into the central canal of the lumbar region of the spinal cord. The injection volume is controlled by mouth. The maximal injection volume is achieved when the *blue* solution reaches both the ventricle and the tail (*arrowheads*). For unilateral electroporation the electrodes are placed laterally in parallel to the body axis of the embryo. Make sure not to touch major blood vessels to avoid fatal bleeding. (**f**) For dorsal or ventral electroporation HH20 embryos are easy to handle because their body is more detached from the egg yolk and slightly tilted to the side compared to the younger embryo shown in (**e**). Note that the lumbar region is tilted to the side and therefore you see more lateral parts of the spinal cord. Use the forceps to pull the extraembryonic membranes sidewards to position the embryo and stabilize it in the desired position for injection and electroporation with anode and cathode positioned dorsally and ventrally, respectively. For dorsal electroporation the polarity would be reversed. Scale bar: 2 mm

6. Inject the mix containing miRNA plasmids (*see* **Notes 11** and **12**) or long dsRNA (*see* **Notes 11** and **13**) into the central canal of the neural tube and control injection volume by mouth (Fig. 2e, f). The maximum volume is reached when the blue solution extends from the brain ventricle to the tail of the embryo.

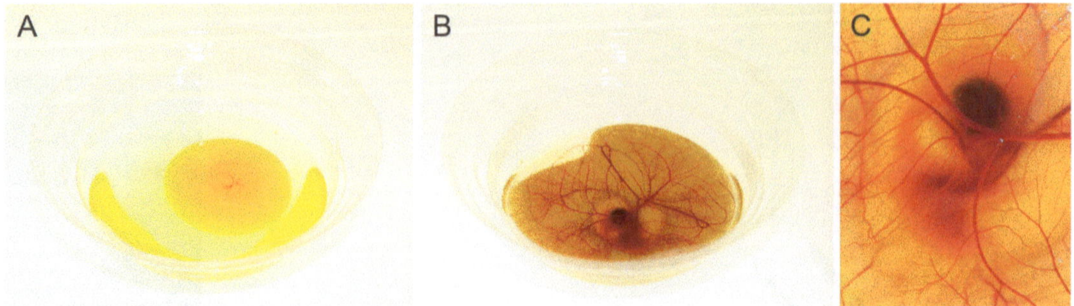

Fig. 3 Ex ovo culture. (**a**) The embryo is transferred from the egg into a domed dish at E2.5 and kept at 38.5 °C. (**b** and **c**) Ex ovo culture of a HH34 embryo (E8) ready for cerebellar injection and electroporation

7. Before applying an electric field, add a few drops of PBS to the embryo to prevent overheating and high electrical resistance during subsequent electroporation.
8. Place the electrodes parallel to the anterior-posterior axis of the spinal cord. Do not touch major blood vessels while applying current to avoid bleeding. Use 5 pulses of 50 ms duration at 18 V for E2 or at 25 V for E3 embryos for electroporation (Fig. 2, *see* **Notes 14–16**).
9. Put the resealed egg back into the incubator until the embryo reaches the desired developmental stage for analysis (*see* **Note 4**).

3.3 Ex Ovo Culture of Chicken Embryos

1. Incubate fertilized eggs for 2.5 days at 38.5 °C and at least 45 % humidity (*see* **Notes 4** and **17**).
2. Position the eggs on the side for 20 min to allow for the embryo to position on top of the egg yolk.
3. Wipe the egg with 70 % ethanol and crack it on a sharp edge. Transfer the whole egg content into the domed dish without destroying the egg yolk (Fig. 3).
4. Cover the culture with a lid from a petri dish to minimize evaporation and keep the ex ovo cultures in the incubator until further use.

3.4 Ex Ovo Injection and Electroporation

Here we describe injection and electroporation of the developing cerebellum of embryos at E8 as an example and as previously described [3]. Other areas of the brain can be targeted with the same parameters as a starting point.

1. Use autoclaved tools and clean the work space with 70 % ethanol.
2. Stage the embryos according to Hamburger and Hamilton [14].

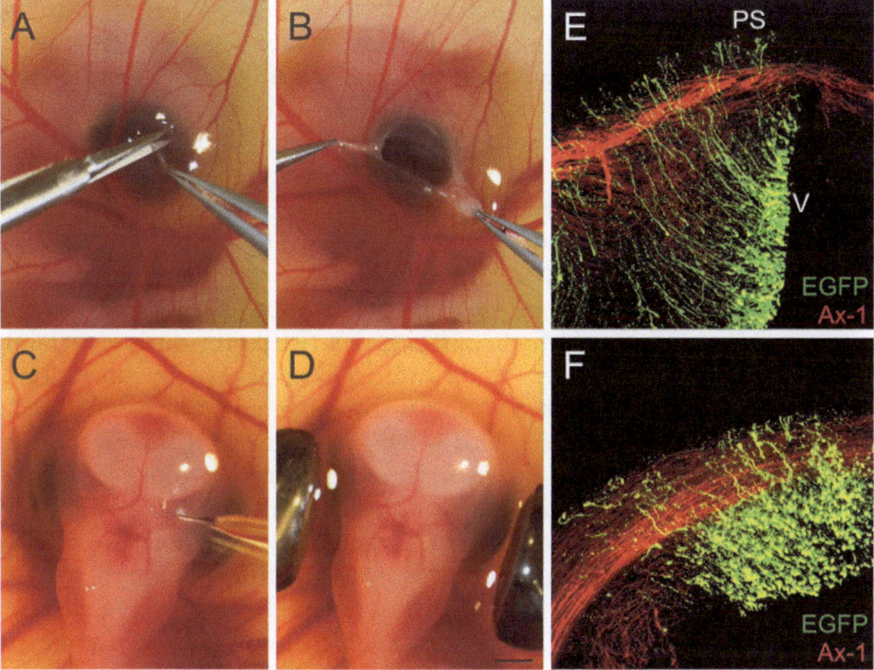

Fig. 4 Ex ovo electroporation. (**a** and **b**) Using spring scissors and forceps, a small hole is cut into the extraembryonic membranes in order to get access to the desired injection site. In the example shown here, the injection would be planned into the eye. (**c**) For injections into the developing cerebellum, the blood vessels on the back of the head can be used as landmarks. (**d**) The tweezer electrodes are placed parallel to the head for electroporation. Importantly, electrodes should not touch the embryo. Scale bar in (**d**): 2 mm. (**e**) and (**f**) Two examples of successful electroporation of the cerebellum. Coronal sections are shown with EGFP-positive cells (*green*) indicating successful transfection. Axonin-1/contactin-2 is a marker for parallel fibers, axons of granule cells (*red*). *V* ventricle, *PS* pial surface

3. Remove the lid and cut a small hole of 3–4 mm diameter into the extraembryonic membranes above the head, where the injection is planned (Fig. 4).
4. Depending on the age of the embryo and the injection site, it may be required to fix the head of the embryo with a spatula bent to a hook by placing it underneath the neck.
5. Use a glass capillary connected to a piece of tubing to inject the mix (*see* **Notes 10–13**) into the cerebellum (Fig. 4c). To target all the cerebellar layers, insert the glass capillary first into the ventricular zone and apply constant pressure while pulling the needle out (*see* **Note 18**).
6. Put a few drops of sterile PBS on the injection site and place the platelet electrodes on either side of the head of the embryo. Avoid touching embryonic tissue with the electrodes to prevent damage. Use six pulses of 40 V and 99 ms duration with 1 s interpulse intervals (Fig. 4d; *see* **Note 16**).
7. Cover the dish and put it back to the incubator until the desired stage for analysis is reached.

3.5 Verification of Gene Silencing

First, electroporation efficiency has to be high to get strong downregulation of the gene of interest. The best way to monitor successful targeting of the tissue is to check for the expression of fluorescent proteins (Figs. 2d and 4e, f; *see* **Note 19**). A first impression can be obtained by simply looking at the embryo in ovo or ex ovo under a fluorescent stereomicroscope. The embryos expressing the reporter protein can then be further analyzed for verification of gene knockdown. The efficacy of electroporation at late stages in ex ovo cultures has to be verified after dissection [3].

There are several possibilities to analyze gene silencing. Immunohistochemistry as described previously is certainly the best way to verify the downregulation of protein expression [1, 4, 10, 15]. Lysates from spinal cord or brain tissue can be used for analysis by Western blot [15]. Alternatively, if there are no antibodies available, in situ hybridization on cryosections [4, 10] or maybe on whole-mount embryos at very young stages [16, 17] can monitor expression levels of the targeted mRNA. Also RT-PCR can serve to demonstrate efficient downregulation, although spatial information is lost [18]. The nontargeted half of the spinal cord or the cerebellum may serve as internal control when compared to the electroporated hemisegment (*see* **Note 20**).

3.6 Phenotype Analysis

Depending on the target gene and its function, the methods of phenotype analysis will vary. It is not within the scope of this chapter to list methods for immunohistochemistry, axonal tracing [1, 19], or analysis of cell migration. No matter what kind of analysis you will use, keep in mind that phenotypes can only be assessed by comparison of your experimental group with *two* control groups (*see* **Note 20**).

4 Notes

1. Eggs have to be windowed no later than the third day of incubation even when the injection and electroporation are made at later stages. After 3 days of development, the blood vessels no longer detach from the eggshell and will therefore be damaged during windowing.
2. Eggs can be stored at 15 °C for up to 1 week before incubation. Longer storage will impair development and viability of the embryos.
3. After windowing, the eggs need to be sealed properly. Loss of humidity from the egg will strongly decrease the viability of the embryo. Furthermore, high humidity in the incubator is crucial (~45 %). Place a tray of water containing 0.1 g/l of copper sulfate into the incubator. Copper sulfate prevents contamination of the water.

4. We routinely use two incubators, one to incubate eggs before they are windowed or used for ex ovo cultures and another one for windowed eggs or dishes. Avoid opening incubators too many times during an experiment, as temperature fluctuations, especially going up to ~40 °C and more, will have negative effects on embryo development and strongly decrease survival rates.
5. Staging of the embryos at the beginning and at the time of analysis is very important. Careful comparison of the developmental progress between experimental and control embryos that have not been handled can give important information about potential interference of embryo handling and/or injection and electroporation procedure with normal development.
6. To reduce the risk of contamination, always clean your work space with 70 % ethanol and use autoclaved tools and sterile solutions. Additionally, keep the time during which the egg is unsealed to a minimum. Working in a laminar flow hood is not necessary, however.
7. Drilling a hole into the corner of the intended window is necessary for allowing airflow during removal of albumen which will result in the detachment of the embryo from the eggshell.
8. Damaging the yolk will result in death of the embryo and therefore has to be avoided for both in ovo and ex ovo development.
9. If the eggs need to be reopened several times, sealing with a coverslip is advantageous since the paraffin can be quickly melted by placing a soldering iron on the coverslip. No matter whether you seal the egg with Scotch tape or a coverslip, make sure the window is properly closed, because dehydration will dramatically lower survival rate.
10. The diameter of the injection needle should be kept at 5–7 μm to minimize tissue damage and to prevent leakage of the injection mix upon retraction of the needle. It is important to have no leakage in order to get efficient and reproducible gene silencing.
11. Make sure that salt concentrations and pH of the injection mix are in the physiological range. Plasmids used in the mix should be purified carefully by making sure that there is no alcohol remaining from previous precipitation steps. Tris buffers should be avoided as they tend to cause unspecific, toxic effects.
12. DNA injection mix: RNAi plasmid at a concentration that must be determined by the user and trypan blue diluted in PBS (for detailed designing of RNAi plasmids and concentration of

ingredients of the injection mix, *see* [4]). We design artificial miRNAs using GenScript's siRNA Target Finder http://www.genscript.com/siRNA_target_finder.html.

13. dsRNA injection mix: sterile PBS containing dsRNA derived from the gene of interest (200–400 ng/μL), EGFP reporter plasmid (20 ng/μL), and 0.04 % (vol/vol) trypan blue (for detailed description of dsRNA synthesis and troubleshooting advice, *see* [1, 5]). Long dsRNA is easily produced by in vitro transcription from a cDNA template. Expressed sequence tags (ESTs) are available from ARK Genomics or Source BioScience Life Sciences.

14. For targeting dorsal or ventral cell types, electrodes can be positioned accordingly along the body of the embryo (*see* Fig. 2c). Make sure not to touch any major blood vessels during electroporation because this will lead to fatal bleeding. Keep the electrodes away from the heart.

15. For in ovo spinal cord electroporation, we use custom made wire electrodes. There are also commercially available ones from BTX, Harvard Apparatus (http://www.btxonline.com/genetrodes/).

16. Make sure to clean the electrodes from denatured proteins *after each electroporation* to maintain a proper electric field between the electrodes for the following embryos.

17. For best survival rate the eggs should be cracked at E2.5. E2 works also but older stages will not give good survival rates.

18. Since injection depth and volume cannot be seen easily at older stages of development, we recommend using injection of a mix of a dye (DiI, CFSE (carboxyfluorescein succinimidyl ester)) or fluorescent dextran beads together with an EGFP reporter plasmid to establish the landmarks that can be used to guide injection. For instance, for the cerebellar anlage, the blood vessels on the back of the head can be used as landmarks [3].

19. If you have low transfection rates, consider the following points:

    - Make sure the electrodes are positioned close enough to the embryo without actually touching it.
    - The injection volume should be maximized but kept small enough to avoid tissue damage or leakage. It is absolutely important to prevent leakage of the injection mix before electroporation in order to get reproducible and effective gene silencing.
    - Take into account that in highly proliferating cells, the electroporated plasmid will be diluted strongly over time. In this case, reconsider the time point of injection and electroporation.

- Finally, injection and electroporation require extensive practice to obtain adequate manual skills for an efficient downregulation of the targeted mRNA.

20. Note depending on time of electroporation and position of the electrodes, it is not possible to compare the two halves of the embryo, as both sides would be affected by the manipulation. Therefore it is important to have adequate controls. *At least two control groups are always required.* One group consists of untreated embryos taken out of unwindowed eggs at the time of analysis. The second control group consists of mock-treated embryos. These are handled exactly the same way as the experimental group(s) but without dsRNA derived from the gene of interest. In these embryos either only a reporter plasmid or dsRNA from a gene that is not expressed in the nervous system is injected. The best control would be dsRNA- or miRNA-based constructs targeting a family member of the gene of interest. This is of course not always available. The treated control group and the embryos from the untreated group have to be indistinguishable to make sure that the handling of the embryos did not cause any artifacts that may be mistaken as a phenotype caused by silencing the target gene. The comparison between the control-treated and the experimental group will provide the desired result indicating the function off the gene of interest.

Acknowledgments

We thank Evelyn Avilés and Nicole Wilson for their help in Figs. 2 and 4. Work in the laboratory of E.S. is supported by a grant from the Swiss National Science Foundation.

References

1. Pekarik V, Bourikas D, Miglino N, Joset P, Preiswerk S, Stoeckli ET (2003) Screening for gene function in chicken embryo using RNAi and electroporation. Nat Biotechnol 21:93–96
2. Chesnutt C, Niswander L (2004) Plasmid-based short-hairpin RNA interference in the chicken embryo. Genesis 39:73–78
3. Baeriswyl T, Stoeckli ET (2008) Axonin-1/TAG-1 is required for pathfinding of granule cell axons in the developing cerebellum. Neural Dev 3:7
4. Wilson NH, Stoeckli ET (2011) Cell type specific, traceable gene silencing for functional gene analysis during vertebrate neural development. Nucleic Acids Res 39:e133
5. Baeriswyl T, Mauti O, Stoeckli ET (2008) Temporal control of gene silencing by in ovo electroporation. Methods Mol Biol 442:231–244
6. Mauti O, Baeriswyl T, Stoeckli ET (2008) Gene silencing by injection and electroporation of dsRNA in Avian embryos. CSH Protoc 2008:pdb.prot5094
7. Luo J, Redies C (2005) Ex ovo electroporation for gene transfer into older chicken embryos. Dev Dyn 233:1470–1477
8. Mauti O, Sadhu R, Gemayel J, Gesemann M, Stoeckli ET (2006) Expression patterns of plexins and neuropilins are consistent with cooperative and separate functions during neural development. BMC Dev Biol 6:32

9. Das RM, van Hateren NJ, Howell GR, Farrell ER, Bangs FK, Porteous VC, Manning EM, McGrew MJ, Ohyama K, Sacco MA et al (2006) A robust system for RNA interference in the chicken using a modified microRNA operon. Dev Biol 294:554–563
10. Bourikas D, Pekarik V, Baeriswyl T, Grunditz A, Sadhu R, Nardó M, Stoeckli ET (2005) Sonic hedgehog guides commissural axons along the longitudinal axis of the spinal cord. Nat Neurosci 8:297–304
11. Domanitskaya E, Wacker A, Mauti O, Baeriswyl T, Esteve P, Bovolenta P, Stoeckli ET (2010) Sonic hedgehog guides post-crossing commissural axons both directly and indirectly by regulating Wnt activity. J Neurosci 30:11167–11176
12. Wilson NH, Stoeckli ET (2013) Sonic hedgehog regulates its own receptor on post-crossing commissural axons in a Glypican-dependent manner. Neuron in press
13. Niederkofler V, Baeriswyl T, Ott R, Stoeckli ET (2010) Nectin-like molecules/SynCAMs are required for post-crossing commissural axon guidance. Development 137:427–435
14. Hamburger V, Hamilton HL (1992) A series of normal stages in the development of the chick embryo. 1951. Dev Dyn 195:231–272
15. Rao M, Baraban JH, Rajaii F, Sockanathan S (2004) In vivo comparative study of RNAi methodologies by in ovo electroporation in the chick embryo. Dev Dyn 231:592–600
15. Stepanek L, Stoker AW, Stoeckli E, Bixby JL (2005) Receptor tyrosine phosphatases guide vertebrate motor axons during development. J Neurosci 25:3813–3823
16. Katahira T, Nakamura H (2003) Gene silencing in chick embryos with a vector-based small interfering RNA system. Dev Growth Differ 45:361–367
17. Dai F, Yusuf F, Farjah GH, Brand-Saberi B (2005) RNAi-induced targeted silencing of developmental control genes during chicken embryogenesis. Dev Biol 285:80–90
18. Sato F, Nakagawa T, Ito M, Kitagawa Y, Hattori M-A (2004) Application of RNA interference to chicken embryos using small interfering RNA. J Exp Zool A Comp Exp Biol 301:820–827
19. Perrin FE, Stoeckli ET (2000) Use of lipophilic dyes in studies of axonal pathfinding in vivo. Microsc Res Tech 48:25–31

Part VII

Mouse Protocols

Chapter 18

Immunohistochemistry and RNA In Situ Hybridization in Mouse Brain Development

Jinling Liu and Aimin Liu

Abstract

During development, the mouse brain is progressively divided into functionally distinct compartments. Numerous neuronal and glial cell types are subsequently generated in response to various inductive signals. Each cell expresses a unique combination of genes encoding proteins from transcription factors to neurotransmitters that define its role in brain function. To understand these important and highly sophisticated processes, it is critical to accurately locate the various proteins and cells that produce them. In this chapter, we introduce the techniques of immunohistochemistry, which detects the localization of specific proteins, and RNA in situ hybridization, which enables the visualization of specific mRNAs.

Key words Immunohistochemistry, RNA in situ hybridization, Cryosection, Antibody, Digoxigenin, Fluorescent

1 Introduction

The mouse brain consists of multiple divisions (cerebrum, epithalamus, thalamus, hypothalamus, cerebellum, and brain stem) and close to 100 million neurons and glia [1]. Extrinsic inductive signals and intrinsic cellular programs both play key roles in the compartmentalization of the brain as well as cellular behaviors such as proliferation, differentiation, migration, and cell death. To better understand the developmental processes involved in mouse brain development, it is important to obtain information regarding the spatial and temporal patterns of gene expression. In this chapter, we will describe methods for the detection of protein (immunohistochemistry, IHC) and mRNA (RNA in situ hybridization, RISH) in brain sections.

IHC detects particular proteins present in tissues. The principle underlying this technique is the specific bindings between antibodies and antigens. To visualize the antibody-antigen interaction, an antibody is tagged with a fluorophore, which can be conveniently detected with a fluorescent microscope. Alternatively,

Simon G. Sprecher (ed.), *Brain Development: Methods and Protocols*, Methods in Molecular Biology, vol. 1082,
DOI 10.1007/978-1-62703-655-9_18,

the antibody can be conjugated to an enzyme that catalyzes a color-producing reaction, which can be visualized under a regular microscope. In this chapter we will describe the method of IHC with the fluorophore-labeled antibody. The procedure comprises tissue preparation, blocking, primary and secondary antibody incubation, mounting, and visualization.

The success of IHC heavily depends on the availability of high-quality antibodies. In addition, secreted signaling proteins as well as proteins of the extracellular matrix are not restricted to the cells producing them, preventing a high-resolution identification of the signaling centers. On the other hand, RISH allows the detection of the expression of virtually all genes in the cells, even genes that do not encode proteins, providing more flexibility compared with the antibody-based IHC method. Traditionally, RISH depends on the hybridization of the specific RNA sequence in situ to radiolabeled probes [2]. Currently, Digoxigenin-labeled probes are more commonly used in RISH, which can be recognized with antibodies coupled with fluorophore or enzymes such as alkaline phosphatase or peroxidase [3].

RISH can be performed on both frozen sections and paraffin sections, with frozen sections allowing more sensitive detection of weak signals [4]. The RISH method we introduce in this chapter uses Digoxigenin-labeled riboprobes (complementary RNA probes) to detect specific mRNA on frozen brain sections. The procedure includes synthesis of riboprobes, hybridization of sections with Digoxigenin-labeled riboprobes, post-hybridization washes, incubation with alkaline phosphatase (AP)-conjugated anti-Digoxigenin antibody, and a color reaction using the phosphatase substrate BM purple solution.

2 Materials

The materials should be stored at room temperature unless otherwise specified.

2.1 Cryosection Preparation

1. Dissection tools: student quality iris scissors (Fine Science Tools), Dumont forceps (Fine Science Tools), and spoon (Fine Science Tools).
2. 6 cm petri dishes.
3. 24-well tissue culture plates (*see* **Note 1**).
4. Stereomicroscope, such as Nikon SMZ645.
5. 4 % paraformaldehyde (PFA): Prepare 16 % stock by adding 32 g PFA powder and 100 μl 5 N NaOH in 150 ml pre-warmed distilled, deionized water (ddH_2O, at ~65 °C) (*see* **Note 2**). Once PFA is dissolved, add 20 ml 10× PBS and adjust the volume to 200 ml with ddH_2O. Filter through Whatman

paper and make 10 ml aliquot in 50 ml centrifuge tubes. Store at −20 °C. Thaw one tube at ~65 °C and dilute to 4 % PFA with PBS before use. 4 % PFA can be stored at 4 °C for up to 1 week (*see* **Note 3**).

6. Phosphate-buffered saline (PBS): Prepare 10× stock by dissolving 80 g NaCl, 2 g KCl, 14.4 g Na_2HPO_4 and 2.4 g KH_2PO_4 in 1 l of ddH_2O. Adjust pH to 7.4 and autoclave at 121 °C for 25 min. Dilute to 1× solution with ddH_2O for use.
7. Nutator mixer (VWR).
8. 30 % sucrose: Dissolve 12 g sucrose powder in 40 ml PBS. Store at 4 °C.
9. O.C.T. compound (VWR): Store at 4 °C.
10. Disposable embedding molds, such as Polysciences.
11. Dry ice or liquid nitrogen in appropriate containers.
12. Cryostat, such as Leica CM1900.
13. Superfrost Plus microscopic slides, such as VWR (*see* **Note 4**).
14. Slide boxes, such as VWR.
15. Micro slide trays, such as VWR.

2.2 Immunohistochemistry

1. Coplin jars, such as Wheaton.
2. Blocking buffer: 1 % normal goat serum, 0.1 % Triton X-100, in PBS. Keep at 4 °C (*see* **Note 5**).
3. A humidified slide incubation chamber (Fig. 1): Cut two 5 ml serological pipettes and tape them to the bottom of a flat-bottom plastic box with lid, such that the two pipettes are parallel and 5 cm apart. Place paper towels soaked with ddH_2O on the bottom to keep a moist environment.
4. Paper towels.
5. Fluorophore-conjugated secondary antibodies: Store at −20 °C in 50 % glycerol.
6. Micro slide trays, such as VWR.
7. Coverslips, such as VWR, 25 × 50 mm.
8. Forceps: Dumont forceps (Fine Science Tools).
9. Dabco: Aldrich. Store at 4 °C (*see* **Note 6**).
10. Nail polish.

2.3 Synthesis of RNA Probes

1. A plasmid with the promoters for viral RNA polymerases (T3, T7, and Sp6) flanking the multiple cloning sites. Store at −20 °C.
2. Restriction endonucleases: Store at −20 °C.
3. Horizontal DNA electrophoresis apparatus and power supply.
4. Agarose.

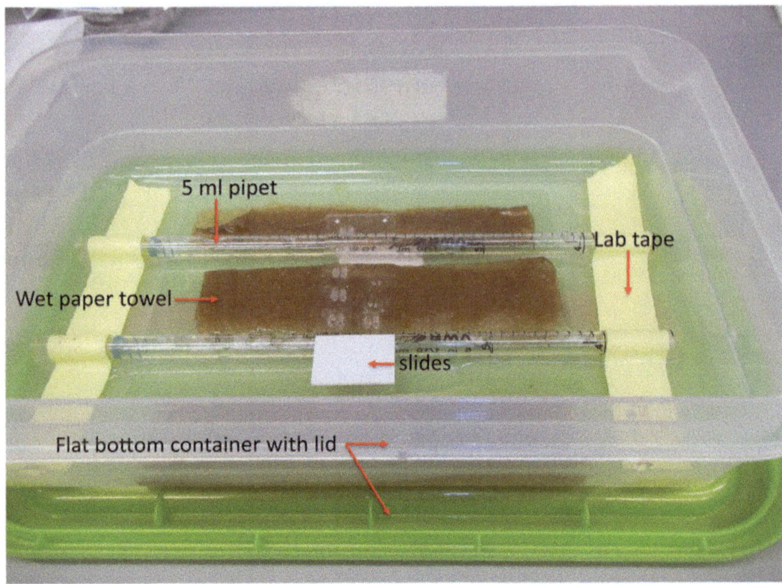

Fig. 1 A homemade humidified slide incubation chamber. This chamber is made by taping two 5 ml serological pipettes, broken to appropriate length, to the bottom of a plastic container with lid. Paper towels soaked in ddH_2O provide humidity during the incubation. For RISH, a microwavable box or a container that can withstand 55 °C temperature is needed

5. 10 mg/ml ethidium bromide solution: Keep away from light (*see* **Note 7**).
6. TBE buffer: Dissolve 54 g Tris base, 27.5 g boric acid in ddH_2O, add 20 ml 0.5 M EDTA (pH 8.0), and bring the volume to 5 l with ddH_2O.
7. 1-kb DNA ladder.
8. 3 M sodium acetate (NaOAc), pH 5.2: Dissolve 408.3 g of $NaOAc{\cdot}3H_2O$ in 800 ml ddH_2O. Adjust pH to 5.2 with acetic acid and adjust the volume to 1 l with ddH_2O. Aliquot and autoclave.
9. Phenol saturated with Tris–HCl, pH 8.0: Store at 4 °C.
10. Chloroform.
11. 70 % and 100 % ethanol.
12. Tabletop microcentrifuge, such as Eppendorf 5415D.
13. Magnetic stir plate.
14. Diethylpyrocarbonate (DEPC): Store at 4 °C (*see* **Note 8**).
15. DEPC-H_2O: Add 0.1 % v/v DEPC to ddH_2O and mix o/n on a magnetic stir plate at room temp; autoclave at 121 °C for 25 min (*see* **Note 9**).
16. 10× transcription buffer (Roche): Store at −20 °C.

17. 10× DIG-labeling mix (Roche): Store at –20 °C (*see* **Note 10**).
18. RNase inhibitor (Promega, 40 U/μl): Store at –20 °C.
19. RNA polymerase (Roche T3, T7, or Sp6, 20 U/μl): Store at –20 °C.
20. DNaseI (RNase-free): Store at –20 °C.

2.4 RNA In Situ Hybridization

The amount of reagents in this section is enough for processing five slides in a 50 ml Coplin jar and should be adjusted according to the number of slides and size of the container used.

2.5 RNA In Situ Hybridization Day 1: Hybridization of Cryosections on the Slides

All containers and reagents need to be RNase-free on day 1.

1. Coplin jars: 50 ml.
2. Diethylpyrocarbonate (DEPC): Store at 4 °C (*see* **Note 8**).
3. Magnetic stir plate.
4. DEPC-H_2O (400 ml): Add 400 μl DEPC to 400 ml ddH_2O. Mix o/n on a magnetic stir plate at room temperature and autoclave at 121 °C for 25 min (*see* **Note 9**).
5. DEPC-PBS (400 ml): Add 400 μl DEPC to 400 ml PBS. Mix o/n on a magnetic stir plate at room temperature and autoclave at 121 °C for 25 min (*see* **Note 9**).
6. 4 % PFA in DEPC-PBS (80 ml): Thaw two 10 ml aliquots of 16 % PFA (*see* Subheading 2.1 for 16 % PFA preparation) and dilute each with 30 ml DEPC-PBS. Store at 4 °C.
7. 0.25 % acetic anhydride in 0.1 M TEA-HCl (40 ml): Add 0.742 g triethanolamine-HCl and 360 μl 5 N NaOH to 39.8 ml DEPC-H_2O to make TEA-HCl. Add 100 μl acetic anhydride right before use (*see* **Note 11**).
8. 20 μg/ml proteinase K (proK; Roche #03115879001): Dissolve one vial of proK (100 mg) in 5 ml DEPC-H_2O to make 20 mg/ml stock. Aliquot and store at –20 °C. Thaw an aliquot before use and add 40 μl into 40 ml DEPC-PBS (*see* **Note 12**).
9. 70 % ethanol in DEPC-H_2O (40 ml): 28 ml 100 % ethanol, 12 ml DEPC-H_2O.
10. 95 % ethanol in DEPC-H_2O (40 ml): 38 ml 100 % ethanol, 2 ml DEPC-H_2O.
11. Hybridization solution (40 ml): Mix the following in a clean 50 ml tube, aliquot, and store at –20 °C (*see* **Note 13**). 20 ml formamide (deionized, aliquot in 50 ml tubes, and store at –20 °C); 8 ml 50 % dextran sulfate; 400 μl 100× Denhardt's (VWR, aliquot in 1.5 ml tubes and store at –20 °C); 1 ml yeast tRNA (10 mg/ml, aliquot in 1.5 ml tubes, store at –20 °C); 2.4 ml 5 M NaCl; 800 μl 1 M Tris–HCl (pH 8.0); 400 μl 0.5 M EDTA; 400 μl 1 M $NaPO_4$; 4 ml 10 % sarkosyl; 2.6 ml DEPC-H_2O.

12. Parafilm (cut into 20 mm × 50 mm pieces) or RNase-free plastic coverslips.
13. A hybridization oven.
14. A humidified slide incubation chamber: *see* Subheading 2.2.

2.6 RNA In Situ Hybridization Day 2: Post-hybridization Washes and Antibody Incubation

Starting from day 2, reagents do not need to be RNase-free.

1. 20× saline-sodium citrate (SSC) buffer: Mix the following in an appropriate container: 800 ml ddH_2O; 175.3 g NaCl; 88.2 g sodium citrate. Adjust pH to 7.0 with a few drops of 12 N HCl. Adjust the volume to 1 l with ddH_2O. Aliquot and autoclave at 121 °C for 25 min.
2. 5× SSC (40 ml): 10 ml 20× SSC, 30 ml ddH_2O.
3. 2× SSC (40 ml): 4 ml 20× SSC, 36 ml ddH_2O.
4. 0.1× SSC (40 ml): 0.2 ml 20× SSC, 39.8 ml ddH_2O.
5. High-stringency wash buffer (120 ml) (make fresh): 60 ml formamide, 12 ml 20× SSC, 48 ml ddH_2O.
6. PBT (350 ml, enough for both days 2 and 3): Add 350 μl Tween 20 to 350 ml PBS in a 500 ml bottle. Mix well by shaking vigorously. Alternatively, 10 % Tween 20 stock can be made in ddH_2O in advance and stored at room temperature.
7. RNase buffer (400 ml) (*see* **Note 14**): 40 ml 5 M NaCl; 4 ml 1 M Tris–HCl, pH 7.5; 4 ml 0.5 M EDTA, pH 8.0; 352 ml ddH_2O.
8. 10 mg/ml RNase A stock (DNase-free): Add 10 ml 0.01 M NaOAc (pH 5.2) to a vial of 100 mg RNase A. Heat at 100 °C for 15 min. Cool to room temperature. Add 1 ml 1 M Tris–HCl (pH 7.5) and aliquot. Store at −20 °C (*see* **Note 15**). On the day of experiment, thaw one aliquot and make 20 μg/ml RNase A in RNase buffer by adding 80 μl 10 mg/ml RNase A stock to 40 ml RNase buffer.
9. Alkaline phosphatase-conjugated anti-Digoxigenin antibody (Fab fragments from sheep; Roche). Store at 4 °C.
10. A humidified slide incubation chamber: *See* Subheading 2.2.

2.7 RNA In Situ Hybridization Day 3: Color Reaction

1. NTMT (80 ml) (make fresh; *see* **Note 16**): 8 ml 1 M Tris–HCl 9.5; 4 ml 1 M $MgCl_2$; 1.6 ml 5 M NaCl; 800 μl 10 % Tween 20; 40 mg levamisole; adjust to 80 ml with ddH_2O.
2. Levamisole: make 50 mg/ml stock in ddH_2O, aliquot, and store in 1.5 ml tubes at −20 °C.
3. BM purple (Roche). Store at 4 °C.
4. Water-based mounting medium, such as Mount Quick aq. (SPI). Store at 4 °C.
5. Coverslips: VWR micro cover glass, 25 × 50 mm.

6. Nuclear fast red solution (100 ml): Dissolve 5 g aluminum sulfate in 100 ml ddH_2O, then add 0.1 g nuclear fast red. Boil and stir on a heated magnetic stir plate to dissolve nuclear fast red. Filter the solution right before use.

3 Methods

Conduct all procedures at room temperature unless otherwise specified.

3.1 Mouse Brain Cryosection Preparation

1. Dissect embryos in ice-cold PBS in a 6 cm dish (*see* **Note 17**).
2. Cut the embryos at the shoulder level and transfer the heads into a 24-well plate with a spoon.
3. Fix the embryos in 4 % PFA on a nutator at 4 °C for 1 h for IHC or o/n for RISH.
4. Rinse the embryos with PBS and then wash the embryos in PBS on a nutator o/n at 4 °C.
5. Immerse the embryos in 30 % sucrose o/n at 4 °C, on a nutator (*see* **Note 18**).
6. Change the 30 % sucrose and incubate for another 2–3 h for further infiltration at 4 °C, on a nutator.
7. Transfer the embryos into a disposable embedding mold and immerse the embryos in O.C.T. compound for 1 h at 4 °C (*see* **Note 19**).
8. Position the samples at desired orientation and freeze them on dry ice or, alternatively, in a paperboard box floating on the surface of liquid nitrogen. Wait for 5 min (*see* **Note 20**).
9. Transfer the frozen O.C.T. block containing the brain samples to the cryostat and wait at least 1 h so that the temperature of the block can reach the optimal cutting temperature (*see* **Note 21**).
10. Cut 10 μm sections using a cryostat (*see* **Note 22**). Collect sections on Superfrost Plus slides.
11. Dry the slides in a micro slide holder for 1 h at room temperature. Store them in a slide box at –80 °C (*see* **Note 23**).

3.2 Immunostaining on Mouse Brain Sections

1. Remove sections from –80 °C freezer and dry slides in a micro slide holder at room temperature for about 45 min (*see* **Note 24**).
2. Place the slides in a Coplin jar and incubate the sections with blocking buffer for 1 h.
3. Take slides out of blocking buffer; wipe the backside (the one without sections) and edges with paper towel. Place them with the front side (the one with sections) up on the pipettes of the humidified slide incubation chamber (Fig. 1; *see* **Note 25**).

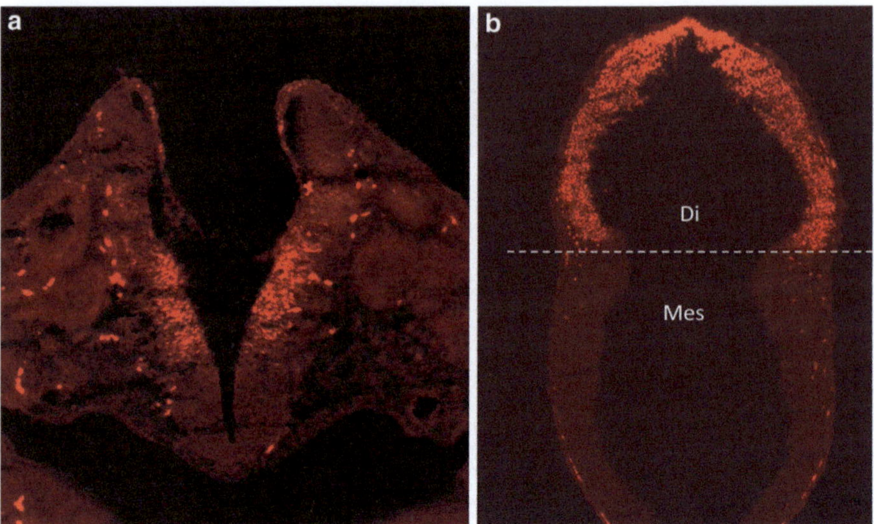

Fig. 2 Immunostaining of brain sections. Shown are two coronal sections of an E12.5 brain incubated with a monoclonal antibody against Pax6 and a Cy3-conjugated secondary antibody. (**a**) Pax6 is weakly expressed in the dorsal region and strongly expressed in the intermediate region in the hindbrain. (**b**) Pax6 is expressed in the diencephalon (Di), but not the mesencephalon (mes). *Dashed line* demarcates the boundary between the diencephalon and mesencephalon

4. Apply primary antibodies to the sections (diluted in blocking buffer, 300 μl per slide) and incubate o/n at 4 °C (*see* **Note 26**).
5. Pour primary antibody onto a paper towel and place the slides in a Coplin jar. Wash the slides with blocking buffer, 3 × 10 min.
6. Similar to **step 3**, place slides in the humidified chamber and incubate in the dark with appropriate fluorophore-conjugated secondary antibody (diluted in blocking buffer, 300 μl per slide) for 2 h (*see* **Note 27**).
7. Pour secondary antibody onto a paper towel and place the slides in a Coplin jar. Wash with blocking buffer, 3 × 10 min.
8. Take slides out of blocking buffer and wipe the backside and edges with paper towel. Place them into a micro slide tray.
9. Apply 45 μl Dabco evenly onto each slide and mount with a coverslip (*see* **Notes 28** and **29**); seal the slides with nail polish.
10. Observe the sections under fluorescent microscope and take photos with a cooled CCD camera. Some examples are shown in Fig. 2.
11. Slides can be stored at 4 °C protected from light (*see* **Note 30**).

3.3 Synthesis of RNA Probes for RISH

1. Clone the cDNA (or part of the cDNA if the full-length cDNA is longer than 1,500 bps; *see* **Note 31**) of the gene of interest into a plasmid containing the promoters for the viral RNA polymerases (T3, T7, and Sp6; such as pBluescript).

2. Linearize the template:
   Choose a unique restriction site in the multiple cloning sites on the 5′ end of the cDNA. Cut ~10 μg DNA in 20–50 μl reaction with the corresponding restriction endonuclease o/n using at least 20 IU enzyme. Run small amount (0.1–0.5 μg) in a 0.8 % agarose gel to check the efficiency of restriction enzyme digestion.
3. Purify the linearized DNA:
   Bring the volume of above reaction to 200 μl with ddH_2O, add 20 μl 3 M NaOAc (pH 5.2), and mix well. Perform standard phenol/chloroform and chloroform extraction. Add 600 μl 100 % ethanol to the purified DNA, mix well, and leave at –80 °C for 30 min. Centrifuge at top speed (>14,000 × *g*) in a tabletop centrifuge for 10 min at 4 °C. Rinse the pellet with 70 % ethanol without disturbing it. Dry the pellet briefly and dissolve it in 10 μl DEPC-H_2O. The linearized template can be stored at –20 °C.
4. In vitro transcription:

2 μl	10× transcription buffer
3 μl	DIG-labeling mix
2 μl	Linearized template DNA
0.5 μl	RNase inhibitor
2 μl	RNA polymerase (T3, T7, or Sp6; use the one whose promoter is at the 3′ end of the cDNA)
10.5 μl	DEPC-H_2O

Incubate at 37 °C for 2 h.

5. Run 1 μl in 1 % agarose gel for 20 min to 1 h to check the yield of the probe.
6. DNase I treatment (optional):
   Add 1 μl DNaseI and 1 μl RNase inhibitor to the RNA probe and incubate at 37 °C for 15 min to remove the template.
7. Add 180 μl DEPC-H_2O, 20 μl 3 M NaOAc, and mix well. Then add 600 μl 100 % ethanol, mix well, and leave at –80 °C for 30 min.
8. Centrifuge at top speed (>14,000 × *g*) in a tabletop centrifuge for 15 min at 4 °C (*see* **Note 32**).
9. Discard the supernatant and rinse the pellet with 70 % ethanol once without disturbing the pellet.
10. Discard 70 % ethanol and dry the pellet for 5 min. Dissolve the RNA probe in 40–50 μl DEPC-H_2O and store at –80°C.

3.4 RNA In Situ Hybridization of Mouse Brain Sections

Day 1: *hybridization of cryosections* (all steps are carried out in Coplin jars unless otherwise specified; avoid RNase contamination)

1. Dry slides at room temperature for about 45 min (*see* **Note 24**).
2. Postfix slides in 4 % PFA in DEPC-PBS for 10 min (*see* **Note 33**).
3. Wash with DEPC-PBS, 2 × 5 min.
4. Drain excess DEPC-PBS and incubate for 6 min in 20 μg/ml proteinase K in DEPC-PBS (*see* **Note 34**).
5. Drain and wash with DEPC-PBS for 5 min.
6. Refix in 4 % PFA for 5 min, then wash 5 min in DEPC-PBS (*see* **Note 35**).
7. Acetylate sections with acetic anhydride in 0.1 M TEA-HCl for 10 min (*see* **Note 36**).
8. Wash in DEPC-PBS for 5 min; dehydrate in 70 % ethanol for 5 min and 95 % ethanol for 2 min. Air dry for 30 min to 2 h.
9. Add 2 μl appropriate probe (approx. 1 μg) to 1 ml hybridization solution and heat at 80 °C for 2 min (*see* **Note 37**).
10. Place slides horizontally in a humidified slide incubation chamber. Cover sections with 200 μl of hybridization solution with the probe and lower parafilm coverslips over sections avoiding bubbles (*see* **Note 38**).
11. Seal the slide incubation chamber carefully and hybridize at 55 °C in a hybridization oven overnight (16–18 h).

Day 2: post-hybridization washes and antibody incubation (all steps are carried out in Coplin jars; RNase-free environment is not required)

1. Dip slides gently in a Coplin jar filled with 5× SSC to let the coverslips float off the slides (*see* **Note 39**).
2. Incubate the sections in high-stringency wash at 65 °C in a Coplin jar for 30 min (*see* **Note 40**).
3. Wash in RNase buffer at 37 °C, 3 × 10 min.
4. Wash in RNase buffer with 20 μg/ml RNase A at 37 °C for 30 min (*see* **Note 41**).
5. Wash in RNase buffer at 37 °C for 15 min.
6. Repeat high-stringency wash (as in **step 2**) at 65 °C, 2 × 20 min.
7. Wash in 2× SSC, then 0.1× SSC for 15 min each at 37 °C.
8. Wash with PBT for 15 min.
9. Take the slide out of PBT; wipe the backside and edges of the slide with paper towel. Place slides horizontally in a humidified slide incubation chamber and block for 1 h with 10 % goat serum in PBT (300 μl per slide).

10. Pour the blocking buffer onto paper towels. Wipe the backside and edges of the slide with paper towel. Incubate with AP-conjugated anti-Digoxigenin antibody (diluted 1/5,000 in PBT with 1 % goat serum, 320 μl per slide) at 4 °C overnight in the same humidified chamber.

Day 3: color reaction

1. Pour the antibody onto paper towels. Place slides in a Coplin jar and wash in PBT for five times, 1 h each.
2. Wash 2 × 10 min in freshly prepared NTMT buffer.
3. Wipe the backside and edges of the slide with paper towel. Place the slides horizontally in a humidified slide incubation chamber and incubate o/n to several days in BM purple solution (300 μl per slide) supplemented with 0.5 mg/ml levamisole in the dark (*see* **Note 42**).
4. Observe periodically the progress of the color reaction under a microscope. If the staining is not ready, reapply BM purple solution and incubate for longer time (*see* **Note 43**).
5. When the staining is ready, i.e., when the signal is strong and the background staining just begins to show, place slides back into a Coplin jar. Wash slides in PBS for 2–5 min and dip briefly in ddH_2O (*see* **Note 18**).
6. (Optional) Counterstain the sections with nuclear fast red until the sections turn slightly pink. Usually it takes 2–3 min.
7. Wash the excess nuclear fast red in slow-running tap water.
8. Wipe the backside and edges of the slide with paper towel. Apply mounting medium to the slides and put coverslips on (*see* **Note 44**).
9. Observe the staining under a microscope and take photos with a color camera. Some examples are shown in Fig. 3.
10. The mounted sections can be stored at 4 °C.

4 Notes

1. The 24-well plates are for the convenient storage of individual small sample, such as the brains of E12.5 embryos or younger. Vials of appropriate size should be used for older/bigger brain samples.
2. Distilled, deionized water (ddH_2O) used in this protocol is ultra-purified water with a resistance of 18.2 MΩ.
3. PFA is highly volatile and irritant. Prepare the 16 % PFA in the fume hood.

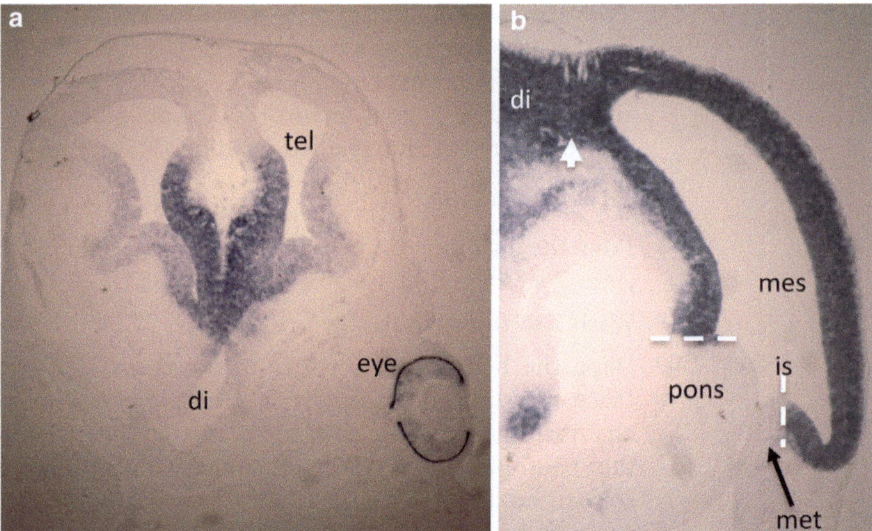

Fig. 3 RNA in situ hybridization of brain sections. Shown are two sections of E12.5 brains hybridized with a Digoxigenin-labeled RNA probe against Otx2. (**a**) A coronal section through the forebrain region shows Otx2 expression in ventral telencephalon (tel) and the retinal pigment epithelium of the eye, but not in ventral diencephalon (di). (**b**) A sagittal section shows that Otx2 is expressed in the mesencephalon (mes) and diencephalon (di), but not in the metencephalon (met) and pons. The *arrowhead* points to the boundary between diencephalon and mesencephalon. *Dashed lines* demarcate the boundary between the midbrain and hindbrain, known as the isthmus (is)

4. It is critical to use Superfrost Plus slides as they contain a special coating to prevent the sections from falling off the slides during the incubation and washing steps.
5. Make fresh blocking buffer for each experiment. Do not store for more than a week.
6. Dabco is irritant. Avoid contact with skin and eyes. Avoid inhalation of vapor or mist.
7. Ethidium is highly carcinogenic. Avoid direct contact and dispose ethidium-containing waste properly.
8. DEPC is toxic. Avoid direct contact.
9. All DEPC-treated solutions need to be autoclaved to degrade DEPC, which can react with RNA.
10. When thawing the DIG-labeling mix, avoid prolonged incubation at 37 °C to reduce the chance of NTP degradation.
11. Keep acetic anhydride from moisture (keep the container tightly closed all the time) and add acetic anhydride immediately before use.
12. Do not refreeze proK for RNA in situ hybridization. The activity of proK should be tested every time a new batch is introduced.

13. To make 50 % dextran sulfate, mix 10 ml dextran sulfate and 10 ml ddH_2O in a 50 ml tube by inverting, shaking, vortexing, and heating at 60 °C. When dextran sulfate starts to get into water, the total volume will decrease. Add more ddH_2O to keep the final volume at 20 ml.
14. Sterile RNase buffer can be stored at room temperature.
15. Take caution to avoid contaminating the bench and other lab materials with RNase A.
16. NTMT should be freshly prepared.
17. At this step, we usually remove the extraembryonic membranes or the tips of the tails to genotype the embryos.
18. The tissue should float on the surface of the 30 % sucrose initially and sink to the bottom of the well after sucrose has fully infiltrated the tissue. Make sure the tissue is completely immersed in the sucrose, as the morphology of tissues staying at the air/liquid interface can be distorted by surface tension.
19. The time the embryos immersed in O.C.T. depends on the size and density of the tissue. For large and/or dense tissue, longer time is needed.
20. Do not freeze samples by directly placing the embedding mold in liquid nitrogen.
21. If not cutting sections right away, wrap the embedding mold with parafilm and keep it at −80 °C for up to 1 week.
22. See manufacturer's manual for how to cut cryosections.
23. Cryosections can be stored for months at −80 °C without noticeable degradation of proteins and mRNAs.
24. Dried sections do not fall off the slides in subsequent experiments. In addition, drying creates holes in the subcellular structure, permeabilizing the cells for further experiments.
25. Take caution not to let the sections dry completely. If necessary, process slides one at a time. This applies to all steps that involve taking slides out of the solution.
26. The antibody solution stays on the top of slides only if the edges and the bottom of the slides are dry. Therefore, add the antibody solution to the center of the slide and avoid moving the chamber once the antibody is added. Place a sign on the chamber to warn others not to move it during incubation.
27. Cover the humidified slide incubation chamber with foil or perform the incubation in a cabinet.
28. Cut the pipette tip to make a large orifice because Dabco is sticky.
29. To avoid bubbles, lower the coverslip slowly.
30. The fluorescent signal decreases over time, so try to observe the fluorescence as early as you can.

31. In general, shorter probes (<500 bps) tend to yield weaker signals, whereas very long probes (>1,500 bps) have difficulty penetrating the cell membranes.
32. Take caution when removing the supernatant because the pellet may not be visible.
33. Postfixation ensures the tissue is fixed equally with cross-linked RNA molecules. It also improves the retention of the tissue on the slide.
34. To drain excess DEPC-PBS, hold the slides and tilt the Coplin jar onto the paper towels. Proteinase K treatment improves the signal intensity by allowing greater access of the target mRNA for the probes.
35. Refixation improves the section stability after proteolytic digestion.
36. Acetylation chemically modifies proteins and reduces their nonspecific bindings.
37. Preheat the hybridization solution at 80 °C before use.
38. To avoid bubbles, lower the parafilm coverslips slowly from one side to the other. Take caution when transporting the slide chamber to the oven such that the coverslips do not fall off the slides.
39. Do not force the coverslips off the slides with forceps, or the sections may tear.
40. Pre-warm solution for this step and **steps 3–7**. During the high-stringency wash, low salt concentration and high temperature inhibit nonspecific bindings.
41. RNase A digests single-stranded RNA to reduce the background signal.
42. Wrap the slide chamber in foil or place the chamber in a dark cabinet.
43. Before checking the staining status of the sections, prepare BM purple solution with 0.5 mg/ml levamisole in case the staining is not ready and more incubation with BM purple is needed. Otherwise, the sections may become dry before the BM purple solution is ready.
44. The mounting medium is very sticky and solidifies quickly. It is better to apply the mounting medium before the slides dry and put coverslip on immediately.

Acknowledgments

We would like to thank Dr. Simeone for providing the RNA in situ probe for Otx2. The monoclonal antibody against Pax6 developed by Dr. Jessell was obtained from the Developmental Studies

Hybridoma Bank developed under the auspices of the NICHD and maintained by The University of Iowa, Department of Biological Sciences, Iowa City, IA 52242. A. L. is supported by an NSF grant (IOS-0949877) and a Penn State new lab start-up fund.

References

1. Kandel E, Schwartz J, Jessell T (2000) Principles of neural science, 4th edn. McGraw-Hill, Health Professions Division, New York, NY. ISBN 0-8385-7701-6
2. Joseph G. Gall, Mary Lou Pardue (1969) Formation and detection of RNA-DNA hybrid molecules in cytological preparations*. Proc Natl Acad Sci USA 63:378–383
3. Komminoth P, Merk FB, Wolfe HJ, Roth J (1992) Comparison of 35S-and digoxigenin-labeled RNA and oligonucleotide probes for in situ hybridization. Histochemistry 98: 217–228
4. Wilcox JN (1993) Fundamental principles of in situ hybridization. J Histochem Cytochem 41:1725–1733. doi:10.1177/41.12.8245419

Chapter 19

In Utero Electroporation to Study Mouse Brain Development

Emilie Pacary and François Guillemot

Abstract

In utero electroporation is a rapid and powerful technique to study the development of many brain regions. This approach presents several advantages over other methods to study specific steps of brain development in vivo, from proliferation to synaptic integration. Here, we describe in detail the individual steps necessary to carry out the technique. We also highlight the variations that can be implemented to target different cerebral structures and to study specific steps of development.

Key words Electroporation, In utero, Brain, Embryo, Neuronal development, Gain and loss of function, Transfection, Mouse

1 Introduction

In utero electroporation is a simple and rapid approach to overexpress or silence genes in the developing mouse brain, representing a major alternative to time-consuming knockout, knock-in, and transgenic strategies to alter the mouse genome. Indeed, injection of plasmid DNA solution into the embryonic brain and application of an electric current are sufficient to allow area- and time-dependent transfection of brain cells. Since DNA is negatively charged, the injected plasmid moves towards the positive pole of the electrode and is thus selectively introduced into a specific brain region. Consequently, distinct areas can be targeted (Table 1), depending on the position and/or angle of the positive electrode. This is an important advantage over viral infections where the infected area is usually large and dependent on the titer of the viral suspension used. The region of interest can thus be individually targeted without affecting the rest of the embryo. The size of the transfected area can also be adjusted by using paddle-type electrodes of different diameter or by using wire-type electrodes (microelectrodes). In addition to the orientation and type of

Simon G. Sprecher (ed.), *Brain Development: Methods and Protocols*, Methods in Molecular Biology, vol. 1082,
DOI 10.1007/978-1-62703-655-9_19,

Table 1
The different brain structures that can be targeted by in utero electroporation

Brain structures	References
Somatosensory cortex	[5–7]
Prefrontal cortex	[8]
Piriform cortex	[9]
Amygdala	[9–11]
Hippocampus	[3, 12, 13]
Retina	[14–16]
Ganglionic eminences	[17–19]
Preoptic area	[20]
Thalamus	[21–23]
Cerebellum	[24–26]
Midbrain	[4, 27, 28]

electrodes, the location of DNA injection is another parameter that can be altered to target specific cell populations. For instance, while injection into the ventricles allows the targeting of a large region because of the diffusion of DNA, DNA can be introduced directly into a specific structure to better control the size of the transfected area. Finally, the ability to choose a particular developmental time point for the electroporation extends the range of brain areas that can be targeted. Moreover, depending on this time point, different developmental events can be studied including cell proliferation, cell migration, axon growth, dendrite development, or synaptogenesis (Table 2). This technique can also be used to electroporate plasmids that provide inducible gene expression or inducible RNAi-mediated silencing, thus refining the temporal control of gene expression. By using this system, the timing of gene targeting can be finely controlled, which allows for a detailed analysis of the cellular functions of the target gene.

Besides the spatial and temporal restrictions that can be achieved in the developing brain, in utero electroporation presents also the advantage to allow introduction of multiple plasmids in the same cells, which is difficult to achieve with viral constructs.

Gain- and loss-of-function studies using in utero electroporation have greatly contributed to the characterization of molecular mechanisms controlling specific steps of brain development. The improvement of this technique, with for instance transposon systems to promote stable gene expression [1, 2], should help in the future to further understand mouse brain development.

Table 2
Developmental events that can be studied using in utero electroporation and corresponding references

Developmental events	Examples of references
Regional patterning	[23, 29]
Proliferation	[18, 30, 31]
Neuronal differentiation	[19, 32, 33]
Neuronal migration	[5, 26, 34]
Dendrite and spine development	[3, 35]
Axon formation and guidance	[21, 36, 37]
Synaptogenesis and synapse maturation	[38, 39]
Astrogliogenesis	[40, 41]

2 Materials

2.1 DNA Injection

1. DNA solution: Purify plasmid DNA with an endotoxin-free Maxi-prep kit. Prepare plasmid DNA solution to desired concentrations in water and add Fast Green (final concentration of 0.05 %; Sigma) (*see* **Note 1**).
2. Needles: Pull borosilicate glass capillaries (1.0 mm O.D. × 0.58 mm I.D.) (Harvard Apparatus) using a micropipette puller.
3. Microloader tips (Eppendorf).

2.2 Surgery

1. Phosphate-buffered saline (PBS), pH 7.4: 10 g NaCl, 2 g KCl, 11.5 g Na_2HPO_4 $7H_2O$, and 2 g KH_2PO_4 in 800 ml of ddH_2O. Bring volume to 1 l and adjust pH.
2. Autoclave: Autoclave surgical instruments and PBS.
3. Buprenorphine: Prepare analgesic solution in PBS to a final concentration of 30 μg/ml.
4. Isoflurane.
5. Anesthetic induction chamber.
6. Heating pads.
7. Recovery chamber.
8. Water bath.
9. Surgical instruments: Extra thin Iris scissors (Fine Science Tools), Curved Forceps (Fine Science Tools), Ring forceps (Fine Science Tools), Needle holder (Fine Science Tools), Graefe Forceps (Fine Science Tools).
10. Beaker.
11. Electric razor.

12. Eye gel.
13. Chlorhexidine.
14. Mask and sterile gloves.
15. Scalpel.
16. Sterile drapes.
17. Sterile swabs.
18. Cotton buds.
19. Vicryl absorbable suture (Ethicon Inc.).

2.3 Electroporation

1. Electroporator.
2. Electrodes: Paddle-type or wire-type electrodes.
3. Microinjector (Femtojet Eppendorf).
4. Foot control for Femtojet microinjector (Eppendorf).
5. Capillary holder (Eppendorf).

2.4 Tissue Processing

1. Sodium pentobarbitone.
2. Pump for perfusion.
3. Fluorescent binocular.
4. 4 % Paraformaldehyde (PFA) in PBS.
5. 20 % sucrose in PBS.
6. OCT Compound (VWR International).
7. Molds for embedding (HistoMold, Leica).
8. Vibratome or cryostat.
9. Microscope slides.
10. Coverslips.
11. Mounting medium: Aqua Poly/mount (Polysciences Inc).

3 Methods

3.1 Preparation of the Surgery

1. Weigh a pregnant mouse at appropriate stage of pregnancy and inject subcutaneously 0.1 mg/kg of the analgesic buprenorphine at least 30 min before the surgery.
2. Turn on the heating pads and the recovery chamber. Place sterile PBS in warm water bath and all sterile surgical instruments and materials (scalpel, drape, swabs, cotton buds, and sutures) on a sterile drape. Place the electrodes into a PBS-filled beaker and connect to the electroporator.
3. Fill the needle with the DNA solution using a microloader tip and connect the needle to the capillary holder.

3.2 Surgery Before DNA Injection and Electroporation

1. Anesthetize the pregnant mouse with isoflurane in oxygen carrier (oxygen 2 l/min) using an anesthetic induction chamber. Wait until the animal loses righting reflex.
2. Transfer the animal to a "presurgery" mask. Place a drop of eye gel on each eye to prevent corneal ulceration of the eyes while it is under general anesthesia. Use an electric razor to shave the hair of the abdomen. Clean the shaved area once with chlorhexidine to collect flying hair.
3. Transfer the animal to a second mask in the surgical area. Place the mouse with its back on the heating pad. Start the surgery when the pedal reflex has been lost.
4. Put on mask and sterile gloves. Cover the animal with a sterile drape (with a small hole over the abdomen) to prevent the tissues and instruments from being contaminated by the areas of skin that have not been shaved and disinfected. Clean the shaved area at least three times with chlorhexidine. Use a different sterile cotton bud each time. Use a scalpel to make a vertical incision along the midline (~1 in. long) through the skin. Using scissors, make a similar incision of the muscle of the abdomen along the linea alba (white line composed mostly of collagen connective tissue).
5. Choose the most accessible embryos, place the ring forceps between two embryos, and carefully pull the embryonic chain out of the abdominal cavity (*see* **Note 2**). From this point on, keep the embryos hydrated with sterile pre-warmed PBS.

3.3 Injection of DNA and Electroporation

1. Pinch off the tip of the needle with a forceps.
2. Start with one of the most lateral embryos, making it easier to keep track of which embryos were electroporated. Manipulate the position of the embryo inside the amniotic sac using the ring forceps and stabilize the head of the embryos between the rings. Squeeze gently to push up the embryo closer to the uterine wall.
3. With the other hand, take the capillary holder and insert the needle carefully into the middle of the hemisphere to target the lateral ventricle or into the desired structure (*see* **Note 3**). Press the pedal to inject the DNA solution (*see* **Note 4**). The location of the DNA solution is monitored with Fast Green (*see* **Note 5**).
4. Place the electrodes on the sides of the embryo head to target the desired area (*see* **Note 6**).
5. Apply electrical pulses (*see* **Notes** 7 and **8**).

All the embryos in a pregnant mouse can be electroporated, usually with the same DNA construct to avoid confusion when harvesting electroporated embryos or pups (*see* **Note 9**). However, a long surgery decreases the survival rate of embryos. The abdominal cavity should not remain open longer than 30 min.

3.4 Surgery Post Electroporation

1. After electroporating the embryos, add PBS into the abdominal cavity and use the ring forceps to replace the uterine horn in its original location.
2. Suture the abdomen wall and skin with Vicryl absorbable sutures.
3. Place the animal in a recovery chamber until it wakes up (usually 5–10 min) and then transfer in a cage placed on a heating pad.

3.5 Post-surgery

Check the behavior of the mice to assess pain, suffering, or distress and weigh the animals 24 and 48 h after the surgery. If needed, analgesics can be administered to minimize pain and discomfort.

3.6 Tissue Processing

Collect the electroporated embryos or pups at the embryonic or the postnatal stages required for the experiment.

1. For analysis at embryonic stages (for example to study cell proliferation or migration):

   Euthanize mother via cervical dislocation and collect the embryos. After decapitation, select the brains that have been properly electroporated, as indicated by the amount and location of the fluorescent signal, visualized across the skull using a fluorescent binocular microscope. Dissect the brain out of the skull, fix overnight in 4 % PFA, and then place in 20 % sucrose/PBS overnight. Embed in OCT compound, freeze at –80 °C, and section using a cryostat.
2. For analysis at postnatal stages (for example to study synaptogenesis):

   Anesthetize pups or adult mice with intraperitoneal injection of pentobarbitone (40–60 mg/kg) and perform transcardial perfusion with PBS, followed by 4 % PFA in PBS. Dissect the brains out of the skull and postfix in 4 % PFA overnight. After washings in PBS, section the brains using a vibratome.

In both types of analysis, sections can be mounted in Aqua Poly/mount directly or after immunostaining.

4 Notes

1. The efficiency of electroporation is highly dependent on the DNA concentration. A concentration of 1 μg/μl is generally used. This concentration is usually sufficient to visualize electroporated neurons (when a GFP-expressing plasmid is electroporated) without affecting their development. However, depending on the promoter used (low expression level with cytomegalovirus promoter/enhancer, stronger expression

level with the cytomegalovirus immediate early enhancer and chicken β-actin promoter fusion (CAG) promoter) as well as the size and the stability of the protein expressed, plasmid concentration can be adjusted (0.25–5 μg/μl).

2. Do not pull too much on the embryos as this will increase the risk of hemorrhage.
3. The sharpness of the needle is critical to pierce properly the uterine wall. It is important to minimize the movement of the needle at the surface of the uterine wall and after insertion into the brain because an enlargement of the hole will result in the leakage of amniotic fluid and embryo death. Avoid piercing blood vessels in the uterine wall, as this will result in bleeding and embryo death.
4. It is important not to inject a too large volume of DNA into the lateral ventricle because it will induce hydrocephalus (do not exceed 2 μl at E14.5). The volume of DNA is adjusted for the purpose of the experiment. 1 μl is generally used for most experiments but a smaller volume of DNA may be required (for example to electroporate fewer cells and visualize the dendritic arbor of isolated electroporated neurons).
5. In case of injection into the ventricle, it is worth noting that the size of the electroporated area increases with the time between the injection and the application of currents.
6. Selection of the correct size of electrodes and adequate arrangement of electrodes are of key importance for successful electroporation. For instance, to target the cerebral cortex, 5 mm platinum electrodes are often used and the positive paddle is placed on the dorsolateral side of the injected ventricle [3].
7. Avoid applying current across the placenta as this will result in embryo death.
8. The electric pulses induce the formation of transient pores in cell membranes, thus allowing the uptake of DNA. These electropores last longer when higher voltage and longer pulse duration are applied. However, this also increases the risk of inducing nonreversible cell membrane damage, which will cause cell death.

   For instance, to electroporate the cerebral cortex at E14.5, five 30 V electrical pulses (50 ms duration) at 1-s intervals are often applied.

   See Saito and Nakatsuji [4]: Table 1 “Effect of Voltage and Pulse Numbers on Survival and EYFP-Positive Rate” and Table 2 “Optimal Conditions for Electroporation at Different Stages.”
9. Conditions (size electrodes, voltage …) should be optimized for each electroporation experiment depending on the stage of the embryo and brain region electroporated. See details in references of Tables 1 and 2.

References

1. Chen F, LoTurco J (2012) A method for stable transgenesis of radial glia lineage in rat neocortex by piggyBac mediated transposition. J Neurosci Methods 207(2):172–180
2. Iguchi T, Yagi H, Wang CC, Sato M (2012) A tightly controlled conditional knockdown system using the Tol2 transposon-mediated technique. PLoS One 7(3):e33380
3. Pacary E, Haas MA, Wildner H, Azzarelli R, Bell DM, Abrous DN, Guillemot F (2012) Visualization and genetic manipulation of dendrites and spines in the mouse cerebral cortex and hippocampus using in utero electroporation. J Vis Exp (65):e4163. doi: 10.3791/4163
4. Saito T, Nakatsuji N (2001) Efficient gene transfer into the embryonic mouse brain using in vivo electroporation. Dev Biol 240(1):237–246
5. Pacary E, Heng J, Azzarelli R, Riou P, Castro D, Lebel-Potter M, Parras C, Bell DM, Ridley AJ, Parsons M, Guillemot F (2011) Proneural transcription factors regulate different steps of cortical neuron migration through Rnd-mediated inhibition of RhoA signaling. Neuron 69(6):1069–1084
6. Shimogori T, Ogawa M (2008) Gene application with in utero electroporation in mouse embryonic brain. Dev Growth Differ 50(6):499–506
7. Tabata H, Nakajima K (2001) Efficient in utero gene transfer system to the developing mouse brain using electroporation: visualization of neuronal migration in the developing cortex. Neuroscience 103(4):865–872
8. Niwa M, Kamiya A, Murai R, Kubo K, Gruber AJ, Tomita K, Lu L, Tomisato S, Jaaro-Peled H, Seshadri S, Hiyama H, Huang B, Kohda K, Noda Y, O'Donnell P, Nakajima K, Sawa A, Nabeshima T (2010) Knockdown of DISC1 by in utero gene transfer disturbs postnatal dopaminergic maturation in the frontal cortex and leads to adult behavioral deficits. Neuron 65(4):480–489
9. Bai J, Ramos RL, Paramasivam M, Siddiqi F, Ackman JB, LoTurco JJ (2008) The role of DCX and LIS1 in migration through the lateral cortical stream of developing forebrain. Dev Neurosci 30(1–3):144–156
10. Remedios R, Huilgol D, Saha B, Hari P, Bhatnagar L, Kowalczyk T, Hevner RF, Suda Y, Aizawa S, Ohshima T, Stoykova A, Tole S (2007) A stream of cells migrating from the caudal telencephalon reveals a link between the amygdala and neocortex. Nat Neurosci 10(9):1141–1150
11. Soma M, Aizawa H, Ito Y, Maekawa M, Osumi N, Nakahira E, Okamoto H, Tanaka K, Yuasa S (2009) Development of the mouse amygdala as revealed by enhanced green fluorescent protein gene transfer by means of in utero electroporation. J Comp Neurol 513(1):113–128
12. Nakahira E, Yuasa S (2005) Neuronal generation, migration, and differentiation in the mouse hippocampal primoridium as revealed by enhanced green fluorescent protein gene transfer by means of in utero electroporation. J Comp Neurol 483(3):329–340
13. Navarro-Quiroga I, Chittajallu R, Gallo V, Haydar TF (2007) Long-term, selective gene expression in developing and adult hippocampal pyramidal neurons using focal in utero electroporation. J Neurosci 27(19): 5007–5011
14. Garcia-Frigola C, Carreres MI, Vegar C, Herrera E (2007) Gene delivery into mouse retinal ganglion cells by in utero electroporation. BMC Dev Biol 7:103
15. Petros TJ, Rebsam A, Mason CA (2009) In utero and ex vivo electroporation for gene expression in mouse retinal ganglion cells. J Vis Exp (31):e1333. doi:10.3791/1333
16. Punzo C, Cepko CL (2008) Ultrasound-guided in utero injections allow studies of the development and function of the eye. Dev Dyn 237(4):1034–1042
17. Borrell V, Yoshimura Y, Callaway EM (2005) Targeted gene delivery to telencephalic inhibitory neurons by directional in utero electroporation. J Neurosci Methods 143(2): 151–158
18. Castro DS, Martynoga B, Parras C, Ramesh V, Pacary E, Johnston C, Drechsel D, Lebel-Potter M, Garcia LG, Hunt C, Dolle D, Bithell A, Ettwiller L, Buckley N, Guillemot F (2011) A novel function of the proneural factor Ascl1 in progenitor proliferation identified by genome-wide characterization of its targets. Genes Dev 25(9):930–945
19. Rouaux C, Arlotta P (2010) Fezf2 directs the differentiation of corticofugal neurons from striatal progenitors in vivo. Nat Neurosci 13(11):1345–1347
20. Gelman DM, Martini FJ, Nobrega-Pereira S, Pierani A, Kessaris N, Marin O (2009) The embryonic preoptic area is a novel source of cortical GABAergic interneurons. J Neurosci 29(29):9380–9389
21. Bonnin A, Torii M, Wang L, Rakic P, Levitt P (2007) Serotonin modulates the response of embryonic thalamocortical axons to netrin-1. Nat Neurosci 10(5):588–597
22. Kataoka A, Shimogori T (2008) Fgf8 controls regional identity in the developing thalamus. Development 135(17):2873–2881

23. Vue TY, Bluske K, Alishahi A, Yang LL, Koyano-Nakagawa N, Novitch B, Nakagawa Y (2009) Sonic hedgehog signaling controls thalamic progenitor identity and nuclei specification in mice. J Neurosci 29(14):4484–4497
24. Kawauchi D, Taniguchi H, Watanabe H, Saito T, Murakami F (2006) Direct visualization of nucleogenesis by precerebellar neurons: involvement of ventricle-directed, radial fibre-associated migration. Development 133(6): 1113–1123
25. Nishiyama J, Hayashi Y, Nomura T, Miura E, Kakegawa W, Yuzaki M (2012) Selective and regulated gene expression in murine Purkinje cells by in utero electroporation. Eur J Neurosci 36(7):2867–2876
26. Okada T, Keino-Masu K, Masu M (2007) Migration and nucleogenesis of mouse precerebellar neurons visualized by in utero electroporation of a green fluorescent protein gene. Neurosci Res 57(1):40–49
27. Alvarez-Maya I, Navarro-Quiroga I, Meraz-Rios MA, Aceves J, Martinez-Fong D (2001) In vivo gene transfer to dopamine neurons of rat substantia nigra via the high-affinity neurotensin receptor. Mol Med 7(3):186–192
28. Willett RT, Greene LA (2011) Gata2 is required for migration and differentiation of retinorecipient neurons in the superior colliculus. J Neurosci 31(12):4444–4455
29. Fukuchi-Shimogori T, Grove EA (2001) Neocortex patterning by the secreted signaling molecule FGF8. Science 294(5544): 1071–1074
30. Lange C, Huttner WB, Calegari F (2009) Cdk4/cyclinD1 overexpression in neural stem cells shortens G1, delays neurogenesis, and promotes the generation and expansion of basal progenitors. Cell Stem Cell 5(3): 320–331
31. Weimer JM, Yokota Y, Stanco A, Stumpo DJ, Blackshear PJ, Anton ES (2009) MARCKS modulates radial progenitor placement, proliferation and organization in the developing cerebral cortex. Development 136(17):2965–2975
32. Nguyen L, Besson A, Heng JI, Schuurmans C, Teboul L, Parras C, Philpott A, Roberts JM, Guillemot F (2006) p27kip1 independently promotes neuronal differentiation and migration in the cerebral cortex. Genes Dev 20(11):1511–1524
33. Singh KK, De Rienzo G, Drane L, Mao Y, Flood Z, Madison J, Ferreira M, Bergen S, King C, Sklar P, Sive H, Tsai LH (2011) Common DISC1 polymorphisms disrupt Wnt/GSK3beta signaling and brain development. Neuron 72(4):545–558
34. Elias LA, Wang DD, Kriegstein AR (2007) Gap junction adhesion is necessary for radial migration in the neocortex. Nature 448(7156):901–907
35. Cubelos B, Sebastian-Serrano A, Beccari L, Calcagnotto ME, Cisneros E, Kim S, Dopazo A, Alvarez-Dolado M, Redondo JM, Bovolenta P, Walsh CA, Nieto M (2010) Cux1 and Cux2 regulate dendritic branching, spine morphology, and synapses of the upper layer neurons of the cortex. Neuron 66(4):523–535
36. Fang WQ, Ip JP, Li R, Ng YP, Lin SC, Chen Y, Fu AK, Ip NY (2011) Cdk5-mediated phosphorylation of Axin directs axon formation during cerebral cortex development. J Neurosci 31(38):13613–13624
37. Yi JJ, Barnes AP, Hand R, Polleux F, Ehlers MD (2010) TGF-beta signaling specifies axons during brain development. Cell 142(1):144–157
38. Elias GM, Elias LA, Apostolides PF, Kriegstein AR, Nicoll RA (2008) Differential trafficking of AMPA and NMDA receptors by SAP102 and PSD-95 underlies synapse development. Proc Natl Acad Sci USA 105(52):20953–20958
39. Trimbuch T, Beed P, Vogt J, Schuchmann S, Maier N, Kintscher M, Breustedt J, Schuelke M, Streu N, Kieselmann O, Brunk I, Laube G, Strauss U, Battefeld A, Wende H, Birchmeier C, Wiese S, Sendtner M, Kawabe H, Kishimoto-Suga M, Brose N, Baumgart J, Geist B, Aoki J, Savaskan NE, Brauer AU, Chun J, Ninnemann O, Schmitz D, Nitsch R (2009) Synaptic PRG-1 modulates excitatory transmission via lipid phosphate-mediated signaling. Cell 138(6):1222–1235
40. Garcia-Marques J, Lopez-Mascaraque L (2013) Clonal Identity Determines Astrocyte Cortical Heterogeneity. Cereb Cortex 23(6): 1463–1472
41. Subramanian L, Sarkar A, Shetty AS, Muralidharan B, Padmanabhan H, Piper M, Monuki ES, Bach I, Gronostajski RM, Richards LJ, Tole S (2011) Transcription factor Lhx2 is necessary and sufficient to suppress astrogliogenesis and promote neurogenesis in the developing hippocampus. Proc Natl Acad Sci USA 108(27):E265–E274

Chapter 20

The Cre/*Lox* System to Assess the Development of the Mouse Brain

Claudius F. Kratochwil and Filippo M. Rijli

Abstract

Cre-mediated recombination has become a powerful tool to confine gene deletions (conditional knock-outs) or overexpression of genes (conditional knockin/overexpression). By spatiotemporal restriction of genetic manipulations, major problems of classical knockouts such as embryonic lethality can be circumvented. Furthermore Cre-mediated recombination has broad applicability in the analysis of the cellular behavior of subpopulations and cell types as well as for genetic fate mapping. This chapter will give an overview about applications for the Cre/*LoxP* system and their execution.

Key words Cre recombinase, Transgenesis, Conditional knockout, Conditional knockin, CreER^{T2}, Flpe recombinase, MADM, Split-Cre, Brainbow

1 Introduction

After the first gene knockout (KO) in the mouse was obtained by Thomas and Capecchi using site-directed mutagenesis of the HPRT gene in 1987 [1], the in vivo function of many genes has been analyzed using this technique. Nevertheless the KO approach has two main restrictions. First of all, genes whose inactivation is embryonically lethal cannot be analyzed for their function in late development and adulthood. Secondly, KOs of genes which have a function in multiple tissues and/or cell types are difficult to analyze, since phenotypes might be combinations of multiple distinct defects and therefore quite complex to dissect.

In 1994 Gu et al. described for the first time the use of the Cre/*Lox* recombination system to induce a gene knockout in mice [2]. This technology made it possible to conditionally knockout genes solely in subsets of cells (i.e., in a cell-type- or tissue-specific manner), where *Cre* recombinase is expressed. Two years before, the technique had been already used to conditionally overexpress the SV40 large tumor antigen in mice [3]. The key principle of the

Simon G. Sprecher (ed.), *Brain Development: Methods and Protocols*, Methods in Molecular Biology, vol. 1082, DOI 10.1007/978-1-62703-655-9_20, © Springer Science+Business Media, LLC 2014

Cre-mediated recombination is that the recombinase enzyme can catalyze the deletion or inversion of a genomic fragment depending on the orientation of small recognition sequences, called *Lox* or *LoxP* sites, flanking such fragment.

Since then, Cre-mediated recombination has been used successfully for many applications, tissues [4–9] and model organisms including, e.g., *Drosophila* [10], *Xenopus* [11], zebrafish [12], and plants [13]. Also the recombination mechanism has been elucidated by the analysis of the crystal structure of Cre and the Cre/*LoxP* interface [14].

In mammalian cells, Cre is the predominantly used recombinase for site-specific recombination and has been shown to be more effective than another (in *Drosophila* widely used) recombination system, the Flp/*Frt* system from *Saccharomyces cerevisiae* [15]. The Flp recombinase has a similar ability as Cre to delete or invert (and therefore was named Flp (pronounced "Flip") recombinase) DNA fragments, though recognizing distinct target sequences. Later on a more efficient version, Flpe has been created which works reliably in mice [16, 17] and offers a suitable alternative for some applications.

To further restrict the knockout in time, inducible variants of Cre were designed. The most widely used version is CreER^{T2} [18]. Hereby Cre is fused to a mutated ligand-binding domain of the estrogen receptor (ER) [19]. The fusion protein is normally confined to the cytoplasm, while in the presence of the synthetic ligand tamoxifen or 4-hydroxytamoxifen, it translocates to the nucleus, where it can trigger recombination events (*see* Subheading 3.5). A light-activatable Cre recombinase to control activity in time and space has also been generated [20].

Another idea to restrict Cre or CreER^{T2} further in space came from Hirrlinger et al. [21–23]. By splitting Cre into two inactive fragments, which regain Cre activity when co-expressed, recombination could be restricted to the intersection of two expression domains (by using different enhancers/promoters for the two fragments). A similar system was established by Farago et al. [24], using a combination of the two recombination systems Cre/*Lox* and Flpe/*Frt* (*see* Subheading 3.6). Both systems have been also tested in vivo in transgenic animals.

A powerful tool to construct a fate map of cells as well as to analyze gene function on a single-cell resolution was introduced with the MADM (mosaic analysis with double markers) system [25, 26] an adaptation of the MARCM (mosaic analysis with a repressible cell marker) system from *Drosophila* [27]. MADM uses Cre-mediated interchromosomal recombination. By employing two markers (e.g., GFP and RFP), not only single-cell progenies can be traced, but the combination with mutations makes it possible to distinguish mutant, heterozygous, and wild-type cells by their distinct fluorescent markers [28] (*see* in detail in Subheading 3.7).

Another attempt to achieve single-cell resolution was given by the brainbow system from Livet et al. [29]. Here, recombination leads to a stochastic choice of expression of a fluorescent protein. Multiple integrations result in a combinatorial color code, creating approximately 100 distinguishable colors (*see* Subheading 3.8), which is especially helpful for the analysis of the nervous system.

In this chapter, we will focus on the basic principles and applications of the Cre/*LoxP* and CreER^{T2}/*LoxP* systems and their variations (Split-Cre, MADM, brainbow) and will provide strategies and protocols for their use.

2 Materials

2.1 Reporter Mice

1. ROSA26 LacZ reporter line [30]; available at Jackson Laboratories:

   # 003504 B6.129S4-*Gt(ROSA)26Sor*^{tm1Sor}/J.

2. ROSA26 tdTomato reporter line [31]; available at Jackson Laboratories:

   # 007914 B6.Cg-*Gt(ROSA)26Sor*$^{tm14(CAG\text{-}tdTomato)Hze}$/J.

3. ROSA26 ZsGreen reporter line [31]; available at Jackson Laboratories:

   # 007906 B6.Cg-*Gt(ROSA)26Sor*$^{tm6(CAG\text{-}ZsGreen1)Hze}$/J.

4. Brainbow reporter lines [29]; six transgenic lines available at Jackson Laboratories:

   # 007901 B6.Cg-Tg(Thy1-Brainbow1.0)HLich/J.

   # 007910 B6;CBA-Tg(Thy1-Brainbow1.0)LLich/J.

   # 007911 B6.Cg-Tg(Thy1-Brainbow1.1)MLich/J.

   # 007921 B6.Cg-Tg(Thy1-Brainbow2.1)RLich/J.

   # 013731 STOCK Gt(ROSA)26Sor$^{tm1(CAG\text{-}Brainbow2.1)Cle}$/J.

   # 017492 B6.129P2-Gt(ROSA)26Sor$^{tm1(CAG\text{-}Brainbow2.1)\ Cle}$/J.

5. MADM mice [25, 26, 32, 33]; several transgenic lines are available at Jackson Laboratories, the following are the most widely used:

   # 013749 STOCK Tg(ACTB-EGFP,-tdTomato)11Luo/J.

   # 013751 STOCK Tg(ACTB-tdTomato,-EGFP)11Luo/J.

   # 017932 STOCK Tg(ACTB-EGFP*)10Luo/J.

   # 017923 STOCK Tg(ACTB-EGFP*,-tdTomato)10Luo/J.

   # 017912 STOCK Gt(ROSA)26Sor$^{tm6(ACTB\text{-}EGFP*,\text{-}tdTomato)Luo}$/J.

   # 017921 STOCK Gt(ROSA)26Sor$^{tm7(ACTB\text{-}EGFP*)Luo}$/J.

2.2 Generation of Genetically Modified Mice

1. Genomic DNA (for amplifying enhancers).
2. Cloning plasmid(s) (for Cre, CreER^{T2}, *Lox* sites, minimal promoter, fluorescent proteins, resistance cassettes (e.g., Addgene)).
3. Standard reagents for molecular biology (restriction enzymes (NEB), ligase (NEB), competent E. coli (e.g., DH5alpha, Top10 (Invitrogen)), primers, antibiotics).
4. Modified recombinogenic bacterial strains for recombination [34] (*see* **Note 1**).

2.3 Tamoxifen Treatment for CreER^{T2}-Mediated Recombination

1. Gavage/feeding needle to administer tamoxifen (Fine Science Tools).
2. Tamoxifen (SIGMA).
3. Corn oil (SIGMA) or sunflower oil.
4. Syringe (graded in 100 μl intervals).

2.4 Genotyping

1. Standard reagents for DNA extraction (1 M Tris–HCl pH 8.5; 5 M NaCl; 0.5 M EDTA; 20 % SDS; proteinase K; isopropanol; ethanol).
2. Standard reagents and primers for polymerase chain reaction (PCR) or GoTaq® Green Master Mix (Promega).
3. Standard reagents for gel electrophoresis.

3 Methods

3.1 Applications of the Cre/LoxP System

To create a conditional knockout, the gene of interest is flanked with recognition sites (*LoxP* sites; **Locus of crossing [*x*-ing]-over of bacteriophage *P1***) for the bacteriophage P1 Cre recombinase [35] (Fig. 1). If Cre recombinase is present, the sequence flanked by two *LoxP* sites (the "floxed" sequence) is excised (if *Lox* sites have the same orientation) or inverted (if *Lox* sites have opposing orientation) (Fig. 1). It is also possible to recombine between two different plasmids or chromosomes (interchromosomal recombination), which is, for example, used in the MADM system (*see* Subheading 3.7). The excision of genomic fragments gives the possibility to perturb gene function either (a) completely, by removing the whole gene or the start codon; (b) partly, by removing certain parts/exons of the gene or by truncating it; or by (c) changing its expression pattern and levels by excising/replacing/modifying promoter or enhancer elements. In a similar way, Cre-mediated recombination can be used to overexpress genes, by excision of an intervening transcriptional termination sequence, flanked by *LoxP* sites, that prevents the transcription of the target gene [3] (Fig. 2b). Since these constructs might result in low levels of leaking readthrough transcription, especially if many copies are present (e.g., when using gene transfer by

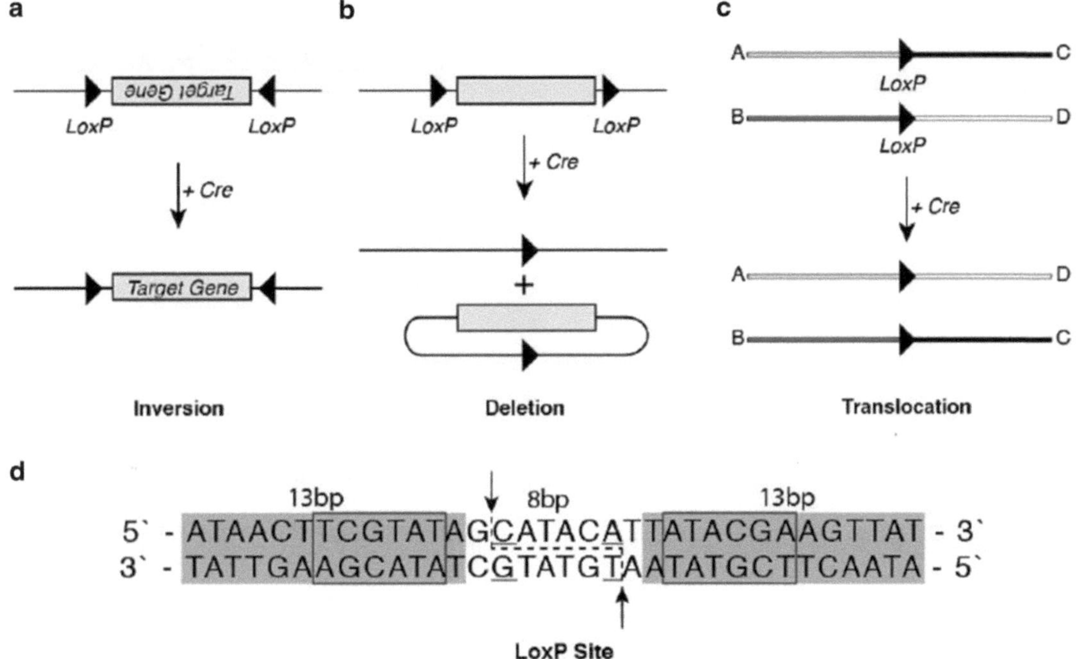

Fig. 1 Distinct genomic outcomes of Cre/*LoxP*-mediated recombination. (**a**) If two *Lox* sites have opposing directionality, the sequence in between is inversed upon Cre-mediated recombination. The recombination is reversible. (**b**) If *Lox* sites have the same orientation, the sequence in between is recombined out, resulting in a circular fragment and the sequence lacking this fragment. (**c**) If *LoxP* sites are present on two chromosomes, recombination can also occur interchromosomally, albeit at a lower frequency than if the *LoxP* sites are on the same chromosome. (**d**) The *LoxP* site is 34 base pair (bp) long. Two palindromic 13 bp sequences, containing the Cre binding site (*boxed*), flank an asymmetric 8 bp sequence. *Arrows* indicate the sites of cleavage during recombination. The *underlined* base pairs are mutated in the most commonly used variants *LoxN* and *Lox2722* (modified from [36])

viruses or electroporation), a different approach (also called flip-excision (FLEX) switch) was created, using *LoxP* sites, which are put in inverted orientation (Fig. 2c), causing an inversion of the intervening DNA (Fig. 1). Thereby the transcription of a gene can be initiated by inverting the open-reading frame. Because here none of the *LoxP* sites is removed, the inversion would continue forth and back. To interfere with that, a second *LoxP*-incompatible site pair is introduced (e.g., *Lox2272*) which results in the termination of the ongoing recombination by cutting out the specific recognition partners of both *Lox* sites (Fig. 2) [37].

It should be mentioned that the efficiency of Cre/*LoxP*-mediated recombination decreases in general with increasing genetic distance, but in principle any desired rearrangement can be made with the Cre/*LoxP* system [38].

3.2 Designing Constructs

Cre/*LoxP* is a binary system. On the one hand, the gene of interest has to be flanked by *LoxP* sites ("floxed") (Fig. 1). On the other hand recombinase expression in the cell or tissue of interest is provided

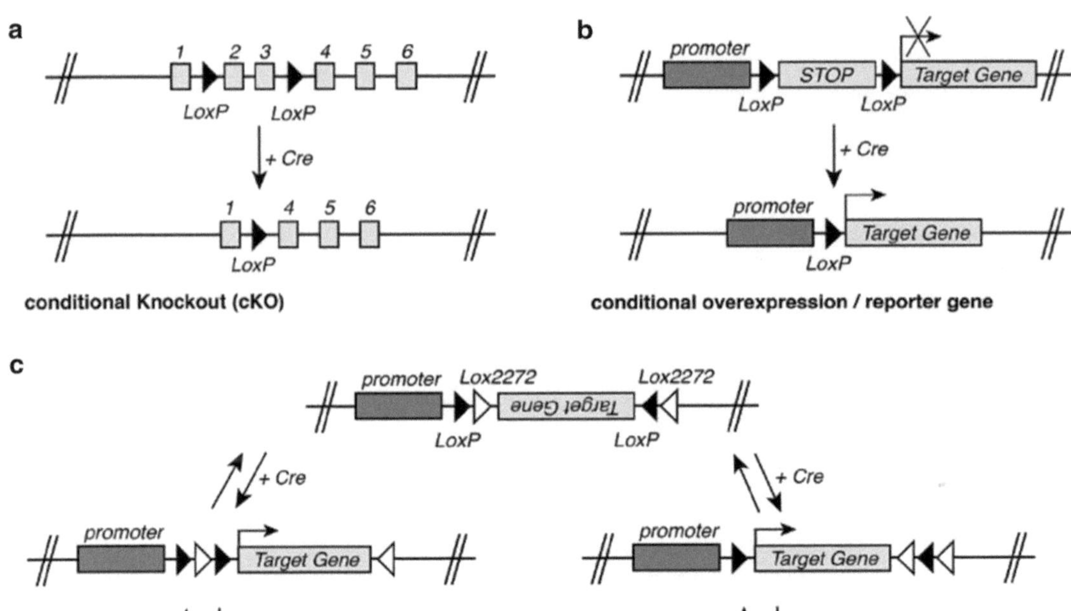

Fig. 2 Three examples for applications using the Cre/*LoxP* system. (**a**) The classic approach to generate a conditional knockout is to place *LoxP* sites in introns either flanking the exon, which contains the ATG, or exons that contain functionally important domains. Using this strategy, the probability to interfere with gene expression before Cre recombination is reasonably smaller than if *LoxP* sites are placed in promoter regions. (**b**) To overexpress genes usually an intervening sequence is placed between promoter and target gene, which blocks transcription. Upon Cre-mediated recombination this sequence is removed, and the gene starts to be expressed under the control of the upstream promoter. (**c**) To reduce the risk of leaking, recently the use of double-inverted *Lox* sites became an alternative approach (flip-excision switch, or FLEX). Here, the gene of interest is placed in inverted or antisense orientation. Cre mediates the inversion of the sequence in the presence of a pair of *Lox* sites in inverted orientation. The use of a second pair of *Lox* sites also in reverse orientation, though incompatible with the first pair, eventually results in an irreversible sense configuration, since both sites lack at the end a partner site, allowing stable transcription

either by (a) a knockin (KI) of *Cre*, (b) a transgene in which *Cre* expression is driven by a specific promoter/enhancer, (c) a bacterial artificial chromosome (BAC) in which *Cre* is inserted in-frame at a specific locus, or (d) gene transfer using, e.g., viruses or *in utero* electroporation (Fig. 3). Many tissue- or cell-type-specific enhancers have been described in the literature. Others can be found in enhancer databases such as the VISTA Enhancer Browser [39] or can be identified by selecting evolutionarily conserved sequences in proximity of genes with expression patterns of interest [40] and by cloning a fragment of a few kilobases of core and proximal promoter in front of the *Cre* recombinase gene. Finally, relevant

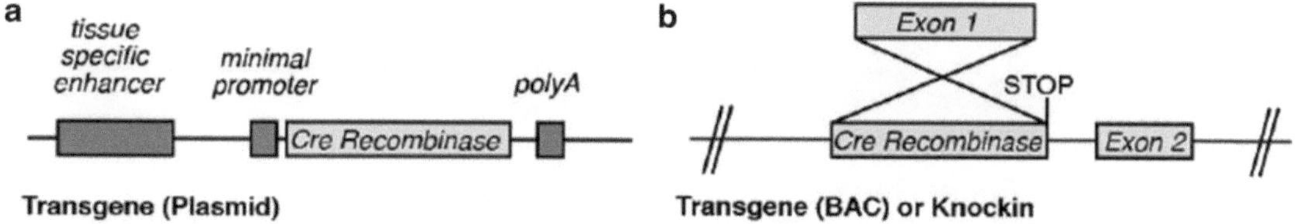

Fig. 3 Two approaches to obtain tissue-/cell-type-specific *Cre* expression. (**a**) A tissue-specific enhancer is combined with a minimal promoter to drive tissue-specific *Cre* expression, where the enhancer is active. (**b**) Another approach is site-directed recombination into a bacterial artificial chromosome (BAC) or into the genome of embryonic stem cells

enhancers can be also identified by genome-wide search, e.g., by ChIP-Seq for the p300 protein [41]. If enhancers are used which lack the core promoter element, they are usually combined with the minimal/core promoter (minP) of the human ß-globin promoter [42], which is unable to drive efficient transcription if no additional enhancer sequences are present proximally.

3.3 Mouse Mating Schemes

Once *Cre* KI (i.e., inserted at specific loci) or *Cre*-expressing transgenic mice have been created, they need to be crossed to mice carrying *LoxP* site-bearing conditional alleles, in order to generate double heterozygotes for *Cre* and the floxed locus. Since most laboratories dispose of multiple *Cre* drivers, it is more space efficient to generate a few double heterozygous males for the *Cre*-expressing line and the conditional allele and mate them to homozygous conditional mutant females, which can be readily maintained as a pool. However, it should be noted that in this type of crossings, double heterozygous *Cre*/conditional allele and/or homozygous conditional mutant specimen in the absence of Cre-mediated excision usually displays a wild-type-like phenotype, thus serving as controls. It may be useful to additionally cross a floxed conditional reporter line into the background of *Cre*/conditional allele double heterozygotes. In many projects this will ease the analysis, as it will allow to (1) have a direct readout of the cells in which the gene has been knocked out, (2) directly analyze cellular behavior between heterozygotes and homozygotes, and (3) select tissue or cells using methods such as fluorescent-activated cell sorting (FACS). Conditional Cre-inducible reporter mice in which the reporter genes are inserted in housekeeping gene loci are available from the Jackson Laboratories. These lines provide stable and constitutive reporter expression once activated by Cre-mediated excision. In particular, three KIs in the ROSA26 locus are the most frequently used reporter lines since they are highly sensitive to Cre-mediated activation, which express either *ß-galactosidase* (*LacZ*) [30], *ZsGreen* or *tdTomato* [31], but additional reporter lines expressing other fluorescent proteins are also available (e.g., [43]). The choice depends on the application. ß-Galactosidase catalyzes the transformation of X-gal into an insoluble blue enzymatic product

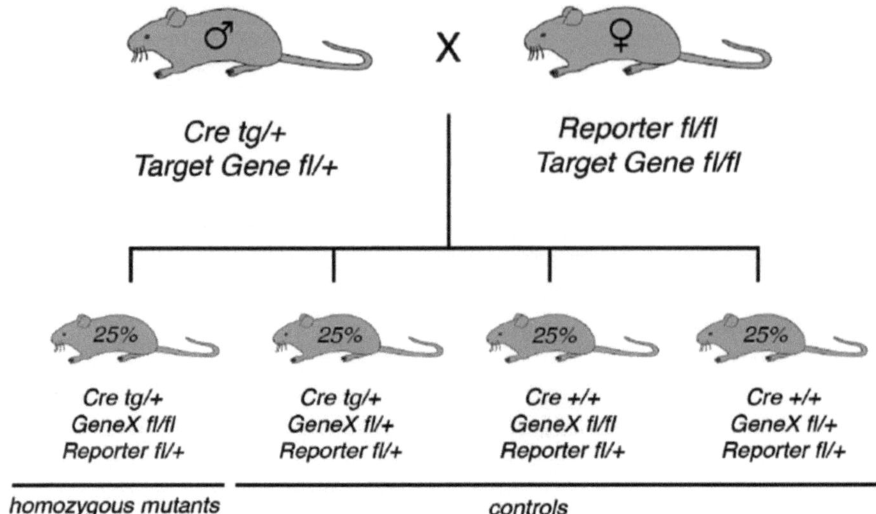

Fig. 4 Mating scheme to analyze gene function by conditional knockout. The use of pools for the conditional knockout (with or without reporter) eases the analysis if multiple *Cre* drivers are used to test gene function in different tissues. The offspring contains animals displaying mutant (*Cre tg/+, Gene fl/fl*) and control (*Cre tg/+, Gene fl/+, and Gene fl/fl*) phenotypes

and is preferentially used for nonfluorescent histochemistry or whole mount stainings of embryos or organs. The strongly fluorescent proteins ZsGreen (*see* **Note 2**) or tdTomato are a better choice for fluorescence histochemistry, live imaging, or cell sorting. The crossing scheme (Fig. 4) will give litters containing all needed controls: 25 % homozygotes (+Cre), 25 % homozygotes (−Cre), and 25 % heterozygotes (−Cre). The controls can be used to exclude an effect of the Cre (heterozygotes +Cre) or of the floxed locus (homozygotes −Cre) when analyzing the phentoype of the homozygous mutants.

3.4 Genotyping

To genotype mice for maintenance and experiments, polymerase chain reaction (PCR) is the method of choice. Primers have to be designed to simultaneously detect (1) the *Cre*, (2) the floxed locus, (3) the locus after Cre-mediated deletion, and (4) the wild-type locus of the respective gene. For (1), specific primers for *Cre* can be used or, alternatively, primers which span the *Cre* extremities and its flanking regions. For (2–4), usually three primers are needed (Fig. 5). One primer pair is chosen on each side of one or the other of the *LoxP* sites to detect the small size difference, as compared to the wild-type fragment, due to the presence of the *LoxP* site (Fig. 5b). Another pair of primers is designed 5′ and 3′ to the two *LoxP* sites, respectively, to be able to detect the Cre-mediated deletion of the locus and confirm the efficiency of the recombination (Fig. 5c). PCRs for genotyping can be done using standard PCR protocols for genotyping on clipped toes, ears, or tails or any other tissue, where recombination has to be tested. We recommend the

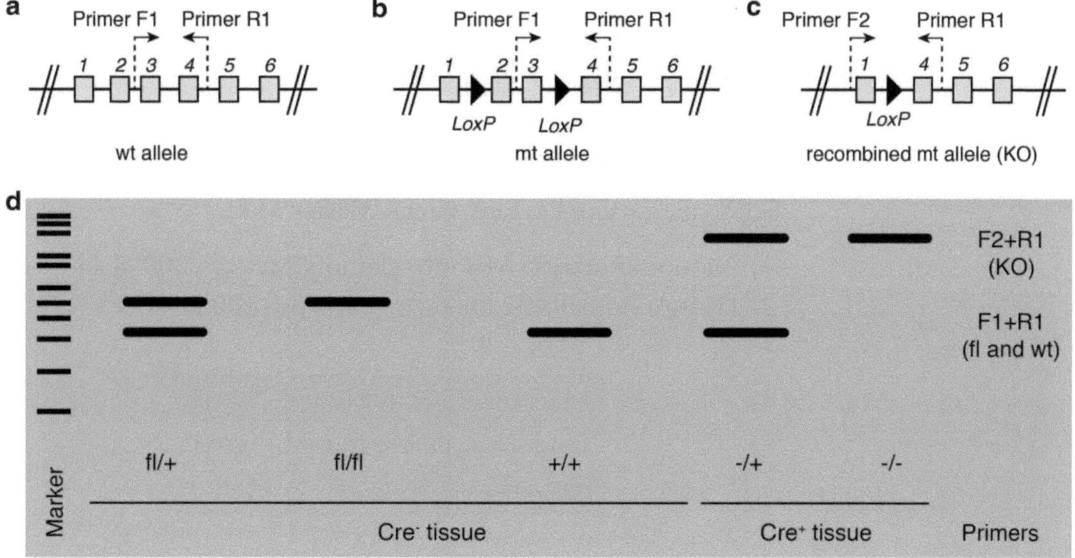

Fig. 5 Genotyping by PCR to detect gene deletion. (**a**) The targeted allele can be detected by primer pair 1 (F1 and R1). (**b**) The floxed allele can be detected by the same primers, yielding a longer amplicon. (**c**) A distinct primer pair (F2 and R1) detects the Cre-mediated recombination. The non-recombined locus usually cannot be detected, as the amplicon will be too long. (**d**) Theoretical result of a PCR with different genotypes in Cre-positive (Cre$^+$) and Cre-negative (Cre$^-$) tissues

use of the GoTaq® Green Master Mix (Promega), because it is faster, reduces pipetting mistakes and contaminations, and increases reproducibility.

DNA extraction:

1. Prepare tail buffer (500 ml stock):

100 mM Tris–HCl pH 8.5	50 ml of 1 M
200 mM NaCl	20 ml of 5 M
5 mM EDTA	5 ml of 0.5 M
0.2 % SDS	5 ml of 20 % SDS
ddH_2O	420 ml

2. Clip the tails or ears and collect them in 1.5 ml Eppendorf tubes.
3. Add 500 μl tail buffer and 10 μl proteinase K (20 mg/ml).
4. Digest overnight in a water incubator at 55 °C or 2 h on a shaking dry incubator at 55 °C.
5. Vortex tubes and centrifuge for 5 min at 12,000–16,000 ×*g*.
6. Collect supernatant in new tube (to remove undigested tissue) with 500 μl isopropanol (2-propanol).

7. Shake vigorously and centrifuge for 10–15 min at 12,000–16,000 × *g*.
8. Remove supernatant, add 500 μl 70 % ethanol to wash pellet, and remove supernatant again.
9. Add 500 μl ddH_2O and vortex vigorously.

Genotyping using GoTaq® Green Master Mix:

1. Aliquot 2× Master Mix into aliquots between 200 μl and 1 ml.
2. Prepare Primer Mix for each genotype (200 μl stock):

Upstream primer, 100 μM	20 μl
Downstream primer, 100 μM	20 μl
ddH_2O	160 μl

   If two up- or downstream primers are needed, use 10 μl of each.

3. Prepare Master Mix for each primer set:

	1 reaction	20 reactions
GoTaq® Green Master Mix, 2×	4 μl	80 μl
Primer Mix, 10 μM for each primer	0.4 μl	8 μl

4. Pipet 4 μl genomic DNA template and 4 μl Master Mix (for many samples a dispenser can be used) per reaction tube. Also half reactions can be done.
5. Annealing conditions should be optimized. It is recommended to design all primers for approximately the same annealing temperature (e.g. 58 °C) because PCRs can be pooled in the same block and different primers can be combined. The following program works for many primer sets:

Initial denaturation 30–35 cycles:	3 min	94 °C
Denaturation	40 s	94 °C
Annealing	40 s	52–62 °C
Extension	1 min	72 °C
Final extension	5 min	72 °C
Soak/refrigeration cycle	–	4 °C

6. Load DNA samples on a 1.5 % agarose gel (one band) 2 % agarose gel (multiple bands with <100 bp difference).

3.5 Inducible Conditional Knockouts

If the gene knockout needs to be induced at a specific developmental stage, a tamoxifen-inducible form of Cre, CreER^{T2} [18], can be used. In this case, Cre activity is only induced if the synthetic hormone tamoxifen is administered. For the activation of CreER^{T2} in embryos, pregnant mice are injected with tamoxifen (dissolved in corn or sunflower oil) using a gavage needle.

Preparation of Tamoxifen (20 mg/ml):

1. Heat 5 ml of corn oil in a tube at 37 °C for 30 min.
2. Put tamoxifen at room temperature (RT).
3. Weigh 100 mg of tamoxifen and let the tamoxifen dissolve on a shaker at 37 °C; vortex regularly. It will take a few hours till the tamoxifen is completely dissolved.
4. Store at 4 °C (1 month) or make aliquots (0.5 ml) and store at −20 °C.

Gavaging:

1. Hold the pregnant female mouse from the back, in a similar fashion as for intraperitoneal injections, and secure such that the head cannot move and the esophagus is straight.
2. Slowly enter the mouth of the animal and proceed along the top of the mouth till the gavage needle enters first the esophagus and then the stomach.
3. Slowly inject the oil with 1–10 mg tamoxifen (*see* **Note 3**) and remove the needle.
4. Minimize the stress of the animal following tamoxifen administration (*see* **Note 4**).

3.6 Intersectional Recombination Systems

Specificity of Cre recombination in distinct cell subpopulations or cell types may be limited by the lack of specific promoters to drive *Cre* expression. Two strategies have been designed to achieve higher cellular specificity of gene recombination, which take advantage of the intersectional activities of two promoters in the same cell. The first approach has been elaborated in the Dymecki laboratory [6, 24] to analyze hindbrain development. Hereby the overlapping expression of two different recombinases, Cre recombinase and Flpe recombinase, driven by two different promoters activates the expression of a reporter gene (*GFP*) solely in cells where both recombinases are present (Fig. 6b). Additionally the construct contains a reporter (*b-galactosidase*), which is activated only if *Cre* though not *Flpe* is expressed (Fig. 6b).

A second approach was created by Hirrlinger et al. [21, 22], where Cre recombinase is split into two inactive parts, N-Cre (the N-terminus) and C-Cre (C-terminus), which are fused to the constitutive GCN4 protein–protein interaction domain, which has been shown to create stable dimers [44]. When co-expressed in the

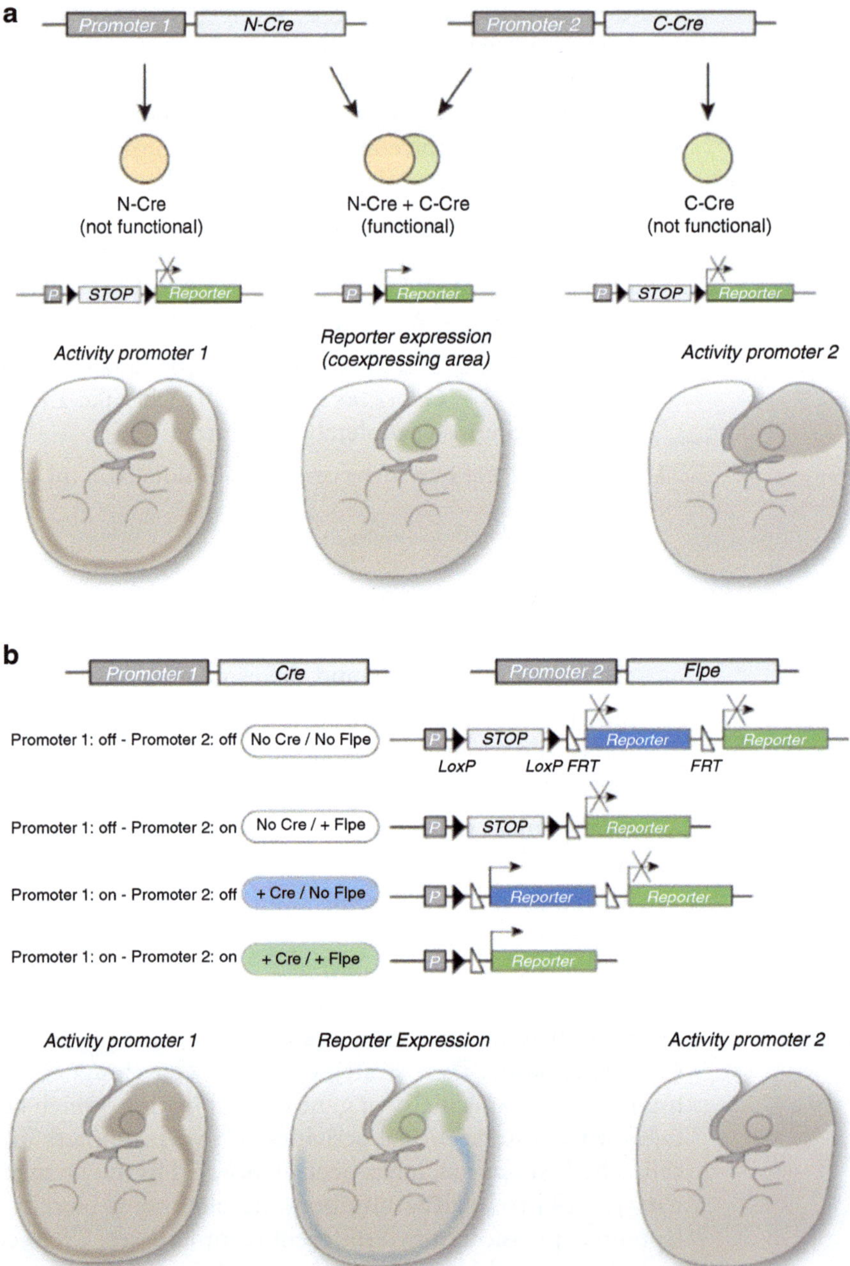

Fig. 6 Tools to generate intersectional recombination. (**a**) In the Split-Cre system, Cre is divided into two inactive fragments expressed under the control of two distinct promoters. In the tissue, or cell type, where both promoters are active, the two fragments are co-expressed resulting in a functional Cre. This triggers a recombination event, e.g., the Cre-mediated activation of a reporter gene. (**b**) The Cre/Flpe system uses two different recombinases recognizing distinct target sites. When both recombinases are active, a specific reporter gene is expressed (e.g., GFP) which is not activated by either one of the recombinases alone

same cell, N-Cre-GCN4 and C-Cre-GCN4 were able to reconstitute a functional Cre recombinase activity, which can induce recombination (Fig. 6a). The system was successfully tested with virus-mediated gene transfer [21, 22] as well as by the use of transgenic mice [23].

3.7 MADM System

The MADM (mosaic analysis with double markers) is an adaptation in mouse of the conceptually similar MARCM (mosaic analysis with a repressible cell marker) used in *Drosophila* [27]. Hereby, rare interchromosomal recombination (Fig. 1c) during mitosis results in sparse labeling of cells with one of two fluorescent reporter genes (e.g., GFP or RFP). In differentiated cells interchromosomal recombination can occur as well but can only result in the expression of both (GFP and RFP) or no reporter (Fig. 7). Because of the low frequency of recombinations during mitosis, this allows (1) to follow the behavior of clonally related (i.e., generated from the same progenitor) cells [45] as well as (2) to sparsely generate mutant cells in the background of a heterozygous mouse labeled by a fluorescent marker (e.g., RFP).

Before recombination no fluorescent marker is expressed in MADM mice, because the N-terminus sequence of either one of the two fluorescent markers (separated by a *LoxP* site) is followed by the C-terminus of the other marker (e.g., N-RFP-LoxP-C-GFP and N-GFP-LoxP-C-RFP, respectively), resulting in nonfunctional fluorescent proteins. After Cre-induced recombination, N-terminus and C-terminus can be reconstituted to functional fluorescent proteins (e.g., RFP and GFP in the old versions of MADM and GFP and tdTomato in the new versions (Fig. 8)). Since recombination can be tightly controlled by the use of tissue- or cell-specific *Cre* drivers as well as by tamoxifen-inducible *CreER*^{T2} drivers for temporal induction, the lineage of single cells can be traced during development and adulthood. Furthermore MADM can be used to selectively knockout genes in labeled cells by introducing (by natural recombination) a mutant allele between the centromere and the MADM transgene, while the second chromosome solely carries the wild-type allele (besides the complementary MADM transgene/knockin). In case of a X-segregation (sister chromatids segregating into different daughter cells) in G2 phase of mitosis, two different cells with different colors (red and green) and genotypes (homozygote mutant and homozygote wild-type allele) are generated, while a Z-segregation would result in two heterozygous sister cells (one unlabeled and one labeled with both fluorescent markers) (Fig. 7).

In contrast to classical conditional knockouts, MADM conditional knockouts have several advantages (Fig. 9). First of all, non-cell-autonomous defects can be almost excluded, because just a small proportion of cells carries the homozygous knockout (which can be titered up or down by the use of a *CreER*^{T2} driver). Secondly it has clear advantages, when the cellular character and behavior (cell shape, cell migration, axon guidance) have to be

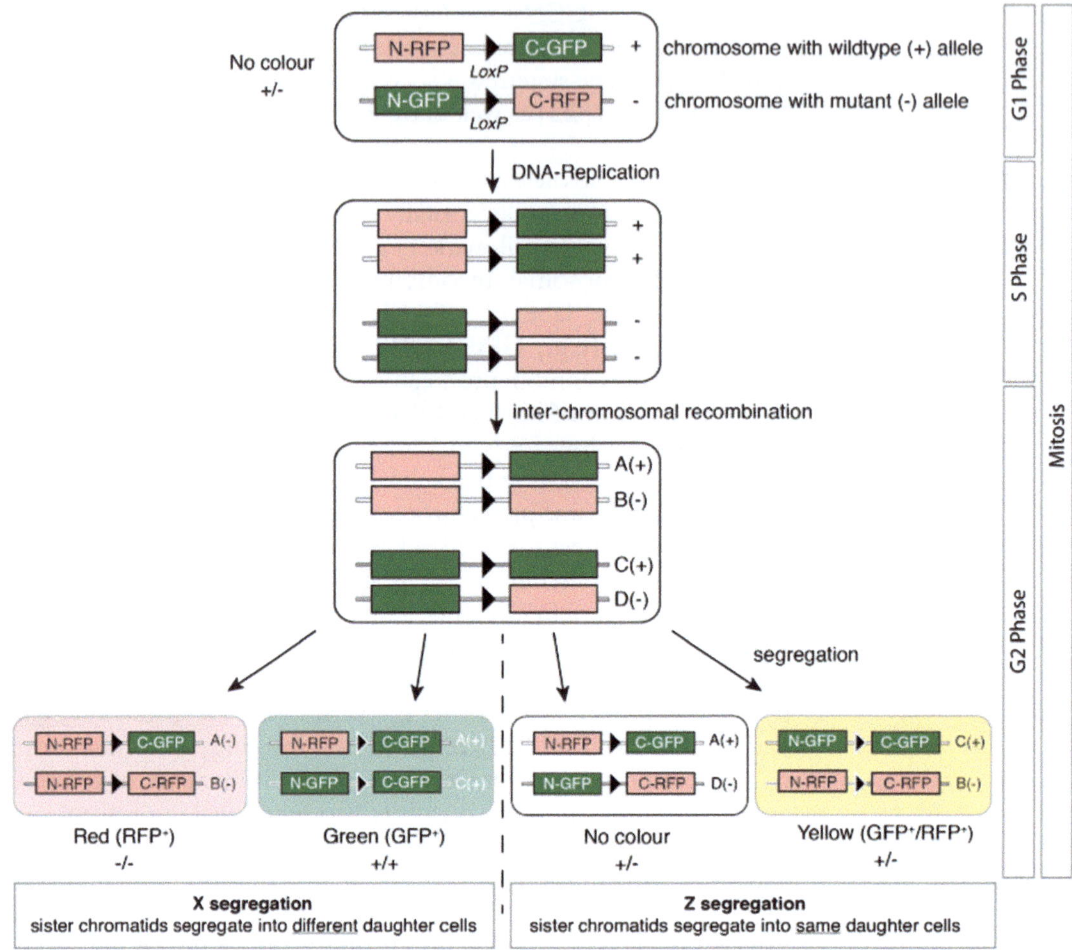

Fig. 7 The MADM system. The MADM system consists of two knockins, which are chimeric genes constituted by the N-terminus and C-terminus of two distinct fluorescent proteins (e.g., RFP and GFP). Before Cre-mediated recombination, the chimeric *N-RFP/C-GFP* or *N-GFP/C-RFP* alleles result in two nonfunctional proteins. During the S phase of the mitosis, sister chromatids duplicate and pair. If an interchromosomal recombination happens, the correspondingly recombined sister chromatids segregate during the G2 phase into different daughter cells (X-segregation). Two cells of different colors (i.e., expressing either *red* or *green* fluorescence) are then generated. In case each sister chromatids segregate into the same cell, a double-labeled and colorless cell is produced. A recombination during G1 or G0 (differentiated cell) phase results as well in a colorless and in a double-labeled cell. If additionally a mutant allele is recombined onto a MADM-bearing chromosome (while the other MADM containing chromosome has the wild-type allele), X-segregation results in the generation of sibling mutant and wild-type cell, labeled with different fluorescent markers, while Z-segregation or recombination in G1 or G0 phase does not change the genotype of the cell

analyzed. Hereby it allows even the direct comparison of cells carrying homozygous mutations and no mutation.

Disadvantages of this technique are that MADM lines are not yet available for all chromosomes, which restricts the use of it at the moment to analyze knockouts of genes located on chr.6, chr.10, and

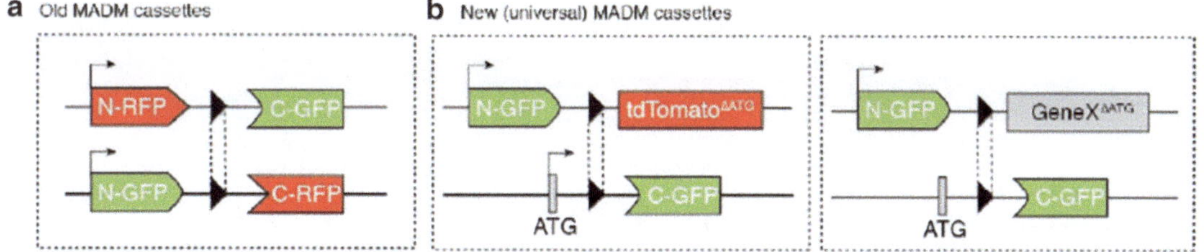

Fig. 8 MADM cassettes. To increase variability in the choice of the second reporter/gene, the old cassettes (**a**) containing *GFP* and *RFP* have been replaced by one cassette containing an ATG and the C-terminus of *GFP* and a second cassette containing *N-GFP* and an ATG-less *tdTomato* (**b**). If another gene except *tdTomato* should be overexpressed (*GeneX*), only one transgenic line has to be created making the system more universal

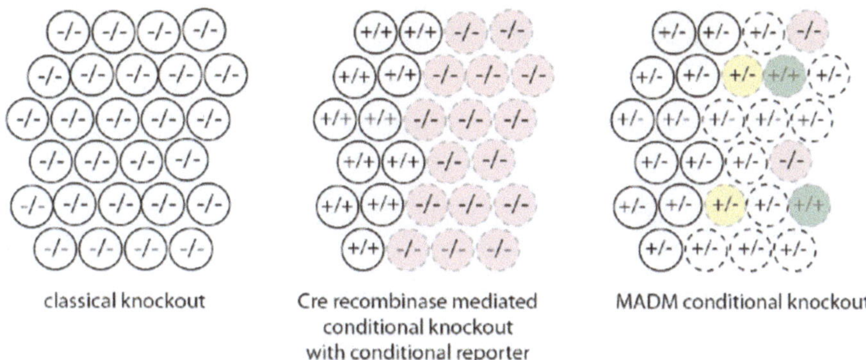

Fig. 9 Advantages of conditional knockouts and MADM conditional knockouts. While classical knockouts give little information about the role of the knockedout gene in different tissues, the use of conditional knockouts allows the analysis of gene function in selected cell subpopulations or tissues. The MADM system allows furthermore to trace the lineage of single cells with different genotypes (homozygous knockout and wild type, respectively) in an environment of heterozygous cells

chr.11 and only between the MADM insertion locus and the centromere.

Recently the system has been further optimized by using constructs with *tdTomato*, a red fluorescent protein brighter than RFP, as well as by making the choice of the second gene to be recombined more flexible (e.g., by substituting a second fluorescent reporter gene with a *GeneX* to be overexpressed) (Fig. 8) [26].

3.8 Brainbow System

An elegant approach to analyze the behavior of single cells in a tissue, especially for the analysis of connectivity patterns in neural circuits, came from the Lichtman group [29]. By the use of different, incompatible *Lox* sites (*LoxN*, *Lox2722*, and *LoxP*), these authors generated mice, which, in a stochastic manner, are induced to express only one fluorescent protein (YFP, CFP, or RFP) as a final recombination event (Figs. 10 and 11).

Four different brainbow versions that stochastically express different kinds of fluorescent proteins exist (Fig. 11a, b), and the corresponding transgenic mouse lines are available from Jackson

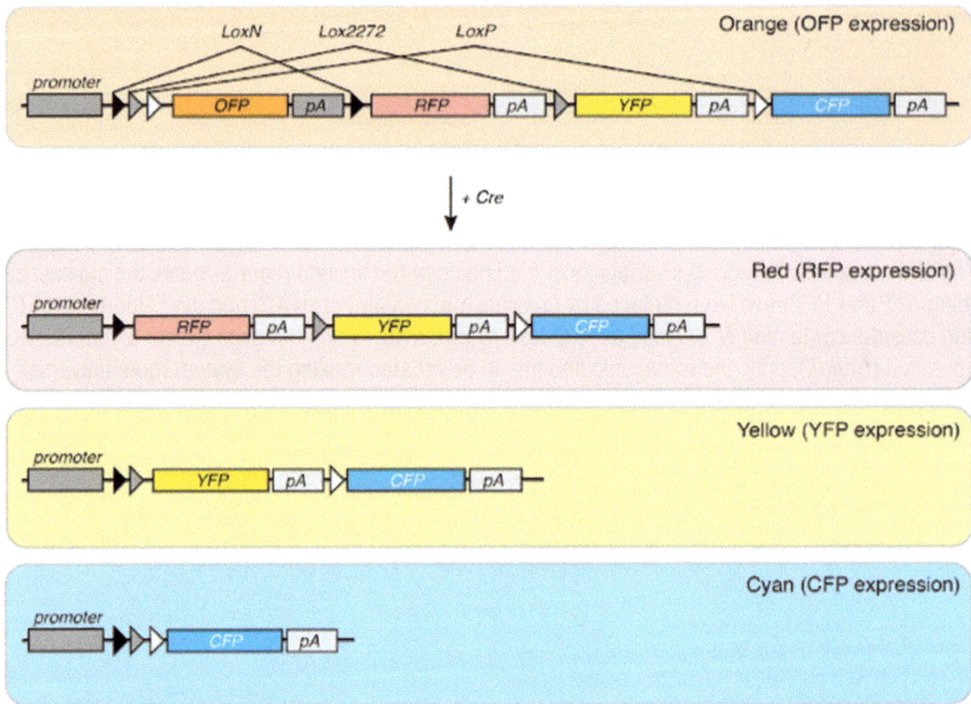

Fig. 10 The brainbow system. Recombination of the Brainbow 1.1 constructs can result in three possible outcomes. A recombination of the *LoxN* site results in the transcriptional activation of *RFP*. Because the other two *Lox* sites (*Lox2722* and *LoxP*) lack a second site because they have been placed 3′ of the *LoxN* site, no further recombination can occur. If the first recombination happens between the *Lox2272* sites, the result is expression of *YFP*; if it happens between the *LoxP* sites, it results instead in expression of *CFP*

Laboratories. Due to the effect that multiple integrations occur as well as they independently result in the stochastic choice of one color, combinations of color lead to almost 100 distinguishable colors (Fig. 11c).

4 Notes

1. Detailed information and troubleshooting about the use of recombination-mediated genetic engineering can be found here.

   http://web.ncifcrf.gov/research/brb/recombineeringInformation.aspx (31/07/13)

2. ZsGreen in contrast to tdTomato is not localizing strongly in cellular extensions like axons or dendrites. For these kinds of analyses, the tdTomato line (# 007914) would be the recommended choice. For localization of cells ZsGreen (# 007906) is better.

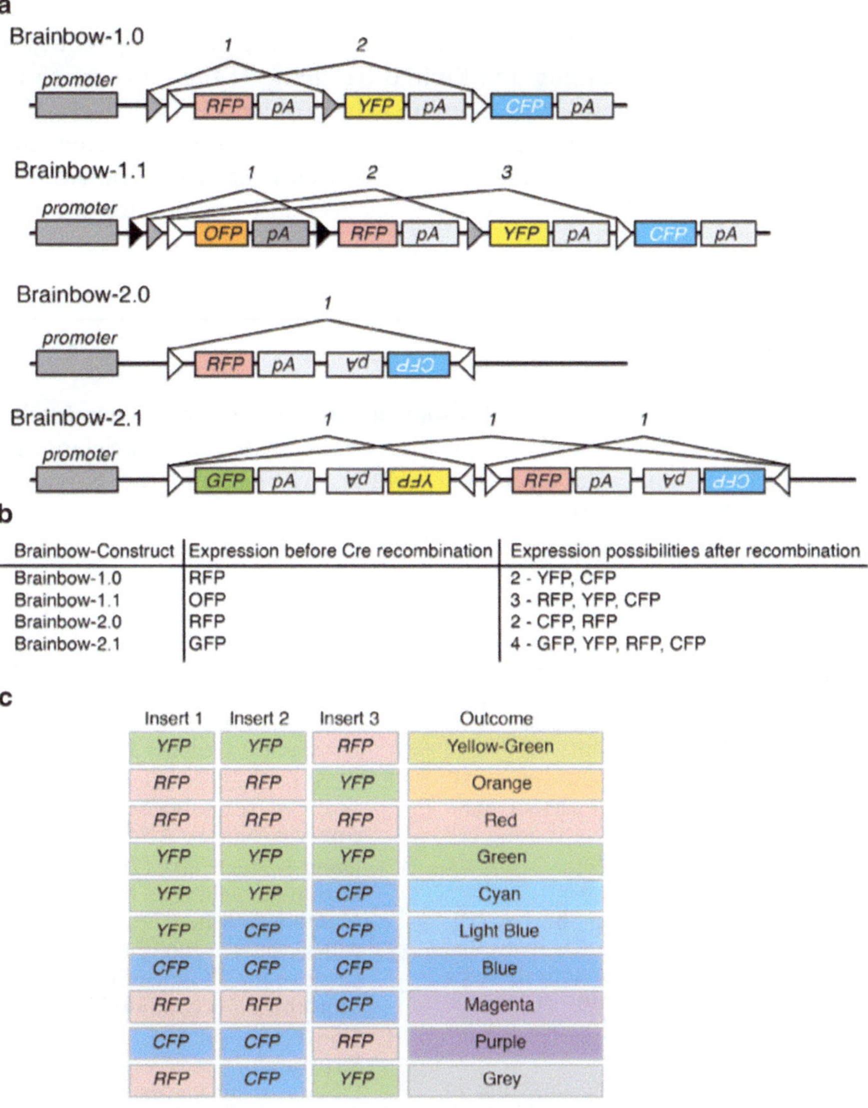

Brainbow-Construct	Expression before Cre recombination	Expression possibilities after recombination
Brainbow-1.0	RFP	2 - YFP, CFP
Brainbow-1.1	OFP	3 - RFP, YFP, CFP
Brainbow-2.0	RFP	2 - CFP, RFP
Brainbow-2.1	GFP	4 - GFP, YFP, RFP, CFP

Insert 1	Insert 2	Insert 3	Outcome
YFP	YFP	RFP	Yellow-Green
RFP	RFP	YFP	Orange
RFP	RFP	RFP	Red
YFP	YFP	YFP	Green
YFP	YFP	CFP	Cyan
YFP	CFP	CFP	Light Blue
CFP	CFP	CFP	Blue
RFP	RFP	CFP	Magenta
CFP	CFP	RFP	Purple
RFP	CFP	YFP	Grey

Fig. 11 Overview of brainbow constructs and principle of "multicolor" brainbow. (**a, b**) Four brainbow constructs have been generated (**a**) with different possible outcomes (**b**). While Brainbows 1.0 and 1.1 recombine only once, Brainbows 2.0 and 2.1 continue to recombine changing the cassette expression pattern after every recombination event. (**c**) Due to multiple insertions more than three different colors are possible. Three inserts would give ten distinguishable colors. Here YFP channel would be set to *green* and CFP channel to *blue*, giving a similar principle as the R(ed)G(reen)B(lue) color model of screen displays

3. For reporter lines and overexpression lines, usually 1–2 mg of tamoxifen per pregnant female is sufficient. For the recombination of two alleles in mutants, 5–10 mg (varies between conditional knockout lines) is necessary. The efficiency of the knockout has to be controlled carefully by in situ hybridization and/or immunohistochemistry.

4. In general, stress of the tamoxifen-treated pregnant female should be kept to a minimum, e.g., by no/reduced change of bedding and limited handling. If the pups are to be analyzed postnatally, there may be problems with delivery, since tamoxifen interferes with the female hormone system in a dosage-dependent manner. To circumvent this problem, the pups can be recovered by C-section and put in adoption with a foster mother.

Acknowledgements

Work in F.M.R. laboratory in Friedrich Miescher Institute (Basel, Switzerland) is supported by the Swiss National Science Foundation (Sinergia CRSI33_127440), ARSEP, and the Novartis Research Foundation.

References

1. Thomas KR, Capecchi MR (1987) Site-directed mutagenesis by gene targeting in mouse embryo-derived stem cells. Cell 51(3):503–512
2. Gu H, Marth JD, Orban PC, Mossmann H, Rajewsky K (1994) Deletion of a DNA polymerase beta gene segment in T cells using cell type-specific gene targeting. Science 265(5168):103–106
3. Lakso M, Sauer B, Mosinger B, Lee EJ, Manning RW, Yu SH, Mulder KL, Westphal H (1992) Targeted oncogene activation by site-specific recombination in transgenic mice. Proc Natl Acad Sci USA 89(14):6232–6236
4. Sauer B (1998) Inducible gene targeting in mice using the Cre/lox system. Methods 14(4):381–392
5. O'Neal KR, Agah R (2007) Conditional targeting: inducible deletion by cre recombinase. Methods Mol Biol 366:309–320
6. Kim JC, Dymecki SM (2009) Genetic fate-mapping approaches: new means to explore the embryonic origins of the cochlear nucleus. Methods Mol Biol 493:65–85
7. Wilson TJ, Kola I (2001) The LoxP/CRE system and genome modification. Methods Mol Biol 158:83–94
8. Brault V, Besson V, Magnol L, Duchon A, Hérault Y (2007) Cre/loxP-mediated chromosome engineering of the mouse genome. Handb Exp Pharmacol 178:29–48
9. Feil S, Valtcheva N, Feil R (2009) Inducible Cre mice. Methods Mol Biol 530:343–363
10. Siegal ML, Hartl DL (1996) Transgene Coplacement and high efficiency site-specific recombination with the Cre/loxP system in Drosophila. Genetics 144(2):715–726
11. Werdien D, Peiler G, Ryffel GU (2001) FLP and Cre recombinase function in Xenopus embryos. Nucleic Acids Res 29(11):E53–3
12. Dong J, Stuart GW (2004) Transgene manipulation in zebrafish by using recombinases. Methods Cell Biol 77:363–379
13. Gilbertson L (2003) Cre-lox recombination: Creative tools for plant biotechnology. Trends Biotechnol 21(12):550–555
14. Guo F, Gopaul DN, van Duyne GD (1997) Structure of Cre recombinase complexed with DNA in a site-specific recombination synapse. Nature 389(6646):40–46
15. Nakano M, Odaka K, Ishimura M, Kondo S, Tachikawa N, Chiba J, Kanegae Y, Saito I (2001) Efficient gene activation in cultured mammalian cells mediated by FLP recombinase-expressing recombinant adenovirus. Nucleic Acids Res 29(7):E40
16. Buchholz F, Angrand P-O, Stewart AF (1998) Improved properties of FLP recombinase evolved by cycling mutagenesis. Nat Biotechnol 16(7):657–662
17. Rodríguez CI, Buchholz F, Galloway J, Sequerra R, Kasper J, Ayala R, Stewart AF, Dymecki SM (2000) High-efficiency deleter mice show that FLPe is an alternative to Cre-loxP. Nat Genet 25(2):139–140
18. Feil R, Wagner J, Metzger D, Chambon P (1997) Regulation of Cre recombinase activity

by mutated estrogen receptor ligand-binding domains. Biochem Biophys Res Commun 237(3):752–757

19. Feil R, Brocard J, Mascrez B, LeMeur M, Metzger D, Chambon P (1996) Ligand-activated site-specific recombination in mice. Proc Natl Acad Sci USA 93(20):10887–10890
20. Edwards WF, Young DD, Deiters A (2009) Light-activated Cre recombinase as a tool for the spatial and temporal control of gene function in mammalian cells. ACS Chem Biol 4(6):441–445
21. Hirrlinger J, Scheller A, Hirrlinger PG et al (2009) Split-cre complementation indicates coincident activity of different genes in vivo. PLoS One 4(1):e4286
22. Hirrlinger J, Requardt RP, Winkler U, Wilhelm F, Schulze C, Hirrlinger PG (2009) Split-CreERT2: temporal control of DNA recombination mediated by split-Cre protein fragment complementation. PLoS One 4(12):e8354
23. Wang P, Chen T, Sakurai K, Han B-X, He Z, Feng G, Wang F (2012) Intersectional cre driver lines generated using split-intein mediated split-cre reconstitution. Sci Rep 2:497
24. Farago AF, Awatramani RB, Dymecki SM (2006) Assembly of the brainstem cochlear nuclear complex is revealed by intersectional and subtractive genetic fate maps. Neuron 50(2):205–218
25. Zong H, Espinosa JS, Su HH, Muzumdar MD, Luo L (2005) Mosaic analysis with double markers in mice. Cell 121(3):479–492
26. Tasic B, Miyamichi K, Hippenmeyer S, Dani VS, Zeng H, Joo W, Zong H, Chen-Tsai Y, Luo L (2012) Extensions of MADM (mosaic analysis with double markers) in mice. PLoS One 7(3):e33332
27. Lee T, Luo L (1999) Mosaic analysis with a repressible cell marker for studies of gene function in neuronal morphogenesis. Neuron 22(3):451–461
28. Hippenmeyer S, Youn YH, Moon HM, Miyamichi K, Zong H, Wynshaw-Boris A, Luo L (2010) Genetic mosaic dissection of Lis1 and Ndel1 in neuronal migration. Neuron 68(4):695–709
29. Livet J, Weissman TA, Kang H, Draft RW, Lu J, Bennis RA, Sanes JR, Lichtman JW (2007) Transgenic strategies for combinatorial expression of fluorescent proteins in the nervous system. Nature 450(7166):56–62
30. Soriano P (1999) Generalized lacZ expression with the ROSA26 Cre reporter strain. Nat Genet 21(1):70–71
31. Madisen L, Zwingman TA, Sunkin SM et al (2010) A robust and high-throughput Cre reporting and characterization system for the whole mouse brain. Nat Neurosci 13(1):133–140
32. Luo L (2007) Fly MARCM and mouse MADM: genetic methods of labeling and manipulating single neurons. Brain Res Rev 55(2):220–227
33. Potter CJ, Tasic B, Russler EV, Liang L, Luo L (2010) The Q system: a repressible binary system for transgene expression, lineage tracing, and mosaic analysis. Cell 141(3):536–548
34. Zhang Y, Buchholz F, Muyrers JP, Stewart AF (1998) A new logic for DNA engineering using recombination in Escherichia coli. Nat Genet 20(2):123–128
35. Sauer B, Henderson N (1988) Site-specific DNA recombination in mammalian cells by the Cre recombinase of bacteriophage P1. Proc Natl Acad Sci USA 85(14):5166–5170
36. Santoro SW, Schultz PG (2002) Directed evolution of the site specificity of Cre recombinase. Proc Natl Acad Sci USA 99(7):4185–4190
37. Schnütgen F, Doerflinger N, Calléja C, Wendling O, Chambon P, Ghyselinck NB (2003) A directional strategy for monitoring Cre-mediated recombination at the cellular level in the mouse. Nat Rev Neurosci 21(5):562–565
38. Zheng B, Sage M, Sheppeard EA, Jurecic V, Bradley A (2000) Engineering mouse chromosomes with Cre-loxP: range, efficiency, and somatic applications. Mol Cell Biol 20(2):648–655
39. Visel A, Minovitsky S, Dubchak I, Pennacchio LA (2007) VISTA Enhancer Browser–a database of tissue-specific human enhancers. Nucleic Acids Res 35:D88–D92
40. Engström PG, Fredman D, Lenhard B (2008) Ancora: a web resource for exploring highly conserved noncoding elements and their association with developmental regulatory genes. Genome Biol 9(2):R34
41. Visel A, Blow MJ, Li Z et al (2009) ChIP-seq accurately predicts tissue-specific activity of enhancers. Nature 457(7231):854–858
42. Yee SP, Rigby PW (1993) The regulation of myogenin gene expression during the embryonic development of the mouse. Genes Dev 7(7A):1277–1289
43. Srinivas S, Watanabe T, Lin CS, William CM, Tanabe Y, Jessell TM, Costantini F (2001) Cre reporter strains produced by targeted insertion of EYFP and ECFP into the ROSA26 locus. BMC Dev Biol 1:4
44. Hope IA, Struhl K (1987) GCN4, a eukaryotic transcriptional activator protein, binds as a dimer to target DNA. EMBO J 6(9):2781–2784
45. Espinosa JS, Luo L (2008) Timing neurogenesis and differentiation: insights from quantitative clonal analyses of cerebellar granule cells. J Neurosci 28(10):2301–2312

Index

Simon G. Sprecher (ed.), *Brain Development: Methods and Protocols*, Methods in Molecular Biology, vol. 1082, DOI 10.1007/978-1-62703-655-9, © Springer Science+Business Media, LLC 2014

Q

R

S

T

V

X

Y

Z

MIX
Papier aus verantwortungsvollen Quellen
Paper from responsible sources
FSC® C105338

If you have any concerns about our products,
you can contact us on
ProductSafety@springernature.com

In case Publisher is established outside the EU,
the EU authorized representative is:
Springer Nature Customer Service Center GmbH
Europaplatz 3, 69115 Heidelberg, Germany

Printed by Libri Plureos GmbH
in Hamburg, Germany